SM358
Science: Level 3

The Quantum World

Book 3

Quantum mechanics of matter

Edited by John Bolton and Stuart Freake

SM358 Course Team

Course Team Chair
John Bolton

Academic Editors
John Bolton, Stuart Freake, Robert Lambourne, Raymond Mackintosh

Authors
Silvia Bergamini, John Bolton, Mark Bowden, David Broadhurst, Jimena Gorfinkiel, Robert Lambourne, Raymond Mackintosh, Nigel Mason, Elaine Moore, Jonathan Underwood

Other Course Team Members
Robert Hasson, Mike Thorpe

Consultant
Derek Capper

Course Manager
Gillian Knight

Course Team Assistant
Yvonne McKay

LTS, Project Manager
Rafael Hidalgo

Editors
Peter Twomey, Alison Cadle

TeX Specialist
Jonathan Fine

Graphic Design Advisors
Mandy Anton, Chris Hough

Graphic Artist
Roger Courthold

Picture Researcher/Copyrights
Lydia Eaton

Software Designers
Fiona Thomson, Will Rawes

Video Producers
Owen Horn, Martin Chiverton

External Assessor
Charles Adams (Durham University)

This publication forms part of an Open University course SM358 The Quantum World. The complete list of texts which make up this course can be found at the back. Details of this and other Open University courses can be obtained from the Student Registration and Enquiry Service, The Open University, PO Box 197, Milton Keynes, MK7 6BJ, United Kingdom: tel. +44 (0)870 333 4340, email general-enquiries@open.ac.uk

Alternatively, you may visit the Open University website at http://www.open.ac.uk where you can learn more about the wide range of courses and packs offered at all levels by The Open University.

To purchase a selection of Open University course materials visit http://www.ouw.co.uk or contact Open University Worldwide, Michael Young Building, Walton Hall, Milton Keynes MK7 6AA, United Kingdom for a brochure. tel. +44 (0)1908 858785; fax +44 (0)1908 858787; email ouwenq@open.ac.uk

The Open University
Walton Hall, Milton Keynes
MK7 6AA

First published 2007, 2009. Copyright © 2007, 2009. The Open University

All rights reserved. No part of this publication may be reproduced, stored in a retrieval system, transmitted or utilised in any form or by any means, electronic, mechanical, photocopying, recording or otherwise, without written permission from the publisher or a licence from the Copyright Licensing Agency Ltd. Details of such licences (for reprographic reproduction) may be obtained from the Copyright Licensing Agency Ltd of 90 Tottenham Court Road, London W1T 4LP.

Open University course materials may also be made available in electronic formats for use by students of the University. All rights, including copyright and related rights and database rights, in electronic course materials and their contents are owned by or licensed to The Open University, or otherwise used by The Open University as permitted by applicable law.

In using electronic course materials and their contents you agree that your use will be solely for the purposes of following an Open University course of study or otherwise as licensed by The Open University or its assigns.

Except as permitted above you undertake not to copy, store in any medium (including electronic storage or use in a website), distribute, transmit or retransmit, broadcast, modify or show in public such electronic materials in whole or in part without the prior written consent of The Open University or in accordance with the Copyright, Designs and Patents Act 1988.

Edited and designed by The Open University.

Typeset at The Open University.

Printed and bound in the United Kingdom by Latimer Trend & Company Ltd, Plymouth.

ISBN 978 07492 2517 9

2.1

QUANTUM MECHANICS OF MATTER

Introduction		7
Chapter 1	**Angular momentum for atomic physics**	9
Introduction		9
1.1	A review of orbital angular momentum	10
1.2	Quantum mechanics in three dimensions	11
1.2.1	The time-independent Schrödinger equation	12
1.2.2	Transforming to spherical coordinates	13
1.2.3	Making the link with angular momentum	15
1.2.4	Energy eigenfunctions and how to label them	17
1.3	Spherical harmonics	18
1.3.1	The form of the spherical harmonics	18
1.3.2	Visualizing spherical harmonics	23
1.4	Particles with spin in a potential energy well	25
1.4.1	Total angular momentum	26
1.4.2	Simultaneous eigenvectors of $\widehat{J}^2$ and $\widehat{J}_z$	28
1.4.3	Energy-level splitting due to spin–orbit interaction	30
1.4.4	Examples from the world of atoms and nuclei	31
Chapter 2	**The hydrogen atom**	35
Introduction		35
2.1	Preliminary thoughts about hydrogen	36
2.2	The time-independent Schrödinger equation	39
2.3	Solutions of the radial equation	42
2.3.1	Limiting behaviour	42
2.3.2	Nodeless solutions of the radial equation	43
2.3.3	General solutions of the radial equation	45
2.4	Putting it together: the complete eigenfunctions	51
2.4.1	The complete solutions	51
2.4.2	Spectroscopic notation	52
2.4.3	Visualizing the complete eigenfunctions	53
2.5	Expectation values and uncertainties	58
2.5.1	Expectation values	58
2.5.2	Uncertainty in electron–proton separation	60
2.5.3	Rydberg states of the hydrogen atom	60
2.5.4	Momentum distribution in the ground state	61
2.5.5	Postscript	64

Chapter 3   Time-independent approximation methods   66

Introduction   66

- 3.1 The variational method   67
  - 3.1.1 The principle of the method   68
  - 3.1.2 Applying the technique   70
  - 3.1.3 A step-by-step recipe for the ground-state energy   74
  - 3.1.4 The variational method and excited states   74
- 3.2 Perturbation methods   75
  - 3.2.1 Basic perturbation theory   75
  - 3.2.2 Taylor expansions and orders of approximation   79
  - 3.2.3 First-order approximation for the energy eigenvalues   80
  - 3.2.4 Higher-order perturbation theory   87

Chapter 4   Hydrogen-like systems   89

Introduction   89

- 4.1 Hydrogen-like atoms   90
  - 4.1.1 Key results for the hydrogen atom   90
  - 4.1.2 The helium ion and similar systems   91
  - 4.1.3 Muonic atoms and their uses   93
  - 4.1.4 Not just muons …   99
  - 4.1.5 Deep within atoms: X-ray spectra   99
- 4.2 The hydrogen spectrum 'under the microscope'   101
  - 4.2.1 Fine structure of the hydrogen spectrum   102
  - 4.2.2 Hyperfine structure of the hydrogen spectrum   107
- 4.3 Putting quantum mechanics and relativity together   108
  - 4.3.1 The Dirac equation   108
  - 4.3.2 The next step: quantum fields   111

Chapter 5   Many-electron atoms   114

Introduction   114

- 5.1 First steps in describing a many-electron atom   114
  - 5.1.1 The Hamiltonian operator   114
  - 5.1.2 The independent-particle model   116
- 5.2 The helium atom   118
  - 5.2.1 The total wave function for two electrons   118
  - 5.2.2 The ground state of helium   120
  - 5.2.3 The excited states of helium   121
- 5.3 Many-electron atoms   123
  - 5.3.1 The central-field approximation   124
  - 5.3.2 Configurations   126
  - 5.3.3 The Periodic Table   128
- 5.4 Terms and levels   131
  - 5.4.1 Good quantum numbers   132
  - 5.4.2 Spin–orbit interaction neglected   133

		5.4.3 Spin–orbit interaction included	136
		5.4.4 Hund's rules	138

Chapter 6 Diatomic molecules 141

Introduction 141

- 6.1 The time-independent Schrödinger equation 142
  - 6.1.1 Setting up the equation 142
  - 6.1.2 The Born–Oppenheimer approximation 143
  - 6.1.3 Three different energy contributions 145
- 6.2 The hydrogen molecule ion 146
  - 6.2.1 Linear combination of atomic orbitals 148
  - 6.2.2 The ground state of the hydrogen molecule ion 150
  - 6.2.3 The ground-state energy curve 153
  - 6.2.4 First excited state of the hydrogen molecule ion 155
- 6.3 Molecular orbitals 157
  - 6.3.1 Spectroscopic notation for molecular orbitals 157
  - 6.3.2 Forming molecular orbitals from atomic orbitals 158
- 6.4 Homonuclear diatomic molecules 161
  - 6.4.1 Molecules of hydrogen and helium 161
  - 6.4.2 Molecules from lithium to neon 163
- 6.5 Beyond the LCAO approximation 165

Chapter 7 Solid state physics 169

Introduction 169

- 7.1 Structure and bonding 169
  - 7.1.1 The arrangement of atoms in solids 169
  - 7.1.2 Types of bonding 171
  - 7.1.3 An LCAO approach to bonding in solids 172
- 7.2 Bloch's theorem and energy bands 175
  - 7.2.1 Bloch's theorem 175
  - 7.2.2 The tight-binding method 180
  - 7.2.3 Band structures in practice 184
- 7.3 Conduction of electricity in solids 186
  - 7.3.1 Metals, insulators and semiconductors 186
  - 7.3.2 Semiconductors 188

Chapter 8 Light and matter 195

Introduction 195

- 8.1 The interaction of light with matter 196
  - 8.1.1 Three types of radiative transition 196
  - 8.1.2 The electric dipole approximation 199
- 8.2 Time-dependent perturbation theory 202
  - 8.2.1 Solving Schrödinger's equation 202
  - 8.2.2 The probability of making a transition 206

		8.2.3	Transitions induced by monochromatic light	207
	8.3		Selection rules for radiative transitions	208
	8.4		Absorption and stimulated emission	211
	8.5		Spontaneous emission and the Einstein coefficients	215
	At the end of the quantum journey			218

Acknowledgements 222

Solutions 223

Index 237

Introduction

Around 1900, physicists had good reason to feel pleased with themselves. The nineteenth century had seen huge advances in mechanics, optics, thermodynamics, fluid mechanics and electromagnetism. But the microscopic world of atoms remained at the boundaries of knowledge. The electron had just been discovered, but the atomic nucleus was still unknown, and the spectral lines emitted by atoms could not be properly explained. Many properties of materials had been measured and catalogued: copper, for example, was characterized by its colour, hardness, melting temperature, electrical conductivity, thermal conductivity, and so on. But no-one knew *why* copper behaved differently from iron. A major spur in the creation of quantum mechanics was the need to understand the structure and properties of atoms, molecules and solids.

Today, the ambitions of physics have changed utterly. Using only the fact that copper atoms have atomic number 29, together with quantum mechanics and powerful computers, it is possible to give quantitative explanations of many of the measured properties of copper. Moreover, quantum mechanics is often used to find an arrangement of atoms that has a desired set of properties. Quantum mechanics is taking the lead in the production of tailor-made materials such as semiconductors for transistors, with huge economic implications. For example, in 2002 it was estimated that the world produced about 10^{18} transistors each year, mainly for use in computer memory; this was about 40 times greater than the annual production of grains of rice.

This book will describe how quantum mechanics is used to explain the structure and properties of matter. Most of the discussion will be about atoms, molecules and solids, but we shall also mention some aspects of nuclear and sub-atomic physics.

When we deal with real systems, such as atoms or molecules, we often run into a quagmire of mathematical complexity. Even a hydrogen atom is not as simple as it might seem; there are small influences, such as those due to the finite size of the proton, the spin of the electron and relativity, that prevent us from obtaining exact solutions. The skill then lies in choosing a suitable model— one that captures the essence of the true situation, but which is still mathematically tractable. We will find systematic ways of taking small effects into account, not exactly, but in an approximate way. Approximation is essential because a vast universe of systems and behaviour cry out for explanation, but very few quantum-mechanical problems can be solved exactly. Systematic approximations allow quantum mechanics to be used in the real world that Nature provides.

The first two chapters begin by considering the hydrogen atom, treating the electron and proton as point charges; this is called the Coulomb model. Chapter 1 shows how the energy eigenfunctions of a hydrogen atom are labelled with angular momentum quantum numbers. Chapter 2 then completes the solution of the Coulomb model of hydrogen, showing how discrete energy levels arise from the time-independent Schrödinger equation. We also discuss how the electron probability density is distributed in various states of the atom.

To go beyond the Coulomb model for hydrogen atoms, we must make approximations. Chapter 3 introduces two different approximation techniques; the variational method and perturbation theory, both of which are used in later chapters. Chapter 4 discusses small corrections to the hydrogen-atom energy levels that arise from a variety of different influences, including relativity and quantum field theory. Over the years, the hydrogen atom has proved to be a good test-bed for physical theories because it is simple enough for very precise theoretical predictions to be compared with equally precise experimental data. Chapter 4 also examines systems that behave very much like hydrogen atoms. These include atoms in which an electron is replaced by a muon, and positronium — a bound system comprising an electron and its antiparticle, the positron.

Chapter 5 discusses atoms beyond hydrogen. As a first step, perturbation theory is used to analyze the helium atom. A helium atom contains two indistinguishable electrons, described by an antisymmetric total wave function. Using this fact, and treating electron–electron repulsion by perturbation theory, the ground-state and low-lying energy levels of helium are calculated. We go on to describe other atoms, outlining the structure of the Periodic Table of the elements in terms of the filling of available quantum states.

Chapter 6 considers simple molecules. The main issue here is the mechanism that binds atoms together in stable molecules. We explore this issue by examining the simplest molecule, the hydrogen molecule ion, which consists of two protons and one electron. By taking combinations of atomic orbitals and using the variational method, we show why this system is stable. Similar principles are applied to other diatomic molecules, and you will see why the nitrogen molecule is very strongly bound, and why the oxygen molecule displays magnetic properties.

Chapter 7 discusses solids. To some extent, these can be thought of as giant molecules, but due account must be taken of the regular spacing of atoms in crystals. The energy levels in a solid form bands separated by gaps, and we will discuss how the filling of energy bands, and the spacing between energy bands, determines the electrical behaviour of conductors, insulators and semiconductors. Technological applications of semiconductors rely on the controlled addition of impurities. You will see how these modify the behaviour of materials, and how devices such as quantum dots function.

Finally, Chapter 8 describes the processes by which light is absorbed or emitted by matter. It explains why some spectral lines are stronger than others, and why some lines that one might expect to observe are absent, or are too faint to be detected. This brings us full-circle in our journey through quantum mechanics. Patterns of spectral lines provide a wealth of data that contributed greatly to the development of quantum mechanics. Now, at the very end of the course, you will see in some detail how quantum mechanics is able to account for the observed patterns of spectral lines.

Chapter 1 Angular momentum for atomic physics

Introduction

It is the proud claim of modern physics that it can, in principle, give an account of the structure and properties of the matter around us, including the matter in stars and galaxies. Underlying this are two great theories: the electromagnetism of Maxwell and quantum theory. The first great triumph of quantum theory was the explanation of the spectrum and other properties of hydrogen atoms; a door that opened upon a previously unimaginable vista of discovery. Soon, Heisenberg had done what had seemed impossible and used quantum mechanics to explain the structure of helium. A rich stream of new developments followed, including the explanations of molecular bonding and electrical conduction in semiconductors. However, it was hydrogen, the simplest element, that provided the starting point.

In many ways, hydrogen atoms have remained at the centre of interest. In the course of refining the understanding of hydrogen, profound new discoveries have been made. The hydrogen atom was the first system for which quantum theory and special relativity were put together — with unexpected results including the prediction of antiparticles. Precise measurements of spectral lines in hydrogen atoms also provided early evidence which confirmed the success of quantum electrodynamics. Moreover, hydrogen-like systems are found in many branches of physics, including particle physics, where a quark and an antiquark bind together to form something that has many similarities with a hydrogen atom.

Much of the quantum mechanics we have developed so far applies in one dimension. This was done to clarify essential ideas, but it is clear that a hydrogen atom is a three-dimensional system, and any realistic description of it must be three-dimensional. An essential concept needed to describe the states of atoms in three dimensions is angular momentum. This chapter will therefore be devoted to extending the theory of angular momentum, previously developed in Book 2, stressing the concepts that are needed for atomic physics.

Section 1.1 begins by listing some properties of orbital angular momentum that were covered in Chapter 2 of Book 2; there is no new material here and the list is given for ease of reference. Section 1.2 then introduces the time-independent Schrödinger equation for a particle in a spherically-symmetric Coulomb potential energy well — a reasonable model of a hydrogen atom. We stress the advantages of using spherical coordinates, and show how orbital angular momentum enters into the time-independent Schrödinger equation. Section 1.3 develops the theory of angular momentum by introducing functions called spherical harmonics, which are eigenfunctions of the magnitude and the z-component of orbital angular momentum; these functions will play an important role in the description of atoms in succeeding chapters. Finally, Section 1.4 combines the orbital and spin angular momenta of a particle to produce the *total angular momentum*, and concludes with brief discussions of applications to atoms and atomic nuclei.

Chapter 1 Angular momentum for atomic physics

1.1 A review of orbital angular momentum

Before getting started on the main subject matter of this chapter, we list some key points about orbital angular momentum that have been discussed earlier in the course (Chapter 2 of Book 2). We shall not elaborate on these points here, but just collect them together for ease of reference. Some of them will be used and developed further in this chapter.

Key points about orbital angular momentum

1. In classical physics, the orbital angular momentum of a particle relative to a fixed origin is defined by

$$\mathbf{L} = \mathbf{r} \times \mathbf{p} = (yp_z - zp_y)\,\mathbf{e}_x + (zp_x - xp_z)\,\mathbf{e}_y + (xp_y - yp_x)\,\mathbf{e}_z,$$

where $\mathbf{r}$ and $\mathbf{p}$ are the instantaneous position and momentum of the particle.

2. In quantum mechanics, L_x, L_y and L_z are represented by linear Hermitian operators, obtained by making the usual operator substitutions, $p_x \Longrightarrow \widehat{p}_x = -i\hbar\partial/\partial x$, etc. This gives

$$\widehat{L}_x = -i\hbar\left[y\frac{\partial}{\partial z} - z\frac{\partial}{\partial y}\right] \tag{1.1a}$$

$$\widehat{L}_y = -i\hbar\left[z\frac{\partial}{\partial x} - x\frac{\partial}{\partial z}\right] \tag{1.1b}$$

$$\widehat{L}_z = -i\hbar\left[x\frac{\partial}{\partial y} - y\frac{\partial}{\partial x}\right]. \tag{1.1c}$$

3. In spherical coordinates, $\widehat{L}_z$ can be expressed as

$$\widehat{L}_z = -i\hbar\frac{\partial}{\partial \phi}. \tag{1.2}$$

The eigenvalues and normalized eigenfunctions of $\widehat{L}_z$ are given by

$$\widehat{L}_z \frac{1}{\sqrt{2\pi}}e^{im\phi} = -i\hbar\frac{\partial}{\partial \phi}\frac{1}{\sqrt{2\pi}}e^{im\phi} = m\hbar\frac{1}{\sqrt{2\pi}}e^{im\phi}, \tag{1.3}$$

where m is the **magnetic quantum number**, which takes integer values (positive, negative or zero): $m = 0, \pm 1, \pm 2, \ldots$.

The magnetic quantum number is sometimes called the *azimuthal quantum number*.

4. The square of the magnitude of the orbital angular momentum is represented by the operator

$$\widehat{L}^2 = \widehat{L}_x^2 + \widehat{L}_y^2 + \widehat{L}_z^2. \tag{1.4}$$

This operator has eigenvalues $l(l+1)\hbar^2$, where l is the **orbital angular momentum quantum number**, which takes non-negative integer values: $l = 0, 1, 2, \ldots$.

5. The orbital angular momentum operators satisfy the commutation relations:

$$\left[\widehat{L}_x, \widehat{L}_y\right] = i\hbar\widehat{L}_z, \quad \left[\widehat{L}_y, \widehat{L}_z\right] = i\hbar\widehat{L}_x, \quad \left[\widehat{L}_z, \widehat{L}_x\right] = i\hbar\widehat{L}_y \tag{1.5}$$

and so do not commute with one another. This makes it impossible to find a set of simultaneous eigenfunctions, with non-zero eigenvalues, for any pair of $\widehat{L}_x$, $\widehat{L}_y$ and $\widehat{L}_z$. It is therefore impossible to find a state in which any pair of L_x, L_y and L_z have definite non-zero values, and the observables L_x, L_y and L_z are said to be *incompatible*.

6. The operator $\widehat{L}^2$ commutes with $\widehat{L}_x$, $\widehat{L}_y$ and $\widehat{L}_z$, so it is possible to find functions that are simultaneous eigenfunctions of $\widehat{L}^2$ and any *one* of $\widehat{L}_x$, $\widehat{L}_y$ and $\widehat{L}_z$. Normally we consider simultaneous eigenfunctions of $\widehat{L}^2$ and $\widehat{L}_z$. In Dirac notation, the eigenvectors are generally labelled by the quantum numbers l and m and obey the eigenvalue equations

$$\widehat{L}^2 |l, m\rangle = l(l+1)\hbar^2 |l, m\rangle \tag{1.6}$$

$$\widehat{L}_z |l, m\rangle = m\hbar |l, m\rangle. \tag{1.7}$$

These eigenvectors describe states in which both L^2 and L_z have definite values, so the observables L^2 and L_z are said to be *compatible*. For a fixed value of l, the possible values of m are $0, \pm 1, \ldots \pm l$, giving $(2l+1)$ values of m for each value of l.

7. When a particle interacts only via a spherically-symmetric potential energy function, $V(r)$, its Hamiltonian operator $\widehat{H}$, and the orbital angular momentum operators $\widehat{L}^2$ and $\widehat{L}_z$, form a mutually-commuting set. It is then possible to find a set of functions that are simultaneous eigenfunctions of all three operators. These eigenfunctions describe states in which the system has definite values of energy, the magnitude of orbital angular momentum and the z-component of orbital angular momentum.

1.2 Quantum mechanics in three dimensions

A simple model of a hydrogen atom treats the proton as being at rest, attracting the electron according to Coulomb's electrostatic force law. The electrostatic potential energy associated with this force is

$$V = -\frac{1}{4\pi\varepsilon_0} \frac{e^2}{r}, \tag{1.8}$$

ε_0 is a fundamental physical constant called the *permittivity of free space*.

where r is the distance between the electron and the proton, and the energy zero has been chosen to be at $r = \infty$, corresponding to the electron and proton being infinitely-far apart.

In this section, we shall develop a form of the time-independent Schrödinger equation that describes this situation. We shall, however, make a generalization which will be useful in later chapters. Rather than considering the Coulomb interaction between an electron and a proton, we shall consider the potential energy function

$$V = \frac{1}{4\pi\varepsilon_0} \frac{qQ}{r}, \tag{1.9}$$

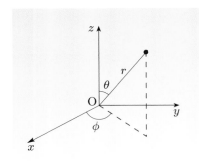

Figure 1.1 Representing the position of a point in spherical coordinates. Here, r is the **radial coordinate**, θ is the **polar angle** and ϕ is the **azimuthal angle**. Both θ and ϕ are measured in radians and they lie in the ranges $0 \le \theta \le \pi$ and $0 \le \phi \le 2\pi$.

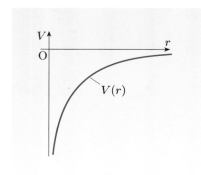

Figure 1.2 The attractive Coulomb potential energy well is infinitely deep at $r = 0$ and does not depend on the angles θ and ϕ.

which describes the Coulomb interaction between charges Q and q. We shall regard the charge Q as being fixed at the origin, while a particle of mass m and charge q moves subject to the potential energy function in Equation 1.9. Clearly, if we put $Q = e$ and $q = -e$ in this equation, we recover the potential energy function for a hydrogen atom.

The potential energy function in Equation 1.9 describes a potential energy well in three dimensions. The three-dimensional nature of the well becomes clear if we remember that $r = \sqrt{x^2 + y^2 + z^2}$ is the distance of a point from the origin, so V is a function of x, y and z, and could be written as $V(x, y, z)$. For most purposes, however, it is better to use the **spherical coordinates** r, θ and ϕ shown in Figure 1.1. In this case, the potential energy function depends only on the radial coordinate, r, and we simply write $V(r)$. Because this potential energy function is independent of the angular coordinates θ and ϕ, it is said to be **spherically-symmetric**. Figure 1.2 shows how $V(r)$ depends on r.

● In classical physics, it can be shown that any spherically-symmetric potential energy function corresponds to a central force. What is a central force, and what significance do central forces have for angular momentum?

○ A central force is one that points directly towards, or directly away from a fixed point (in this case, the origin) and it has a magnitude that only depends on the distance from the fixed point. Any particle subject to a central force has an angular momentum that is conserved.

Referring back to Point 7 in Section 1.1, we note that the spherically-symmetric nature of $V(r)$ is highly significant in quantum mechanics. It tells us that the energy eigenfunctions can be chosen so that they are simultaneously eigenfunctions of $\widehat{L}^2$ and $\widehat{L}_z$. This means that the stationary states of the system can be labelled by their energy, and also by the quantum numbers l and m that determine the values of L^2 and L_z. Much of this chapter is devoted to showing explicitly how this comes about, by writing down the time-independent Schrödinger equation and investigating the general form of its solutions. The argument will be similar to that given in Section 2.5.4 of Book 2, which considered a two-dimensional model of an 'atom'; here we use a more realistic three-dimensional model.

1.2.1 The time-independent Schrödinger equation

We now write down the time-independent Schrödinger equation for a particle of mass m moving in three dimensions subject to the potential energy function of Equation 1.9. Following the general procedure of Book 1, Chapter 2, we first write down the Hamiltonian function of the system in Cartesian coordinates. This is the sum of kinetic energy and potential energy terms:

$$H = \frac{1}{2m}(p_x^2 + p_y^2 + p_z^2) + V(x, y, z). \tag{1.10}$$

We then derive the Hamiltonian operator $\widehat{H}$ for the system by substituting $p_x \Longrightarrow \widehat{p}_x = -i\hbar \, \partial/\partial x$, and making similar substitutions for p_y and p_z, to get

$$\widehat{H} = -\frac{\hbar^2}{2m}\left[\frac{\partial^2}{\partial x^2} + \frac{\partial^2}{\partial y^2} + \frac{\partial^2}{\partial z^2}\right] + V(x, y, z). \tag{1.11}$$

The time-independent Schrödinger equation is obtained by substituting this Hamiltonian operator into the energy eigenvalue equation

$$\widehat{H}\psi(x,y,z) = E\psi(x,y,z).$$

The energy eigenvalues, E, are the allowed energies of the system, and the eigenfunctions describe states of definite energy. The eigenfunctions are subject to conditions of continuity and finiteness. They are interpreted, as usual, according to Born's rule: $|\psi(x,y,z)|^2 \, \delta x \, \delta y \, \delta z$ is the probability of finding the particle in a small volume element $\delta x \, \delta y \, \delta z$, centred on the point (x,y,z).

This may sound straightforward, but let's consider what is involved. Writing out the time-independent Schrödinger equation in Cartesian coordinates, and using Equation 1.9, we obtain

$$-\frac{\hbar^2}{2m}\left[\frac{\partial^2 \psi(x,y,z)}{\partial x^2} + \frac{\partial^2 \psi(x,y,z)}{\partial y^2} + \frac{\partial^2 \psi(x,y,z)}{\partial z^2}\right]$$
$$+ \frac{1}{4\pi\varepsilon_0}\frac{qQ}{\sqrt{x^2+y^2+z^2}}\psi(x,y,z) = E\,\psi(x,y,z). \tag{1.12}$$

The key point to note is that, unlike the case of the three-dimensional infinite square well, this equation does *not* separate: we *cannot* write $\psi(x,y,z) = X(x)Y(y)Z(z)$ and solve three separate one-dimensional equations. The problem is that the potential energy term contains the expression $\sqrt{x^2+y^2+z^2}$, which involves all three coordinates and prevents such a separation. This last expression is, of course, simply r, suggesting that we should turn to spherical coordinates. Note that Cartesian coordinates are the essential starting point for *writing down* the time-independent Schrödinger equation through the transformation from p_x to $\widehat{p}_x$. However, now that we have established the equation in Cartesian coordinates, we are free to change coordinates to obtain an equation that is easier to solve.

1.2.2 Transforming to spherical coordinates

We first introduce a convenient definition: the **Laplacian operator** ∇^2 is defined as

$$\nabla^2 = \frac{\partial^2}{\partial x^2} + \frac{\partial^2}{\partial y^2} + \frac{\partial^2}{\partial z^2}. \tag{1.13}$$

In terms of this operator, the time-independent Schrödinger equation becomes

$$-\frac{\hbar^2}{2m}\nabla^2\psi + V\psi = E\psi, \tag{1.14}$$

where we have omitted the arguments of ψ and V for simplicity (and also because they are about to change). Since the potential energy is most naturally expressed in spherical coordinates, it is sensible to recast Equation 1.14 in terms of these coordinates. The main task is that of expressing the Laplacian operator ∇^2 in terms of r, θ and ϕ. This task was carried out by mathematicians long before quantum mechanics; we simply quote the result. When the Laplacian operator acts on any scalar function $f(r,\theta,\phi)$, the following identity applies:

$$\nabla^2 f = \frac{1}{r^2}\frac{\partial}{\partial r}\left(r^2 \frac{\partial f}{\partial r}\right) + \frac{1}{r^2 \sin\theta}\frac{\partial}{\partial \theta}\left(\sin\theta \frac{\partial f}{\partial \theta}\right) + \frac{1}{r^2 \sin^2\theta}\frac{\partial^2 f}{\partial \phi^2}. \tag{1.15}$$

You are not expected to remember the form of ∇^2 in spherical coordinates.

This equation is not as complicated as it might appear, since a wonderful coincidence will soon be revealed. It turns out that the last two terms have a special significance, and we shall denote them by $\widehat{X}f$ where $\widehat{X}$ is the operator:

$$\widehat{X} = \frac{1}{r^2 \sin\theta} \left[\frac{\partial}{\partial \theta} \left(\sin\theta \frac{\partial}{\partial \theta} \right) + \frac{1}{\sin\theta} \frac{\partial^2}{\partial \phi^2} \right]. \tag{1.16}$$

The Laplacian operator then becomes

$$\nabla^2 = \frac{1}{r^2} \frac{\partial}{\partial r} \left(r^2 \frac{\partial}{\partial r} \right) + \widehat{X}, \tag{1.17}$$

and the time-independent Schrödinger equation for a particle of mass m, charge q, in the field of a charge Q at the origin, can be written as

$$-\frac{\hbar^2}{2m} \left[\frac{1}{r^2} \frac{\partial}{\partial r} \left(r^2 \frac{\partial \psi}{\partial r} \right) + \widehat{X}\psi \right] + \frac{1}{4\pi\varepsilon_0} \frac{qQ}{r} \psi = E\psi. \tag{1.18}$$

Comparing Equation 1.18 with Equation 1.12, you can see that the conversion to spherical coordinates has simplified the potential energy term at the expense of complicating the kinetic energy term. You might wonder whether any progress has been made. You will soon see that it has, because Equation 1.18 is *separable*, and this will allow us to obtain separate differential equations for the r, θ and ϕ-dependences.

Born's rule and normalization in spherical coordinates

The solutions of Equation 1.18 are the energy eigenfunctions $\psi(r, \theta, \phi)$, expressed in spherical coordinates. As usual, these eigenfunctions are interpreted using Born's rule: the probability of finding the particle in a small volume element δV, centred on a point with spherical coordinates (r, θ, ϕ), is $|\psi(r, \theta, \phi)|^2 \, \delta V$. To make this result more useful, we need to express the small volume element δV in spherical coordinates. Using the geometric argument shown in Figure 1.3, it can be shown that

$$\delta V = r^2 \sin\theta \, \delta r \, \delta\theta \, \delta\phi.$$

So the probability of finding the particle in a small volume element in which the spherical coordinates range from (r, θ, ϕ) to $(r + \delta r, \theta + \delta\theta, \phi + \delta\phi)$ is

$$\text{probability} = |\psi(r, \theta, \phi)|^2 \, r^2 \sin\theta \, \delta r \, \delta\theta \, \delta\phi. \tag{1.19}$$

Since the particle is certain to be found somewhere, the expression in Equation 1.19, when integrated over all space, must be equal to 1. The whole of space is mapped out when r varies between zero and infinity, θ varies between 0 and π, and ϕ varies between 0 and 2π. So, in spherical coordinates, the normalization condition is expressed as

$$\int_{\phi=0}^{\phi=2\pi} \int_{\theta=0}^{\theta=\pi} \int_{r=0}^{r=\infty} |\psi(r, \theta, \phi)|^2 \, r^2 \sin\theta \, \mathrm{d}r \, \mathrm{d}\theta \, \mathrm{d}\phi = 1. \tag{1.20}$$

Here, we have attached the variables θ, ϕ and r to the limits on the integral signs, but we shall not always do this. The general rule is that we work from the inside outwards; the inner integral is over r when the inner infinitesimal element of integration is $\mathrm{d}r$, and so on.

1.2 Quantum mechanics in three dimensions

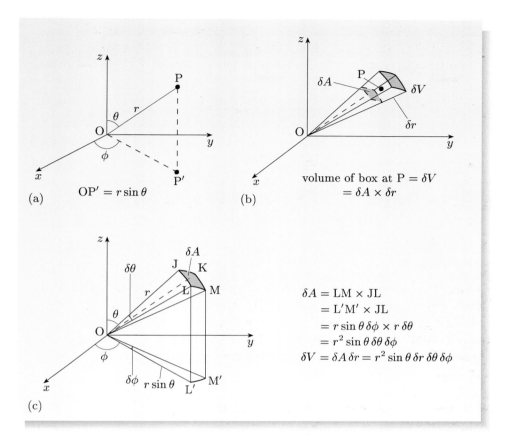

Figure 1.3 A small volume element around a point P expressed in spherical coordinates. (a) The point P has spherical coordinates (r, θ, ϕ). (b) The volume element δV between the two shaded surfaces is the product of the area δA and the length δr. (c) The area δA is the product of two lengths: $\text{LM} = r \sin\theta\, \delta\phi$ and $\text{JL} = r\, \delta\theta$. The final working for the volume element δV is shown in the bottom right of the figure.

Exercise 1.1

(a) What are the dimensions of $\psi(r, \theta, \phi)$?

(b) What are the dimensions of the Laplacian operator and of the operator $\widehat{X}$? ■

1.2.3 Making the link with angular momentum

Apart from a multiplicative factor, it turns out that the operator $\widehat{X}$ is none other than the operator $\widehat{L}^2 = \widehat{L}_x^2 + \widehat{L}_y^2 + \widehat{L}_z^2$ that represents the square of the orbital angular momentum. We shall not give a detailed derivation of this fact, but just sketch the reasoning that is involved. First, recall that $\widehat{L}_z$ can be expressed in spherical coordinates as

$$\widehat{L}_z = -i\hbar \frac{\partial}{\partial \phi}. \tag{1.21}$$

We derived this formula in Chapter 2 of Book 2. With more effort we can also express $\widehat{L}_x$ and $\widehat{L}_y$ in spherical coordinates. The results are:

$$\widehat{L}_x = -i\hbar \left[-\sin\phi \frac{\partial}{\partial \theta} - \cot\theta \cos\phi \frac{\partial}{\partial \phi} \right], \tag{1.22a}$$

$$\widehat{L}_y = -i\hbar \left[\cos\phi \frac{\partial}{\partial \theta} - \cot\theta \sin\phi \frac{\partial}{\partial \phi} \right]. \tag{1.22b}$$

You are not expected to remember these equations.

Now we can use these results to express $\widehat{L}^2$ in spherical coordinates. First, note that $\widehat{L}^2$ is an operator which acts on an arbitrary function $f(\theta, \phi)$. We therefore consider the identity

$$\widehat{L}^2 f = \widehat{L}_x^2 f + \widehat{L}_y^2 f + \widehat{L}_z^2 f,$$

and substitute Equations 1.21 and 1.22 into the right-hand side. Some care is needed because, for example, the formula for $\widehat{L}_x$ involves $\cot\theta$ which must itself be differentiated when we evaluate $\widehat{L}_x^2 f = \widehat{L}_x \widehat{L}_x f$. After some work, it turns out that

$$\widehat{L}^2 f = -\hbar^2 \left[\frac{1}{\sin\theta} \frac{\partial}{\partial\theta} \left(\sin\theta \frac{\partial f}{\partial\theta} \right) + \frac{1}{\sin^2\theta} \frac{\partial^2 f}{\partial\phi^2} \right]. \qquad (1.23)$$

Since this equation is true for arbitrary f, we can write it as a relationship between operators:

$$\widehat{L}^2 = -\hbar^2 \left[\frac{1}{\sin\theta} \frac{\partial}{\partial\theta} \left(\sin\theta \frac{\partial}{\partial\theta} \right) + \frac{1}{\sin^2\theta} \frac{\partial^2}{\partial\phi^2} \right]. \qquad (1.24)$$

Comparing this expression with Equation 1.16 we now see that

$$\widehat{X} = -\frac{\widehat{L}^2}{\hbar^2 r^2}.$$

So the time-independent Schrödinger equation (Equation 1.18) can finally be written in the form

$$-\frac{\hbar^2}{2m} \frac{1}{r^2} \frac{\partial}{\partial r} \left(r^2 \frac{\partial \psi}{\partial r} \right) + \frac{\widehat{L}^2}{2mr^2} \psi + V(r)\psi = E\psi, \qquad (1.25)$$

where we are especially interested in the case $V(r) = qQ/4\pi\varepsilon_0 r$. Equation 1.25 can also be written as $\widehat{H}\psi = E\psi$, where the Hamiltonian operator in spherical coordinates is

$$\widehat{H} = -\frac{\hbar^2}{2m} \frac{1}{r^2} \frac{\partial}{\partial r} \left(r^2 \frac{\partial}{\partial r} \right) + \frac{\widehat{L}^2}{2mr^2} + V(r). \qquad (1.26)$$

We said earlier that a wonderful coincidence would occur, and here it is: the Hamiltonian operator involves the magnitude of the orbital angular momentum in the simple way revealed in Equation 1.26. It is perhaps naive to talk about coincidences in physics. In classical mechanics, it can be shown that

$$\frac{1}{2m}(p_x^2 + p_y^2 + p_z^2) = \frac{p_r^2}{2m} + \frac{L^2}{2mr^2}, \qquad (1.27)$$

where p_r is the radial component of the momentum and L is the magnitude of the orbital angular momentum, so Equation 1.26 may not be so coincidental after all!

The classical expression in Equation 1.27 is so like the first two terms in Equation 1.26 that you might wonder why we started with Cartesian coordinates in the first place. The reason is that Cartesian coordinates provide the only safe way of converting a classical Hamiltonian function into a quantum Hamiltonian operator. The radial momentum p_r is not expressed in Cartesian coordinates, and there are no simple rules for finding the corresponding operator in quantum mechanics.

1.2.4 Energy eigenfunctions and how to label them

It is a general principle that, when a particle is subject only to a spherically-symmetric potential energy function $V(r)$, then its Hamiltonian operator $\widehat{H}$, and the orbital angular momentum operators $\widehat{L}^2$ and $\widehat{L}_z$, form a mutually-commuting set. Using the explicit operators given above, we can now verify that this statement is true.

Exercise 1.2 Use Equations 1.21, 1.24 and 1.26 to verify that each pair of $\widehat{L}_z$, $\widehat{L}^2$ and $\widehat{H}$ commute. ∎

Given a mutually-commuting set of operators, we can find a set of functions that are simultaneous eigenfunctions of all the operators in the set. In the present context, this means that we can arrange for the eigenfunctions of energy to be also eigenfunctions of $\widehat{L}^2$ and $\widehat{L}_z$. Let us go back to the time-independent Schrödinger equation in spherical coordinates (Equation 1.25) and see exactly how this works.

The key feature of Equation 1.25 is that it is separable. This means that we can look for solutions in the product form:

$$\psi(r, \theta, \phi) = R(r) Y(\theta, \phi).$$

When such a function is substituted into Equation 1.25, and the method of separation of variables is used, we get two equations, one for $R(r)$ and the other for $Y(\theta, \phi)$, linked by a separation constant K:

$$-\frac{\hbar^2}{2m} \frac{1}{r^2} \frac{d}{dr}\left(r^2 \frac{dR(r)}{dr}\right) + \frac{K}{2mr^2} R(r) + V(r) R(r) = E R(r) \quad (1.28)$$

$$\widehat{L}^2 Y(\theta, \phi) = K Y(\theta, \phi). \quad (1.29)$$

Now, the second of these equations is just the eigenvalue equation for $\widehat{L}^2$. This means that $Y(\theta, \phi)$ is an eigenfunction of $\widehat{L}^2$, and K is the corresponding eigenvalue.

- What are the allowed values of K?
- According to Point 4 in Section 1.1, the allowed values of K are $l(l+1)\hbar^2$, where $l = 0, 1, 2, \ldots$ is the orbital angular momentum quantum number.

Substituting these allowed values of K back into Equation 1.28, we obtain

$$-\frac{\hbar^2}{2m} \frac{1}{r^2} \frac{d}{dr}\left(r^2 \frac{dR(r)}{dr}\right) + \frac{l(l+1)\hbar^2}{2mr^2} R(r) + V(r) R(r) = E R(r). \quad (1.30)$$

The solutions of this equation give $R(r)$, the radial part of the energy eigenfunction. We therefore see that $R(r)$ depends on the orbital angular momentum quantum number, l. For a given value of l, there may be several different solutions, and we can distinguish these with a second index, n. So the radial part of the wave function can be written as $R_{nl}(r)$, with corresponding energy eigenvalue E_{nl}.

In some special cases, states with the same value of n and different values of l may be degenerate, as you will see in Chapter 2.

The term $l(l+1)\hbar^2/2mr^2$ in Equation 1.30 arises from the kinetic energy terms in the Hamiltonian operator, but it plays the role of an effective potential energy.

It is sometimes called the **centrifugal barrier** since it has the effect of repelling particles away from $r = 0$, with a strength that increases with increasing angular momentum. This term plays an important role in atoms, as it tends to keep electrons with high angular momenta away from the nucleus.

Returning to Equation 1.29, different eigenvalues have different eigenfunctions, so $Y(\theta, \phi)$ must depend on l. That is not all, however. We know that $\widehat{L}^2$ and $\widehat{L}_z$ commute with one another, so we can choose each $Y(\theta, \phi)$ to be an eigenfunction of $\widehat{L}_z$. The eigenvalues of L_z are given by $m\hbar$, where $m = 0, \pm 1, \ldots, \pm l$ is the magnetic quantum number. So, a complete specification of $Y(\theta, \phi)$ involves two quantum numbers, l and m, and the angular part of the eigenfunction is written as $Y_{lm}(\theta, \phi)$. Functions of this type are called **spherical harmonics**; we shall discuss them further in the next section.

Putting the radial and angular parts together, the energy eigenfunctions, which are also eigenfunctions of the magnitude, and the z-component, of the orbital angular momentum, are written as

$$\psi_{nlm}(r, \theta, \phi) = R_{nl}(r) Y_{lm}(\theta, \phi).$$

This function describes a state in which the energy, the magnitude of orbital angular momentum and the z-component of orbital angular momentum all have definite values, given by the eigenvalues in the equations:

$$\widehat{H}\psi_{nlm}(r, \theta, \phi) = E_{nl}\,\psi_{nlm}(r, \theta, \phi)$$

$$\widehat{L}^2 \psi_{nlm}(r, \theta, \phi) = l(l+1)\hbar^2\,\psi_{nlm}(r, \theta, \phi)$$

$$\widehat{L}_z \psi_{nlm}(r, \theta, \phi) = m\hbar\,\psi_{nlm}(r, \theta, \phi).$$

In general, the fact that L_z has a definite value means that L_x and L_y have indefinite values, with one exceptional case that will be discussed shortly.

It is worth noting that the radial part of the solution, $R_{nl}(r)$, depends on the precise potential energy function, $V(r)$, but the angular part does not. Whenever we have a spherically-symmetric potential energy function, and L^2 and L_z have given values, the angular part of the solution is always the same.

Exercise 1.3 Write all combinations of possible quantum numbers l and m corresponding to eigenvalues of $\widehat{L}^2$ that are less than or equal to $20\hbar^2$. ■

1.3 Spherical harmonics

1.3.1 The form of the spherical harmonics

Spherical harmonics are well-behaved functions of θ and ϕ that are eigenfunctions of both $\widehat{L}^2$ and $\widehat{L}_z$, with eigenvalues $l(l+1)\hbar^2$ and $m\hbar$, respectively. We denoted these functions by $|l, m\rangle$ in Book 2 and Section 1.1, and by $Y_{lm}(\theta, \phi)$ in Section 1.2.4. Now that we have operators for $\widehat{L}^2$ and $\widehat{L}_z$ in spherical coordinates, we can investigate the explicit forms of these functions, starting from scratch and making no special assumptions about the form of the eigenfunctions or eigenvalues.

In this context, 'well-behaved' means single-valued and finite.

Combining Equations 1.24 and 1.29, the eigenvalue equation for $\widehat{L}^2$ can be written as

$$-\left[\frac{1}{\sin\theta}\frac{\partial}{\partial\theta}\left(\sin\theta\frac{\partial}{\partial\theta}\right) + \frac{1}{\sin^2\theta}\frac{\partial^2}{\partial\phi^2}\right]Y(\theta,\phi) = \frac{K}{\hbar^2}Y(\theta,\phi).$$

We can look for solutions in the form of a product of a function of θ and a function of ϕ:

$$Y(\theta,\phi) = \Theta(\theta)F(\phi). \tag{1.31}$$

Θ is the Greek capital letter theta.

The usual method of separation of variables then leads to two ordinary differential equations for $\Theta(\theta)$ and $F(\phi)$. The differential equation for $F(\phi)$ is

$$\frac{d^2 F(\phi)}{d\phi^2} = -m^2 F(\phi), \tag{1.32}$$

where m^2 is a separation constant (written in this form for reasons that will soon become apparent). Equation 1.32 is satisfied by functions of the form

$$F(\phi) = e^{im\phi}, \tag{1.33}$$

where m could, *at this stage*, be any real or complex number. However, we can insist that $F(\phi)$ should return to the same value when ϕ increases by 2π. This single-valuedness condition leads to the requirement that

$$e^{im(\phi+2\pi)} = e^{im\phi}e^{im2\pi} = e^{im\phi},$$

and this tells us that m must be real and must be an integer: $m = 0, \pm 1, 2, \ldots$. So, the method of separation of variables automatically produces eigenfunctions of $\widehat{L}_z$:

$$\widehat{L}_z e^{im\phi} = -i\hbar\frac{\partial}{\partial\phi}e^{im\phi} = m\hbar e^{im\phi},$$

with the expected eigenvalues, $0, \pm\hbar, \pm 2\hbar$, etc.

Now let us turn to the equation for $\Theta(\theta)$, which takes the form

$$\left[-\frac{1}{\sin\theta}\frac{d}{d\theta}\left(\sin\theta\frac{d}{d\theta}\right) + \frac{m^2}{\sin^2\theta}\right]\Theta(\theta) = l(l+1)\Theta(\theta), \tag{1.34}$$

where m is the integer that appears in the solution for $F(\phi)$, and we have written $K/\hbar^2$ as $l(l+1)$; there is no loss of generality in doing this, provided we defer making any special assumptions about l.

It would be a lengthy detour to derive the general solutions of Equation 1.34, and we shall not attempt to do so. Instead, we rely on the fact that the solutions of this equation were extensively studied by mathematicians long before the advent of quantum mechanics. The main points to note are:

- $\Theta(\theta)$ diverges as $\theta \to 0$ and $\theta \to \pi$ *unless* l is a positive integer greater than or equal to the integer $|m|$.
- For each value of $l = 0, 1, 2, \ldots$ and $|m| = 0, 1, 2, \ldots l$, we get a solution $\Theta_{lm}(\theta)$ which remains finite for all values of θ.
- $\Theta_{lm}(\theta)$ is a sum of integer powers of $\sin\theta$ and $\cos\theta$. The precise combination depends on l and $|m|$, but does not depend on the sign of m.

Apart from normalization factors, $\Theta_{lm}(\theta)$ is what mathematicians call an *associated Legendre function* with argument $\cos\theta$.

In physical terms, l and m are the orbital angular momentum quantum number and the magnetic quantum number. These mathematical results show, among other things, why l is a non-negative integer, and why the magnitude of m is less than or equal to l.

It is instructive to see how the first few solutions of Equation 1.34 arise. Let us start by supposing that there is a solution that is a constant. We can test this by substituting $\Theta(\theta) = \text{constant}$ into Equation 1.34. We then get the equation

$$\frac{m^2}{\sin^2\theta} = l(l+1). \tag{1.35}$$

The only way of satisfying this equation for all θ is to set $l = m = 0$. So a constant solution is possible *if and only* if $l = m = 0$. Apart from a normalizing constant, we have found the $l = 0$ and $m = 0$ spherical harmonic; we shall see in the next chapter that this constant function leads to a very simple shape for the ground-state eigenfunction in a hydrogen atom.

Since Equation 1.34 contains trigonometrical factors, we might guess that another possible solution would be $\Theta_{lm} = \sin\theta$. We substitute this in Equation 1.34 and find

$$-\frac{1}{\sin\theta}\left(\cos^2\theta - \sin^2\theta\right) + \frac{m^2}{\sin\theta} = l(l+1)\sin\theta.$$

In other words,

$$-\cos^2\theta + \sin^2\theta + m^2 = l(l+1)\sin^2\theta,$$

which, using $\cos^2\theta + \sin^2\theta = 1$, can be rewritten as

$$2\sin^2\theta - 1 + m^2 = l(l+1)\sin^2\theta. \tag{1.36}$$

This last equation holds for all values of θ if and only if $l = 1$ and $m^2 = 1$. So we have found (apart from the normalizing factors) the solutions for $l = 1, m = -1$ and $l = 1, m = +1$. These will turn up later in the description of the first excited states of a hydrogen atom.

Exercise 1.4 For what values of l and m is $\cos\theta$ a solution of Equation 1.34? ∎

Combining Equations 1.31 and 1.33, we see that the complete spherical harmonics are of the form

$$Y_{lm}(\theta, \phi) = \Theta_{lm}(\theta)\, e^{im\phi}. \tag{1.37}$$

Although we write Y_{lm} for the general case, it is helpful to separate the orbital and magnetic quantum numbers with a comma if l and m are specific numbers, e.g. $Y_{3,-2}$ or $Y_{10,9}$.

Table 1.1 displays all the spherical harmonics for $l \leq 3$; there is no upper limit to the value of l for which spherical harmonics exist. The numerical prefactors are normalization constants; we shall discuss the normalization of spherical harmonics shortly.

Table 1.1 Normalized spherical harmonics for $l \leq 3$. Note that all the ϕ-dependence is in the factor $e^{im\phi}$.

l	m	$Y_{lm}(\theta, \phi)$
0	0	$\frac{1}{\sqrt{4\pi}}$
1	0	$\sqrt{\frac{3}{4\pi}} \cos\theta$
1	-1	$\sqrt{\frac{3}{8\pi}} \sin\theta \, e^{-i\phi}$
1	$+1$	$-\sqrt{\frac{3}{8\pi}} \sin\theta \, e^{+i\phi}$
2	0	$\sqrt{\frac{5}{16\pi}} (3\cos^2\theta - 1)$
2	-1	$\sqrt{\frac{15}{8\pi}} \cos\theta \sin\theta \, e^{-i\phi}$
2	$+1$	$-\sqrt{\frac{15}{8\pi}} \cos\theta \sin\theta \, e^{+i\phi}$
2	-2	$\sqrt{\frac{15}{32\pi}} \sin^2\theta \, e^{-2i\phi}$
2	$+2$	$\sqrt{\frac{15}{32\pi}} \sin^2\theta \, e^{+2i\phi}$
3	0	$\sqrt{\frac{7}{16\pi}} (2\cos^3\theta - 3\cos\theta \sin^2\theta)$
3	-1	$\sqrt{\frac{21}{64\pi}} (4\cos^2\theta \sin\theta - \sin^3\theta) \, e^{-i\phi}$
3	$+1$	$-\sqrt{\frac{21}{64\pi}} (4\cos^2\theta \sin\theta - \sin^3\theta) \, e^{+i\phi}$
3	-2	$\sqrt{\frac{105}{32\pi}} \cos\theta \sin^2\theta \, e^{-2i\phi}$
3	$+2$	$\sqrt{\frac{105}{32\pi}} \cos\theta \sin^2\theta \, e^{+2i\phi}$
3	-3	$\sqrt{\frac{35}{64\pi}} \sin^3\theta \, e^{-3i\phi}$
3	$+3$	$-\sqrt{\frac{35}{64\pi}} \sin^3\theta \, e^{+3i\phi}$

Exercise 1.5 (a) Verify that the first two columns of Table 1.1 contain all the allowed combinations of quantum numbers for spherical harmonics with $l \leq 3$.

(b) Identify in Table 1.1 those spherical harmonics that contain as a factor the functions of θ we studied above.

(c) Write out the spherical harmonic that corresponds to $L^2 = 6\hbar^2$ and $L_z = -\hbar$. ∎

The exceptional case

The exceptional case in which $\widehat{L}_x$, $\widehat{L}_y$, $\widehat{L}_z$ and $\widehat{L}^2$ have a simultaneous eigenfunction, despite the fact that they do not all mutually commute, can now be revealed. The eigenfunction is $Y_{0,0}$, which, being independent of θ and ϕ, gives zero when operated on by the operators of Equations 1.21, 1.22, and 1.24. Hence $Y_{0,0}$ is an eigenfunction of all four operators, with eigenvalue 0 for each operator.

The orthonormality of spherical harmonics

To ensure that an eigenfunction $\psi_{nlm}(r, \theta, \phi) = R_{nl}(r) Y_{lm}(\theta, \phi)$ is normalized, we impose the normalization condition of Equation 1.20:

$$\int_0^{2\pi} \int_0^{\pi} \int_0^{\infty} |\psi_{nlm}(r, \theta, \phi)|^2 \, r^2 \sin\theta \, \mathrm{d}r \, \mathrm{d}\theta \, \mathrm{d}\phi = 1,$$

> Remember that multiple integrals are done 'from the inside out', so the integral with limits 0 and ∞ is over r.

and this implies that

$$\int_0^{\infty} |R_{nl}(r)|^2 \, r^2 \, \mathrm{d}r \times \int_0^{2\pi} \int_0^{\pi} |Y_{lm}(\theta, \phi)|^2 \sin\theta \, \mathrm{d}\theta \, \mathrm{d}\phi = 1.$$

We adopt a convention that requires each of the integrals in the above expression to be separately normalized. Thus, our normalization condition for spherical harmonics is

$$\int_0^{2\pi} \int_0^{\pi} |Y_{lm}(\theta, \phi)|^2 \sin\theta \, \mathrm{d}\theta \, \mathrm{d}\phi = 1, \qquad (1.38)$$

for any allowed values of l and m.

As well as being normalized it turns out that any spherical harmonics with different values of l or different values of m are *orthogonal* in the sense that

$$\int_0^{2\pi} \int_0^{\pi} Y_{l_1 m_1}^*(\theta, \phi) Y_{l_2 m_2}(\theta, \phi) \sin\theta \, \mathrm{d}\theta \, \mathrm{d}\phi = 0, \quad \text{if } l_1 \neq l_2 \text{ or } m_1 \neq m_2. \qquad (1.39)$$

Combining Equations 1.38 and 1.39, we have

$$\int_0^{2\pi} \int_0^{\pi} Y_{l_1 m_1}^*(\theta, \phi) Y_{l_2 m_2}(\theta, \phi) \sin\theta \, \mathrm{d}\theta \, \mathrm{d}\phi = \delta_{l_1 l_2} \, \delta_{m_1 m_2}, \qquad (1.40)$$

and we say that the spherical harmonics are *orthonormal*.

Exercise 1.6 Show explicitly, by integration, that:

(a) the spherical harmonics for $l = 0$, $m = 0$ and for $l = 1$, $m = 1$, as given in Table 1.1, are normalized;

(b) the same two spherical harmonics are orthogonal to each other. ∎

The parity of spherical harmonics

> The 1957 Nobel prize in physics was awarded to T-D. Lee and C. N. Yang for their important discoveries relating to the parity properties of fundamental particles.

You will recall from Book 1 that a function $\psi(x)$ is said to be *odd* if $\psi(-x) = -\psi(x)$ for all x, and it is said to be *even* if $\psi(-x) = \psi(x)$. This idea can be generalized in three dimensions. The transformation $x \longrightarrow -x$, $y \longrightarrow -y$, $z \longrightarrow -z$, in which all the coordinates are reflected through the origin, is called **inversion**. A function $f(x, y, z)$ is said to have **odd parity** (or parity -1) if it changes sign under inversion, and it is said to have **even parity** (or parity $+1$) if it is unchanged by inversion. In terms of spherical coordinates (with $r \geq 0$ always), reversing the signs of x, y and z corresponds to the transformation

$$\theta \longrightarrow \pi - \theta \quad \text{and} \quad \phi \longrightarrow \phi + \pi, \tag{1.41}$$

and a function $f(\theta, \phi)$ is said to have odd parity if it changes sign under this transformation, and even parity if it remains unchanged. It turns out that the spherical harmonic $Y_{lm}(\theta, \phi)$ has parity $(-1)^l$. So spherical harmonics with $l = 0, 2, 4, \ldots$ have even parity, and those with $l = 1, 3, 5, \ldots$ have odd parity.

Worked Example 1.1
Show explicitly that the spherical harmonics for $l = 1$ all have odd parity.

Essential skill

Determining the parity of a function of θ and ϕ

Solution
Table 1.1 shows that the three $l = 1$ spherical harmonics involve $\cos\theta$, $\sin\theta$ and $e^{im\phi}$. We have

$$\cos(\pi - \theta) = -\cos\theta$$

$$\sin(\pi - \theta) = \sin\theta$$

$$e^{im(\phi+\pi)} = e^{im\pi} e^{im\phi} = (-1)^m e^{im\phi} = \begin{cases} e^{im\phi} & \text{for } m = 0, \\ -e^{im\phi} & \text{for } m = \pm 1. \end{cases}$$

So, applying the parity transformation to the $l = 1$ spherical harmonics in Table 1.1, we see that each changes sign and therefore has odd parity.

Exercise 1.7 Show that all the $l = 2$ spherical harmonics have even parity.

Exercise 1.8 Show that $|Y_{lm}|^2$ is independent of ϕ for all l and m. ■

1.3.2 Visualizing spherical harmonics

Even for low values of the quantum number l, it is hard to visualize the spherical harmonics from the expressions in Table 1.1. It is easy to see that $Y_{0,0}$ represents a function that is the same for all angles; it becomes harder to visualize $Y_{2,2}$ because it is complex. Fortunately, for many purposes it is the real quantity $|Y_{lm}(\theta, \phi)|^2$ that is important. For $l = 2$ and $m = 2$, this quantity is zero for $\theta = 0$ or $\theta = \pi$, i.e. in the direction of the 'north pole' or the 'south pole', and takes its maximum value for $\theta = \pi/2$, i.e. in the 'equatorial plane' — see Figure 1.4, where $|Y_{2,2}(\theta, \phi)|^2$ is presented in two different ways.

The 'polar diagram' in Figure 1.4b has to be interpreted with care. In such a diagram, the magnitude of a quantity that depends upon angle is proportional to the length of an arrow drawn from the origin at that angle. The line traced out by the tip of the arrow as it covers all the angles shows how the magnitude of the quantity varies with the polar angle θ.

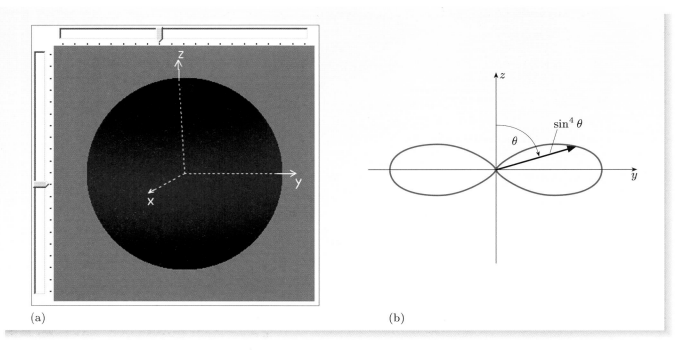

Figure 1.4 (a) A colour-scale representation of $|Y_{2,2}(\theta,\phi)|^2$ taken from the software package *Spherical harmonics*. The brightness of the blue colouring of the surface of the sphere represents $|Y_{2,2}(\theta,\phi)|^2$. (b) A 'polar diagram' in which the length of the arrow represents $|Y_{2,2}(\theta,\phi)|^2$. In this panel, the x-axis is pointing straight out of the paper. This diagram shows that $|Y_{2,2}(\theta,\phi)|^2$ is zero in the directions of the poles ($\theta = 0$ and $\theta = \pi$) and has its maximum value at all points in the equatorial plane, i.e. for $\theta = \pi/2$.

> **Computer simulation: Spherical harmonics**
> Now is the ideal time to study the software package *Spherical harmonics*. This is designed to help you visualize the spatial behaviour of both the complex $Y_{lm}(\theta,\phi)$ and the real $|Y_{lm}(\theta,\phi)|^2$.

Exercise 1.9 (a) All of the spherical harmonics have a modulus that is symmetric about one of the three Cartesian axes. Which axis?

(b) Are the spherical harmonics for $l = 1$ and $m = \pm 1$ equal to zero at the poles or on the equatorial plane? Where are their magnitudes greatest?

(c) All the spherical harmonics in Table 1.1 for which $m = \pm l$ share a common feature. What is it? ■

Another general property of the spherical harmonics should now become apparent, especially after you have studied the computer simulations: they have values of θ for which they are zero (with the exception of $Y_{0,0}$). For $m = \pm l$, the zeros are in the directions of the poles, but as $|m|$ decreases, there is a larger number of values of θ for which the value of Y_{lm} is zero.

1.4 Particles with spin in a potential energy well

It may not have escaped your attention that an electron bound in a potential energy well will have *both* orbital angular momentum *and* spin angular momentum. We will now see that these can couple together in a way that affects the energy levels and the labelling of energy eigenfunctions.

We start by reviewing the situation before spin makes an entrance. For a particle bound by a spherically-symmetric potential energy well, $V(r)$, the time-independent Schrödinger equation is $\widehat{H}\psi = E\psi$, where the Hamiltonian operator is given by

$$\widehat{H} = -\frac{\hbar^2}{2m}\frac{1}{r^2}\frac{\partial}{\partial r}\left(r^2\frac{\partial}{\partial r}\right) + \frac{\widehat{L}^2}{2mr^2} + V(r). \qquad \text{(Eqn 1.26)}$$

We have seen that $\widehat{H}$, $\widehat{L}^2$ and $\widehat{L}_z$ are a mutually-commuting set of operators, and this means that we can choose the energy eigenfunctions to be eigenfunctions of $\widehat{L}^2$ and $\widehat{L}_z$, labelled by the quantum numbers l and m.

It turns out that things are not as simple as we have so far implied. For example, the Coulomb interaction is *not* the only interaction felt by an electron in a hydrogen atom; there is an additional contribution to the energy called the **spin–orbit interaction**. This has the form $V_{\text{so}}(r)\,\mathbf{L}\cdot\mathbf{S}$, where $\mathbf{L}$ is the orbital angular momentum of the electron and $\mathbf{S}$ is its spin angular momentum.

In classical terms, the spin–orbit interaction can be understood as follows: a magnetic field arises as a result of the electron's motion around the nucleus, and the electron behaves like a tiny magnet in this field. The interaction is proportional to $\mathbf{L}\cdot\mathbf{S}$ because the magnetic field is proportional to $\mathbf{L}$ and the magnetic dipole moment of the electron is proportional to $\mathbf{S}$.

Including the spin–orbit interaction term in the Hamiltonian operator for a hydrogen atom gives

$$\widehat{H}_{\text{so}} = -\frac{\hbar^2}{2m}\frac{1}{r^2}\frac{\partial}{\partial r}\left(r^2\frac{\partial}{\partial r}\right) + \frac{\widehat{L}^2}{2mr^2} + V(r) + V_{\text{so}}(r)\,\widehat{\mathbf{L}}\cdot\widehat{\mathbf{S}}, \qquad (1.42)$$

where

$$\widehat{\mathbf{L}}\cdot\widehat{\mathbf{S}} = \widehat{L}_x\widehat{S}_x + \widehat{L}_y\widehat{S}_y + \widehat{L}_z\widehat{S}_z.$$

The operator $\widehat{L}^2$ commutes with this Hamiltonian operator because $\widehat{L}^2$ commutes with any function of r, and with each of the operators $\widehat{L}_x$, $\widehat{L}_y$ and $\widehat{L}_z$, and also with any spin operator. Similarly, $\widehat{S}^2$ commutes with both $\widehat{H}_{\text{so}}$ and with $\widehat{L}^2$. Since $\widehat{H}_{\text{so}}$, $\widehat{L}^2$ and $\widehat{S}^2$ form a mutually-commuting set of operators, we can choose the eigenfunctions of $\widehat{H}_{\text{so}}$ to be eigenfunctions of both $\widehat{L}^2$ and $\widehat{S}^2$, and this means that we can label them with the quantum numbers l and $s = 1/2$, for the orbital angular momentum and the spin of the electron.

By contrast, the operator $\widehat{L}_z$ does *not* commute with $\widehat{H}_{\text{so}}$ because it commutes with neither $\widehat{L}_x$ nor $\widehat{L}_y$. As a consequence, the eigenfunctions of $\widehat{H}_{\text{so}}$ are *not* eigenfunctions of $\widehat{L}_z$, and the energy eigenstates of the atom *cannot* be labelled by the magnetic quantum number, m. By a similar token, $\widehat{S}_z$ does not commute with

> The subscript 'so' stands for spin–orbit. The specific form of $V_{\text{so}}(r)$ will be given in a later chapter; the simple existence of $V_{\text{so}}(r)$ with a significant magnitude is all that matters here.

$\widehat{H}_{so}$, and the energy eigenstates *cannot* be labelled by the spin magnetic quantum number, m_s. As a result, our treatment of angular momentum to this point is incomplete. The new concept that is required is that of *total angular momentum*.

1.4.1 Total angular momentum

We define the **total angular momentum** of a spin-$\frac{1}{2}$ particle by the vector relation $\mathbf{J} = \mathbf{L} + \mathbf{S}$, meaning that the components add: for example, $J_z = L_z + S_z$. The quantum-mechanical operators representing the components of $\mathbf{J}$ also add:

$$\widehat{J}_x = \widehat{L}_x + \widehat{S}_x, \quad \widehat{J}_y = \widehat{L}_y + \widehat{S}_y, \quad \widehat{J}_z = \widehat{L}_z + \widehat{S}_z, \qquad (1.43)$$

and our first task is to see what this means. The operators $\widehat{J}_x$, $\widehat{J}_y$ and $\widehat{J}_z$ are each the sum of a differential operator (such as $\widehat{L}_x$) that acts on functions of θ and ϕ, and a 2×2 matrix (such as $\widehat{S}_x$) that acts on spinors. This may seem strange, but any operator is defined by its actions, so let us see what $\widehat{J}_x$, $\widehat{J}_y$ and $\widehat{J}_z$ do to state vectors.

For a spin-$\frac{1}{2}$ particle, the spin-up and spin-down states relative to the z-axis are

> In this book, $|\uparrow\rangle$ always represents a state that is spin-up with respect to the z-axis.

$$|\uparrow\rangle = \begin{bmatrix} 1 \\ 0 \end{bmatrix} \quad \text{and} \quad |\downarrow\rangle = \begin{bmatrix} 0 \\ 1 \end{bmatrix}, \qquad (1.44)$$

and we can construct state vectors that are products of spherical harmonics and spin state vectors, e.g.

$$Y_{lm}|\uparrow\rangle = \begin{bmatrix} Y_{lm} \\ 0 \end{bmatrix} \quad \text{or} \quad Y_{2,0}|\uparrow\rangle + Y_{2,1}|\downarrow\rangle = \begin{bmatrix} Y_{2,0} \\ Y_{2,1} \end{bmatrix}. \qquad (1.45)$$

When an operator such as $\widehat{L}_z$ acts on such a product, it operates as usual on the spherical harmonic, but ignores the spinor; likewise, a spin operator ignores the spherical harmonic, and any other spatial factor, and acts only on the spinor.

- Given that

$$\widehat{S}_z|\uparrow\rangle = +\tfrac{1}{2}\hbar|\uparrow\rangle, \quad \widehat{S}_z|\downarrow\rangle = -\tfrac{1}{2}\hbar|\downarrow\rangle, \quad \text{and} \quad \widehat{L}_z Y_{lm} = m\hbar Y_{lm},$$

what is the result of operating on $Y_{2,0}|\uparrow\rangle + Y_{2,1}|\downarrow\rangle$ with $\widehat{J}_z = \widehat{L}_z + \widehat{S}_z$?

- We have

$$\widehat{J}_z\bigl(Y_{2,0}|\uparrow\rangle + Y_{2,1}|\downarrow\rangle\bigr) = (\widehat{L}_z + \widehat{S}_z)Y_{2,0}|\uparrow\rangle + (\widehat{L}_z + \widehat{S}_z)Y_{2,1}|\downarrow\rangle$$
$$= (\widehat{L}_z Y_{2,0})|\uparrow\rangle + Y_{2,0}(\widehat{S}_z|\uparrow\rangle)$$
$$+ (\widehat{L}_z Y_{2,1})|\downarrow\rangle + Y_{2,1}(\widehat{S}_z|\downarrow\rangle).$$

So using the eigenvalue equations given in the question, we have

$$\widehat{J}_z\bigl(Y_{2,0}|\uparrow\rangle + Y_{2,1}|\downarrow\rangle\bigr) = 0|\uparrow\rangle + Y_{2,0}\tfrac{1}{2}\hbar|\uparrow\rangle + \hbar Y_{2,1}|\downarrow\rangle - Y_{2,1}\tfrac{1}{2}\hbar|\downarrow\rangle$$
$$= \tfrac{1}{2}\hbar\bigl(Y_{2,0}|\uparrow\rangle + Y_{2,1}|\downarrow\rangle\bigr). \qquad (1.46)$$

Exercise 1.10 Is $\bigl(Y_{2,0}|\uparrow\rangle + Y_{2,1}|\downarrow\rangle\bigr)$ an eigenfunction of $\widehat{J}_z$? If so, what is its eigenvalue? ■

Commutation relations

The operators $\widehat{J}_x$, $\widehat{J}_y$ and $\widehat{J}_z$ do not commute with each other. In fact, they obey commutation relations similar to those for the components of orbital angular momentum or spin:

$$[\widehat{J}_x, \widehat{J}_y] = i\hbar\widehat{J}_z, \quad [\widehat{J}_y, \widehat{J}_z] = i\hbar\widehat{J}_x, \quad [\widehat{J}_z, \widehat{J}_x] = i\hbar\widehat{J}_y. \tag{1.47}$$

We can also introduce the operator

$$\widehat{J}^2 = \widehat{J}_x^2 + \widehat{J}_y^2 + \widehat{J}_z^2,$$

which represents the square of the magnitude of the total angular momentum. As you might expect (by analogy with orbital angular momentum) this operator commutes with all the components of $\widehat{\mathbf{J}}$:

$$[\widehat{J}^2, \widehat{J}_x] = [\widehat{J}^2, \widehat{J}_y] = [\widehat{J}^2, \widehat{J}_z] = 0. \tag{1.48}$$

These commutation relations tell us that it is impossible to find states in which any two of J_x, J_y and J_z have definite values, but it is possible to find states in which both J^2 and any *one* of J_x, J_y and J_z have definite values.

Exercise 1.11 Use the commutation relations for orbital angular momentum (Equation 1.5) and the analogous commutation relations for spin to verify that

$$[\widehat{J}_x, \widehat{J}_y] = i\hbar\widehat{J}_z.$$ ∎

We pointed out earlier that the operators $\widehat{L}_z$ and $\widehat{S}_z$ do not commute with the spin–orbit term in $\widehat{H}_{so}$ (Equation 1.42). The consequence is profound: when the spin–orbit term is included, the energy eigenfunctions are not eigenfunctions of $\widehat{L}_z$ or $\widehat{S}_z$, and they cannot be labelled by the quantum numbers, m or m_s. This is where the total angular momentum operators $\widehat{J}_z$ and $\widehat{J}^2$ become important, because we can show that both these operators commute with $\widehat{H}_{so}$.

The operator $\widehat{J}^2$ involves orbital angular momentum operators that act on functions of θ and ϕ, and spin operators that act on spinors, so it clearly commutes with all functions of r. The real issue is whether it commutes with $\widehat{\mathbf{L}} \cdot \widehat{\mathbf{S}}$. To show that it does, we note that

$$\widehat{J}^2 = (\widehat{\mathbf{L}} + \widehat{\mathbf{S}}) \cdot (\widehat{\mathbf{L}} + \widehat{\mathbf{S}}) = \widehat{L}^2 + \widehat{S}^2 + 2\widehat{\mathbf{L}} \cdot \widehat{\mathbf{S}}. \tag{1.49}$$

and so

$$[\widehat{J}^2, \widehat{\mathbf{L}} \cdot \widehat{\mathbf{S}}] = [\widehat{L}^2, \widehat{\mathbf{L}} \cdot \widehat{\mathbf{S}}] + [\widehat{S}^2, \widehat{\mathbf{L}} \cdot \widehat{\mathbf{S}}] + 2[\widehat{\mathbf{L}} \cdot \widehat{\mathbf{S}}, \widehat{\mathbf{L}} \cdot \widehat{\mathbf{S}}].$$

The first commutator on the right-hand side vanishes because $\widehat{L}^2$ commutes with each of $\widehat{L}_x$, $\widehat{L}_y$ and $\widehat{L}_z$ and, of course, it also commutes with the spin operators $\widehat{S}_x$, $\widehat{S}_y$ and $\widehat{S}_z$. The second commutator vanishes for similar reasons, but applied to $\widehat{S}^2$. Finally, the last commutator vanishes because any operator commutes with itself. We therefore conclude that $\widehat{J}^2$ commutes with $\widehat{\mathbf{L}} \cdot \widehat{\mathbf{S}}$, and hence with the entire Hamiltonian operator $\widehat{H}_{so}$.

Exercise 1.12 Show that $\widehat{J}_z = \widehat{L}_z + \widehat{S}_z$ commutes with $\widehat{\mathbf{L}} \cdot \widehat{\mathbf{S}}$, and hence that it commutes with $\widehat{H}_{so}$. ∎

1.4.2 Simultaneous eigenvectors of $\widehat{J}^2$ and $\widehat{J}_z$

Let us summarize the situation for a spin-$\frac{1}{2}$ particle in a spherically-symmetric potential energy well $V(r)$:

- When the spin–orbit interaction is neglected, the Hamiltonian operator $\widehat{H}$ commutes with $\widehat{L}^2, \widehat{L}_z, \widehat{S}^2$ and $\widehat{S}_z$.
- When the spin–orbit interaction is included, the Hamiltonian operator $\widehat{H}_{so}$ commutes with $\widehat{L}^2$ and $\widehat{S}^2$, but it no longer commutes with $\widehat{L}_z$ or $\widehat{S}_z$; instead, it commutes with the total angular momentum operators $\widehat{J}^2$ and $\widehat{J}_z$.

For any spin-$\frac{1}{2}$ particle, $S^2 = \frac{1}{2}(\frac{1}{2}+1)\hbar^2 = 3\hbar^2/4$, no matter what state the particle is in, so this operator does not help us specify the state of the particle. However, the eigenvalues of $\widehat{J}^2$ and $\widehat{J}_z$ do help us specify states, and we need to know the possible eigenvalues of these operators. The answer is as follows:

> **Eigenvalues of total angular momentum for a spin-$\frac{1}{2}$ particle**
>
> The eigenvalues of $\widehat{J}^2$ are $j(j+1)\hbar^2$, where j is called the **total angular momentum quantum number**. This quantum number is always non-negative. For a spin-$\frac{1}{2}$ particle with $l \neq 0$, it has two possible values: $j = l + \frac{1}{2}$ or $j = l - \frac{1}{2}$. For $l = 0$, it only has the value $j = \frac{1}{2}$.
>
> The eigenvalues of $\widehat{J}_z$ are $m_j \hbar$, where m_j is called the **total magnetic quantum number**. This quantum number can have the values: $m_j = -j, -j+1, \ldots, j-1, j$. So, there are $2j+1$ different values of m_j for each value of j.

Quantum numbers such as l, j and m_j, that can be used to label energy eigenfunctions are sometimes called **good quantum numbers**.

Because $\widehat{H}_{so}$ and $\widehat{L}_z$ do not commute, we cannot label the exact energy eigenfunctions with l and m, but $\widehat{H}_{so}, \widehat{L}^2, \widehat{J}^2$ and $\widehat{J}_z$ do commute, so we *can* label the energy eigenfunctions with l, j and m_j. It is convenient to use ket notation and write $|l, j, m_j\rangle$ for a combination of spherical harmonics and spinors that has these quantum numbers. The fact that $|l, j, m_j\rangle$ is a simultaneous eigenvector of $\widehat{L}^2, \widehat{J}^2$ and $\widehat{J}_z$ is then expressed as

$$\widehat{L}^2 |l, j, m_j\rangle = l(l+1)\hbar^2 |l, j, m_j\rangle \tag{1.50a}$$

$$\widehat{J}^2 |l, j, m_j\rangle = j(j+1)\hbar^2 |l, j, m_j\rangle \tag{1.50b}$$

$$\widehat{J}_z |l, j, m_j\rangle = m_j \hbar |l, j, m_j\rangle. \tag{1.50c}$$

What specific combinations of spherical harmonics and spinors have these eigenvalues? They turn out to be of the form

$$|l, j, m_j\rangle = A\, Y_{l, m_j - \frac{1}{2}} |\uparrow\rangle + B\, Y_{l, m_j + \frac{1}{2}} |\downarrow\rangle,$$

where each Y_{lm} is a function of θ and ϕ and A and B are numerical constants. In general, there are two terms, one spin-up and the other spin-down. The spherical harmonics accompanying these terms both have the same value of l. This ensures that $|l, j, m_j\rangle$ is an eigenvector of $\widehat{L}^2$. However, the spherical harmonic with the spin-up state $|\uparrow\rangle$ has a lower value of m than that with the spin-down state $|\downarrow\rangle$. This means that $|l, j, m_j\rangle$ is not an eigenvector of $\widehat{L}_z$, but is instead an eigenvector of $\widehat{J}_z$. Explicitly, we have

$$\begin{aligned}\widehat{J}_z|l, j, m_j\rangle &= (\widehat{L}_z + \widehat{S}_z)|l, j, m_j\rangle \\ &= A\left(\widehat{L}_z Y_{l, m_j-\frac{1}{2}}\right)|\uparrow\rangle + A Y_{l, m_j-\frac{1}{2}}\left(\widehat{S}_z|\uparrow\rangle\right) \\ &\quad + B\left(\widehat{L}_z Y_{l, m_j+\frac{1}{2}}\right)|\downarrow\rangle + B Y_{l, m_j+\frac{1}{2}}\left(\widehat{S}_z|\downarrow\rangle\right) \\ &= (m_j - \tfrac{1}{2} + \tfrac{1}{2})\hbar\, A Y_{l, m_j-\frac{1}{2}}|\uparrow\rangle + (m_j + \tfrac{1}{2} - \tfrac{1}{2})\hbar\, B Y_{l, m_j+\frac{1}{2}}|\downarrow\rangle \\ &= m_j \hbar |l, j, m_j\rangle.\end{aligned}$$

To ensure that $|l, j, m_j\rangle$ is also an eigenvector of $\widehat{J}^2$, the numerical coefficients A and B must be chosen carefully. For example, the state with $l = 1$, $j = \frac{1}{2}$ and $m_j = \frac{1}{2}$ is

$$|1, \tfrac{1}{2}, \tfrac{1}{2}\rangle = \sqrt{\tfrac{1}{3}} Y_{1,0} |\uparrow\rangle - \sqrt{\tfrac{2}{3}} Y_{1,1} |\downarrow\rangle. \tag{1.51}$$

We could verify that this is an eigenvector of $\widehat{J}^2$ by writing out the spherical harmonics in full and using explicit forms for orbital and spin angular momentum operators that appear in $\widehat{J}^2$. The calculation is straightforward in principle but lengthy in practice, so we shall not carry it out here.

Different choices of l, j and m_j produce different coefficients. Amazingly, the required coefficients were found by the mathematicians Clebsch and Gordan in the nineteenth century long before any application to quantum mechanics had been dreamt of. Table 1.2 lists all the possible $|l, j, m_j\rangle$ eigenvectors for $l = 0$ and $l = 1$.

Table 1.2 Eigenvectors of $\widehat{L}^2$, $\widehat{J}^2$ and $\widehat{J}_z$ for all states with $l = 0$ and $l = 1$.

l	j	m_j	$\|l, j, m_j\rangle$
0	$\frac{1}{2}$	$\frac{1}{2}$	$Y_{0,0}\|\uparrow\rangle$
0	$\frac{1}{2}$	$-\frac{1}{2}$	$Y_{0,0}\|\downarrow\rangle$
1	$\frac{1}{2}$	$\frac{1}{2}$	$\sqrt{\frac{1}{3}} Y_{1,0}\|\uparrow\rangle - \sqrt{\frac{2}{3}} Y_{1,1}\|\downarrow\rangle$
1	$\frac{1}{2}$	$-\frac{1}{2}$	$\sqrt{\frac{2}{3}} Y_{1,-1}\|\uparrow\rangle - \sqrt{\frac{1}{3}} Y_{1,0}\|\downarrow\rangle$
1	$\frac{3}{2}$	$\frac{3}{2}$	$Y_{1,1}\|\uparrow\rangle$
1	$\frac{3}{2}$	$\frac{1}{2}$	$\sqrt{\frac{2}{3}} Y_{1,0}\|\uparrow\rangle + \sqrt{\frac{1}{3}} Y_{1,1}\|\downarrow\rangle$
1	$\frac{3}{2}$	$-\frac{1}{2}$	$\sqrt{\frac{1}{3}} Y_{1,-1}\|\uparrow\rangle + \sqrt{\frac{2}{3}} Y_{1,0}\|\downarrow\rangle$
1	$\frac{3}{2}$	$-\frac{3}{2}$	$Y_{1,-1}\|\downarrow\rangle$

Exercise 1.13 Consider a spin-$\frac{1}{2}$ particle with orbital angular momentum quantum number $l = 3$. What are the possible values of j? How many different

states $|3, j, m_j\rangle$ correspond to $l = 3$?

Exercise 1.14 Verify that the state $|1, \frac{1}{2}, \frac{1}{2}\rangle$ in Equation 1.51 is an eigenfunction of $\widehat{J}_z$ with eigenvalue $\frac{1}{2}\hbar$.

Exercise 1.15 Show that the state $|1, \frac{1}{2}, \frac{1}{2}\rangle$ in Equation 1.51 is normalized and that it is orthogonal to $|1, \frac{1}{2}, -\frac{1}{2}\rangle$. ■

1.4.3 Energy-level splitting due to spin–orbit interaction

Ignoring the spin–orbit interaction, a spin-$\frac{1}{2}$ particle in a spherically-symmetric potential energy well has a series of energy levels. Corresponding to a given solution $R_{nl}(r)$ of the radial time-independent Schrödinger equation, there will be a number of degenerate quantum states with different values of the magnetic quantum number m and the spin magnetic quantum number, m_s. The allowed values of these quantum numbers are $m = 0, \pm 1, \ldots, \pm l$ and $m_s = \pm\frac{1}{2}$, so each solution $R_{nl}(r)$ corresponds to $2(2l + 1)$ different quantum states, all with the same energy.

When the spin–orbit interaction is included, the degeneracy of the energy levels is changed. To see why, we use Equation 1.49 to write

$$\widehat{\mathbf{L}} \cdot \widehat{\mathbf{S}} = \tfrac{1}{2}\left(\widehat{\mathbf{J}}^2 - \widehat{\mathbf{L}}^2 - \widehat{\mathbf{S}}^2\right).$$

For a spin-$\frac{1}{2}$ particle, we then see that

$$\widehat{\mathbf{L}} \cdot \widehat{\mathbf{S}} \,|l, j, m_j\rangle = \tfrac{1}{2}\left(j(j+1) - l(l+1) - \tfrac{1}{2}(\tfrac{1}{2}+1)\right)\hbar^2 \,|l, j, m_j\rangle.$$

So, the presence of the spin–orbit interaction, causes states with the same value of l, but different values of j, to have different energies. Nevertheless, some degeneracy remains because states with the same values of l and j, but different values of m_j have the same energy.

For a spin-$\frac{1}{2}$ particle, each non-zero value of l corresponds to two values of j (namely, $j = l + \frac{1}{2}$ and $j = l - \frac{1}{2}$). So, each energy level with $l \neq 0$ splits into two. One of the resulting levels is labelled by $j = l + \frac{1}{2}$ and consists of $2(l + \frac{1}{2}) + 1 = 2l + 2$ degenerate states with different values of m_j. The other level is labelled by $j = l - \frac{1}{2}$ and consists of $2(l - \frac{1}{2}) + 1 = 2l$ degenerate states with different values of m_j.

It is interesting to note that the total number of states is the same whether we include the spin–orbit interaction or not. Including spin, but ignoring the spin–orbit interaction, there are $2(2l + 1)$ states for each value of l. With the spin–orbit interaction, there are again $2l + 2 + 2l = 2(2l + 1)$ states.

In Chapter 3 of Book 2 we referred to Pauli's discovery that a 'classically non-describable two-valuedness' was required to explain the *Periodic Table of the elements*. This was later understood to be an effect of spin. We now see that the number of combinations of quantum numbers is unaffected by the fact that the significant quantum numbers are l, j and m_j rather than l, m and m_s. When spin is included, there are $2(2l + 1)$ sets of quantum numbers belonging to a particular value of the orbital quantum number l. The following table gives a hint of the connection with chemistry and the Periodic Table.

l	$2l+1$	$\times 2$	Cumulative sum
0	1	2	2
1	3	6	8
2	5	10	18
3	7	14	32

If you are familiar with the Periodic Table, you will know that the numbers in the last column above, i.e. the cumulative sums 2, 8, 18, 32, play a central role — they are the lengths of the periods (or rows) in the Periodic Table. This is just a glimpse of how the quantum mechanics of angular momentum affects the atomic structure of all elements, a story that will be continued later in this book.

1.4.4 Examples from the world of atoms and nuclei

We now present two very different examples of the spin–orbit interaction in operation, one from atomic physics, and one from nuclear physics.

Yellow flame on the gas stove

If you have ever cooked potatoes on a gas stove and the salty water has spilled over, you will doubtless have noticed a yellow colour in the flame (Figure 1.5). This is due to sodium atoms (from the salt) jumping between stationary states and emitting photons of yellow light in the process. A spectroscope would reveal two closely spaced lines: there are actually two different types of 'yellow' photon, with slightly different wavelengths λ and hence different energies $hf = hc/\lambda$. The explanation is as follows (see Figure 1.6). The sodium atom can, for many purposes, be viewed as a single electron, known as a *valence electron*, bound to a rather inert system consisting of a nucleus and ten other electrons (sodium has atomic number 11). The lowest energy level for the valence electron in sodium has $l = 0$ and $j = \frac{1}{2}$ (this consists of two degenerate states with $m_j = \pm\frac{1}{2}$).

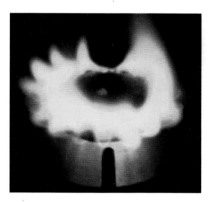

Figure 1.5 Salt in a gas flame reveals the characteristic spectrum of sodium, also seen in some street lighting.

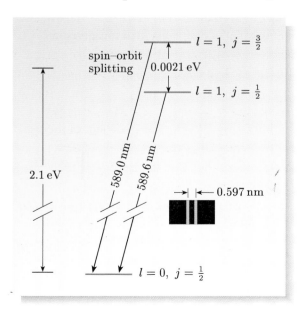

Figure 1.6 The two $l = 1$ states of sodium are about 2.1 eV above the $l = 0$ ground state, and they differ in energy by only 0.0021 eV. Consequently, the two yellow spectral lines, produced by transitions from the $l = 1$ levels to the $l = 0$ level, differ in wavelength by only one part in a thousand. The inset shows their appearance in a spectroscope capable of resolving closely-spaced spectral lines.

- Write out the possible angular momentum eigenvectors for the lowest energy states of the valence electron.
- $|0, \frac{1}{2}, \frac{1}{2}\rangle$ or $|0, \frac{1}{2}, -\frac{1}{2}\rangle$.

There are also two groups of states that, in the absence of a spin–orbit interaction, would have the same energy: the states $|1, \frac{1}{2}, m_j\rangle$ and $|1, \frac{3}{2}, m_j\rangle$ for permitted m_j.

- What values of m_j would be possible for each value of j?
- For $j = \frac{1}{2}$, $m_j = \pm\frac{1}{2}$; for $j = \frac{3}{2}$, $m_j = \pm\frac{1}{2}, \pm\frac{3}{2}$.

The effect of the spin–orbit interaction is to make the states with $j = \frac{1}{2}$ just a little lower in energy than the states with $j = \frac{3}{2}$. As a result, the photon emitted when an electron with $j = \frac{1}{2}$ jumps down to the state with $l = 0$ and $j = \frac{1}{2}$ has a little less energy (and a little longer wavelength) than the photon emitted when an electron with $j = \frac{3}{2}$ jumps down to the state with $l = 0$ and $j = \frac{1}{2}$ (see Figure 1.6). The difference in energy between the $j = \frac{3}{2}$ and $j = \frac{1}{2}$ states, the spin–orbit splitting, is very small, just 0.0021 eV. This shows up when the yellow light is observed through a spectroscope; there is not one yellow line but a pair of closely-spaced yellow lines with wavelengths 589.0 nm and 589.6 nm, i.e. differing in wavelength by about one part in a thousand.

Spin–orbit splitting in atomic nuclei

The spin–orbit effect may be small in atoms, but it can be very large in nuclei, a fact that has a decisive influence on the combinations of protons and neutrons that make the most stable nuclei. This, in turn, has profound effects on the world about us, on the processes taking place in stars, and hence on the history of the Universe. I shall give one example that has some features rather like the sodium atom, but with significant differences. Just as the sodium atom can, for many purposes, be viewed as a single electron bound to a rather inert system of a nucleus and ten other electrons, so the ^{17}O nucleus, an isotope of oxygen with 8 protons and 9 neutrons, often behaves rather as if it were a single neutron (a spin-$\frac{1}{2}$ particle) bound in a potential energy well created by the other 16 particles. The last neutron is governed by a Hamiltonian operator like that in Equation 1.42, but with a very large spin–orbit term.

Figure 1.7 shows the energies of some of the stationary states in a ^{17}O nucleus. The ground state is a state in which the neutron has $l = 2$ and $j = \frac{5}{2}$. There is also a state at an energy 5.09 MeV higher that has $l = 2$ and $j = \frac{3}{2}$. This energy difference is a result of the spin–orbit interaction. Even taking account of the fact that nuclear energies are roughly a million times higher than atomic energies (i.e. measured in MeV rather than eV), this is proportionately very much greater than the 0.0021 eV splitting of the levels in the sodium atom.

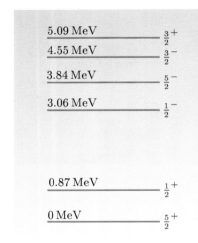

Figure 1.7 The six states of an ^{17}O nucleus with the lowest energies. The energies are shown in MeV, and the total angular momentum j is also marked. The superscript + or − indicates the parity of the state: + if l is zero or an even integer, and − if l is odd. Thus the ground state is at 0 MeV (by definition) and has $j = \frac{5}{2}$ and positive parity; the highest state shown is at an energy of 5.09 MeV and has $j = \frac{3}{2}$ and positive parity.

- From the j and parity indicated for each state in Figure 1.7, deduce the orbital angular momentum quantum number of each state.
- A neutron is a spin-$\frac{1}{2}$ particle, so $j = l \pm \frac{1}{2}$. Using the fact that the parity determines whether l is odd or even, we deduce that the orbital angular momentum quantum number is, for each level (starting with the lowest), 2, 0, 1, 3, 1, 2.

The strong spin–orbit interaction felt by protons and neutrons in nuclei has a decisive influence on which nuclei are particularly stable. These tend to be the nuclei made most copiously in supernovas and the other cosmic furnaces where elements are born. Without the strong nuclear spin–orbit interaction, tin would not be such a common element and there would have been no bronze age, perhaps with drastic consequences for the course of history and the type of civilization we have today.

Summary of Chapter 1

Section 1.1 This section gave a brief review of facts about orbital angular momentum from Chapter 2 of Book 2.

Section 1.2 Ignoring spin, the time-independent Schrödinger equation for a particle in a spherically-symmetric potential energy well is separable in spherical coordinates. It has product solutions $\psi_{nlm}(r,\theta,\phi) = R_{nl}(r)Y_{lm}(\theta,\phi)$, which are also eigenfunctions of $\widehat{L}^2$ and $\widehat{L}_z$:

$$-\frac{\hbar^2}{2m}\frac{1}{r^2}\frac{d}{dr}\left(r^2\frac{dR_{nl}}{dr}\right) + \frac{l(l+1)\hbar^2}{2mr^2}R_{nl}(r) + V(r)R_{nl}(r) = E_{nl}R_{nl}(r)$$

$$\widehat{L}^2 Y_{lm}(\theta,\phi) = l(l+1)\hbar^2\, Y_{lm}(\theta,\phi)$$

$$\widehat{L}_z Y_{lm}(\theta,\phi) = m\hbar\, Y_{lm}(\theta,\phi),$$

where the orbital angular momentum quantum number $l = 0, 1, 2, \ldots$, and the magnetic quantum number $m = -l, -l+1, \ldots, l-1, l$. The ability to label the energy eigenfunctions in this way is consistent with the fact that the Hamiltonian operator $\widehat{H}$, $\widehat{L}^2$ and $\widehat{L}_z$ form a mutually-commuting set. The term $l(l+1)\hbar^2/2mr^2$ is called the centrifugal barrier; it tends to keep particles with high angular momentum away from the origin.

Section 1.3 The functions $Y_{lm}(\theta,\phi)$ are spherical harmonics. They can be written as $Y_{lm}(\theta,\phi) = \Theta_{lm}(\theta)e^{im\phi}$. The restrictions on the quantum numbers m and l arise from the need to ensure that $Y_{lm}(\theta,\phi)$ is single-valued and finite. Spherical harmonics are orthonormal

$$\int_0^{2\pi}\int_0^{\pi} Y^*_{l_1 m_1}(\theta,\phi)\, Y_{l_2 m_2}(\theta,\phi)\, \sin\theta\, d\theta\, d\phi = \delta_{l_1 l_2}\, \delta_{m_1 m_2},$$

and $Y_{lm}(\theta,\phi)$ has parity $(-1)^l$.

Section 1.4 The Hamiltonian operator $\widehat{H}_{so}$ of a spin-$\frac{1}{2}$ particle contains a spin–orbit contribution $V_{so}(r)\widehat{\mathbf{L}}\cdot\widehat{\mathbf{S}}$. If the particle is bound in a spherically-symmetric potential energy well, its Hamiltonian operator commutes with $\widehat{L}^2$ and $\widehat{S}^2$ but does not commute with $\widehat{L}_z$ or $\widehat{S}_z$; instead, it commutes with the total angular momentum operators $\widehat{J}^2$ and $\widehat{J}_z$, where $\widehat{\mathbf{J}} = \widehat{\mathbf{L}} + \widehat{\mathbf{S}}$. The eigenvalues of $\widehat{J}^2$ are $j(j+1)\hbar^2$ where, for a fixed l, we have $j = l+\frac{1}{2}$ or $j = l-\frac{1}{2}$. The corresponding eigenvalues of $\widehat{J}_z$ are $m_j\hbar$, where $m_j = -j, -j+1, \ldots, j-1, j$. Eigenfunctions and eigenvalues of the Hamiltonian operator can be chosen to be simultaneously eigenfunctions of $\widehat{L}^2$, $\widehat{J}^2$ and $\widehat{J}_z$, and so can be labelled by the quantum numbers l, j and m_j. As a result of the spin–orbit interaction, states with the same value of l and different values of j have different energies. The spin–orbit interaction is small for electrons in atoms, but can be large for neutrons and protons in nuclei; this has a decisive effect on their energy-level structure and hence on their relative abundances.

Achievements from Chapter 1

After studying this chapter, you should be able to:

1.1 Explain the meanings of the newly defined (emboldened) terms and symbols, and use them appropriately.

1.2 Write down the time-independent Schrödinger equation for a particle in a spherically-symmetric potential energy well. Given the form of the Laplacian operator in spherical coordinates, rewrite this equation in spherical coordinates.

1.3 Normalize an energy eigenfunction in spherical coordinates.

1.4 For a particle in a spherically-symmetric potential energy well, describe the role played by the operator $\widehat{L}^2$ in the time-independent Schrödinger equation and the role of compatible observables in labelling energy eigenfunctions.

1.5 Give an account of the general properties of the spherical harmonics, including their parity, and demonstrate their orthonormality in simple cases by performing appropriate angular integrals.

1.6 Explain why a spin–orbit interaction implies that $\widehat{L}_z$ and $\widehat{S}_z$ no longer commute with the Hamiltonian operator, and interpret what this means for the labels that can be given to energy eigenfunctions.

1.7 Define the total angular momentum operators, $\widehat{J}^2$ and $\widehat{J}_z$, for a single particle, and write down their eigenvalue equations and commutation relations. Apply these operators to given combinations of spherical harmonics and spinors.

1.8 Explain how the spin–orbit interaction reduces the degeneracy of energy levels.

1.9 Describe atomic or nuclear physics phenomena arising from the spin–orbit interaction.

Chapter 2 The hydrogen atom

Introduction

In this chapter, and in Chapter 4, you will study the quantum theory of the hydrogen atom. Analysis of the hydrogen atom was crucial to the acceptance of Schrödinger's equation. If this simplest atom, with just one electron and one proton, could not be modelled correctly, then it was unlikely that Schrödinger's equation could account for other atoms or molecules.

When hydrogen atoms are heated sufficiently, they emit electromagnetic radiation. If this radiation is passed through a diffraction grating to separate the different wavelengths, it is apparent that only certain wavelengths are emitted, and these wavelengths are characteristic of hydrogen atoms. Figure 2.1 shows the emission spectrum from hydrogen atoms in the visible part of the electromagnetic spectrum. The wavelengths of these spectral lines can be measured with very high accuracy.

The main aim of this chapter is to apply Schrödinger's equation to a simple model of the hydrogen atom and demonstrate that it can account for the wavelengths of the spectral lines shown in Figure 2.1, and other lines in hydrogen's spectrum. The model treats the atom as a pair of point-like particles — an electron and a proton — that are subject to Coulomb's electrostatic interaction; the spins of both particles are ignored. We shall call this the **Coulomb model of a hydrogen atom**. By applying the time-independent Schrödinger equation to this model, we will obtain good approximations to the energy eigenfunctions and energy eigenvalues of a hydrogen atom, and hence be able to explain the observed pattern of spectral lines.

The chapter begins, in Section 2.1, by introducing the Bohr radius and the Rydberg energy, which provide characteristic length and energy scales for the hydrogen atom. We also briefly outline the Bohr model of the hydrogen atom which, although unsatisfactory, provides a reference against which the success of Schrödinger's theory can be judged.

In Section 2.2 we set up the time-independent Schrödinger equation for the hydrogen atom, and show that it reduces to an equation for a single particle with a spherically-symmetric potential energy function. This equation can be separated in spherical coordinates. The angular part of the solution is a spherical harmonic, $Y_{lm}(\theta, \phi)$, which you met in Chapter 1. In Section 2.3 we obtain the radial part of the solution and show how energy quantization arises from the condition that this part should not diverge as r tends to infinity. Then in Section 2.4 we combine the two parts and look at the complete energy eigenfunction for several states of the hydrogen atom. We consider the energy eigenvalues of the various states and link these to the observed atomic spectrum. Finally, in Section 2.5, we use the eigenfunctions to calculate expectation values for the electron–proton separation r, and show that there is a large uncertainty Δr in r, unlike the well-defined orbits of the Bohr model. We also show that the theory can successfully account for the results of experiments that measure the momentum distribution of electrons in hydrogen atoms.

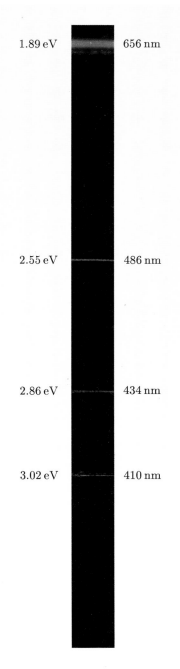

Figure 2.1 The emission spectrum from hydrogen atoms in the visible region of the electromagnetic spectrum comprises four spectral lines. Wavelengths are shown on the right and photon energies on the left.

2.1 Preliminary thoughts about hydrogen

Before attempting to set up and solve the time-independent Schrödinger equation for a hydrogen atom, we shall briefly consider the situation Schrödinger found himself in at the end of 1925. Following de Broglie's insight that matter can be described by waves, Schrödinger tried to find an appropriate wave equation, using the emission spectrum of hydrogen atoms as a crucial test. Hydrogen is simple enough to allow a precise comparison to be made between theory and observation; the theoretical predictions for hydrogen can be carried out with fewer approximations than for other atoms, such as helium, sodium or mercury.

The experimental data for hydrogen were well known, and had been analyzed by various physicists. The pattern of the spectral lines in Figure 2.1 was analyzed by Balmer in 1885, who found that the wavelengths of the lines in this part of the spectrum are given by the formula

$$\lambda = \left(\frac{n^2}{n^2 - 4}\right) 365 \text{ nm}, \tag{2.1}$$

with n taking the values 3, 4, 5 and 6 for the four lines. Balmer also suggested that n might take higher values, corresponding to lines in the ultraviolet region of the spectrum. Other series of spectral lines from hydrogen atoms were soon discovered; these followed similar patterns, but with the constants 4 and 365 nm replaced by other values.

In some ways, the most remarkable feature of the spectrum is that all samples of atomic hydrogen give exactly the same set of spectral lines when they are heated in a similar way. This suggests that all hydrogen atoms are similar to one another, with the same set of energy levels, the same radius, and so on. From the viewpoint of classical physics, this is already a mystery.

Suppose we try to construct a *classical* non-relativistic model of a hydrogen atom. The electron and proton will attract one another according to Coulomb's law, which gives the magnitude of the force on the electron as

$$F_{\text{elec}} = \frac{e^2}{4\pi\varepsilon_0 r^2},$$

where r is the electron–proton separation, e is the proton charge (which is also equal to the magnitude of the electron charge) and ε_0 is the *permittivity of free space*, a fundamental constant that determines the strength of the electrostatic interaction. Now, if we were to calculate the size or energy of a hydrogen atom within a classical model, we would expect the answer to depend on e and ε_0, and also perhaps m_e and m_p, the electron and proton masses. No other constants are relevant to this situation in classical physics. However, it turns out that there is no way of combining these constants to produce a characteristic energy or length. This means that a classical non-relativistic model of the hydrogen atom would allow any energy or size, depending on the initial conditions. In quantum physics, Planck's constant h, or equivalently $\hbar = h/2\pi$, transforms the situation. If we combine this fundamental constant with the charges and masses of the electron and proton, and with ε_0, we can produce combinations with the dimensions of energy and length.

Let us consider how the various constants might be combined. The charge of the proton is e and the charge of the electron is $-e$, so our knowledge of Coulomb's law suggests that these charges will appear in the combination $e^2/4\pi\varepsilon_0$. We can also say that the masses of the electron and proton will not appear independently. When considering diatomic molecules in Section 5.1 of Book 1, we saw that a two-particle system, viewed from its centre of mass, can be treated as a single particle with the reduced mass μ. In the case of the hydrogen atom, the reduced mass is

$$\mu = \frac{m_e m_p}{m_e + m_p}, \tag{2.2}$$

where m_e is the mass of the electron, and m_p is the mass of the proton. Because the mass of a proton is much greater than that of an electron (about 1840 times greater), the reduced mass μ is only very slightly less than the mass of an electron.

Combining the various constants, including $\hbar$, we can define a quantity with the dimensions of length,

$$a_0 = \frac{4\pi\varepsilon_0}{e^2} \frac{\hbar^2}{\mu} = 5.29 \times 10^{-11} \,\text{m}, \tag{2.3}$$

and a quantity with the dimensions of energy,

$$E_R = \left(\frac{e^2}{4\pi\varepsilon_0}\right)^2 \frac{\mu}{2\hbar^2} = 2.18 \times 10^{-18} \,\text{J} = 13.6 \,\text{eV}. \tag{2.4}$$

The length a_0 is called the **Bohr radius** of the hydrogen atom, and the energy E_R is called the **Rydberg energy**. These two constants will turn out to be highly significant in describing the states of a hydrogen atom.

> The factor $\frac{1}{2}$ in the expression for E_R cannot be justified by dimensional arguments alone, but is included here for later convenience.

In 1913, Niels Bohr proposed a model of the hydrogen atom that incorporates Planck's constant in a very simple way. Bohr thought of an un-ionized hydrogen atom as a sort of miniature solar system, with the electron in a circular orbit around the proton. In the **Bohr model**, various stable orbits are allowed, but the key assumption is that the stable orbits have an orbital angular momentum of magnitude $n\hbar$, where $n = 1, 2, 3, \ldots$ is a positive integer. Each stable orbit is called a **Bohr orbit**. Combining this revolutionary idea with classical physics, Bohr showed that the Bohr orbit with quantum number n has radius

$$r_n = n^2 a_0,$$

and that the electron in this orbit has energy

$$E_n = -\frac{E_R}{n^2},$$

where a_0 and E_R are the Bohr radius and Rydberg energy introduced above. The energy levels are negative because the energy zero corresponds to the electron and proton being infinitely far apart (an ionized atom). The orbit with $n = 1$ has the lowest energy, $E_1 = -E_R$, and the smallest radius, $r = a_0$. An infinite number of Bohr orbits have energies above this, but at positive energies the atom becomes ionized. The Rydberg energy E_R is the ionization energy of the ground state.

Bohr assumed that a line in the emission spectrum arises when the electron makes a transition from a higher energy level with $n = n_i$ to a lower level with $n = n_f$. If the emitted photon has energy E_{photon}, conservation of energy requires that

$$E_{\text{photon}} = -\left(\frac{1}{n_i^2} - \frac{1}{n_f^2}\right) E_R = \left(\frac{n_i^2 - n_f^2}{n_i^2 n_f^2}\right) E_R.$$

The wavelength λ of a photon is related to its energy by $E_{\text{photon}} = hc/\lambda$, so the wavelengths of the spectral lines should be given by

$$\lambda = \left(\frac{n_i^2 \, n_f^2}{n_i^2 - n_f^2}\right) \frac{hc}{E_R} = \left(\frac{n_i^2 \, n_f^2}{n_i^2 - n_f^2}\right) 9.11 \times 10^{-8} \text{ m}.$$

● What wavelengths does this formula predict for the radiation emitted when electrons make transitions to the $n = 2$ level from higher levels (with $n > 2$)?

○ For $n_f = 2$,

$$\lambda = \left(\frac{4n_i^2}{n_i^2 - 4}\right) 9.11 \times 10^{-8} \text{ m} = \left(\frac{n_i^2}{n_i^2 - 4}\right) 365 \text{ nm},$$

where $n_i = 3, 4, 5, 6, \ldots$.

As you can see, this is just Balmer's formula, Equation 2.1.

Thus, according to Bohr's theory, the series of lines in Figure 2.1 arises when electrons that have been excited to energy levels with $n_i = 3, 4, 5$ or 6 undergo transitions to the energy level with $n_f = 2$. Since n_f can be any positive integer, Bohr's theory predicts that there should be other series of lines for which electrons undergo transitions from a higher energy level to levels with $n_f = 1$, or $n_f = 3$, etc. Spectral lines have indeed been observed at the predicted wavelengths, and Figure 2.2 shows the lines for transitions to final levels with $n_f = 1$ (Lyman series), $n_f = 3$ (Paschen series), $n_f = 4$ (Brackett series) and $n_f = 5$ (Pfund series). These names are part of the history of the subject, but are unimportant today.

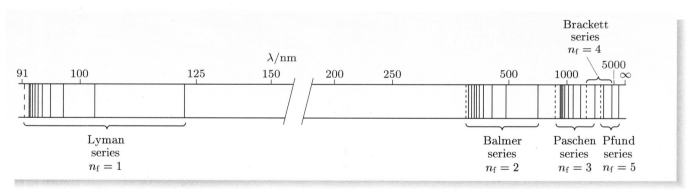

Figure 2.2 The hydrogen atomic spectrum in the ultraviolet, visible and infrared regions of the electromagnetic spectrum.

In spite of this apparent success, Bohr's model fails to account for the energy levels of other atoms, even those of helium. From all you know about quantum mechanics, you will appreciate that Bohr's model cannot be taken too seriously. The Heisenberg uncertainty principle tells us that the electron does not follow a definite trajectory around the proton, and we also know that orbital angular momentum has magnitude $\sqrt{l(l+1)}\hbar$, where $l = 0, 1, 2, \ldots$, rather than $n\hbar$, where $n = 1, 2, 3, \ldots$. As Dirac once said, it is possible to get the right answer for the wrong reasons.

Exercise 2.1 In classical physics, a particle of mass m, travelling with speed v in a circular orbit of radius r, has an orbital angular momentum of magnitude

mvr. Use this result to obtain an expression for the speed of an electron in a Bohr orbit. Evaluate your answer for the $n=1$ orbit. Does this suggest that a hydrogen atom can be dealt with reasonably accurately without using special relativity? ■

2.2 The time-independent Schrödinger equation

The task that confronted Schrödinger was that of finding the energy levels and stationary states of the hydrogen atom in the Coulomb model. As usual, the starting point is the Hamiltonian function of the corresponding classical system: an electron and a proton, treated as two point-particles interacting purely through a Coulomb force.

We shall use centre-of-mass coordinates. Viewed from the centre of mass, the energy of the two-particle system can be written as

$$E = \frac{p^2}{2\mu} - \frac{e^2}{4\pi\varepsilon_0 r}, \tag{2.5}$$

where the first term is the kinetic energy of a particle with the reduced mass μ, and the second term is the electrostatic potential energy of the electron and proton, with the energy zero taken to be at infinity. The right-hand side of Equation 2.5 is the Hamiltonian function for a hydrogen atom in the centre-of-mass frame. The corresponding Hamiltonian operator is

$$\widehat{H} = -\frac{\hbar^2}{2\mu}\nabla^2 - \frac{e^2}{4\pi\varepsilon_0 r}. \tag{2.6}$$

Remember,
$$\nabla^2 = \frac{\partial^2}{\partial x^2} + \frac{\partial^2}{\partial y^2} + \frac{\partial^2}{\partial z^2}.$$

We can now construct Schrödinger's equation

$$i\hbar \frac{\partial}{\partial t}\Psi(\mathbf{r},t) = \widehat{H}\,\Psi(\mathbf{r},t),$$

and find its stationary-state solutions

$$\Psi(\mathbf{r},t) = \psi(\mathbf{r})\,e^{-iEt/\hbar},$$

where $\psi(\mathbf{r})$ satisfies the time-independent Schrödinger equation $\widehat{H}\psi = E\psi$. In more detail, $\psi(\mathbf{r})$ is an energy eigenfunction that satisfies the eigenvalue equation

$$\left(-\frac{\hbar^2}{2\mu}\nabla^2 - \frac{e^2}{4\pi\varepsilon_0 r}\right)\psi(\mathbf{r}) = E\,\psi(\mathbf{r}), \tag{2.7}$$

with eigenvalue E.

The potential energy function $V(r) = -e^2/4\pi\varepsilon_0 r$ is spherically symmetric; in spherical coordinates, it depends on the radial coordinate r, but does not depend on the angular coordinates θ and ϕ. So, following the advice given in Chapter 1 of this book, we shall write the time-independent Schrödinger equation in spherical coordinates. Using a standard result for ∇^2 in spherical coordinates (without attempting to justify it), we then obtain

$$\left(-\frac{\hbar^2}{2\mu}\frac{1}{r^2}\frac{\partial}{\partial r}\left(r^2\frac{\partial}{\partial r}\right) + \frac{1}{2\mu r^2}\widehat{L}^2 - \frac{e^2}{4\pi\varepsilon_0 r}\right)\psi(\mathbf{r}) = E\,\psi(\mathbf{r}), \tag{2.8}$$

where $\widehat{L}$ is the orbital angular momentum operator.

This is simply Equation 1.25 from the previous chapter, with μ substituted for m and $-e^2$ substituted for qQ. Note that the first and third terms in the Hamiltonian operator in Equation 2.8 are functions of r only, and the term involving $\widehat{L}^2$ is the only one that depends on θ and ϕ. This gives us the opportunity to separate the equation further, into one equation in r, and another in θ and ϕ. We achieve this by writing $\psi(\mathbf{r}) = R(r)\,Y(\theta, \phi)$ and using the familiar technique of separation of variables. The two separated equations are

$$\left(-\frac{\hbar^2}{2\mu}\frac{1}{r^2}\frac{\mathrm{d}}{\mathrm{d}r}\left(r^2\frac{\mathrm{d}}{\mathrm{d}r}\right) + \frac{1}{2\mu r^2}K - \frac{e^2}{4\pi\varepsilon_0 r}\right)R(r) = E\,R(r), \quad (2.9)$$

and

$$\widehat{L}^2 Y_{lm}(\theta, \phi) = K\,Y_{lm}(\theta, \phi), \quad (2.10)$$

which are similar to Equations 1.28 and 1.29 of the previous chapter. Equation 2.9 is called the **radial equation**, and its solutions $R(r)$ are called **radial functions**. Equation 2.10 is the eigenvalue equation for the square of the orbital angular momentum, and its solutions, $Y_{lm}(\theta, \phi)$, are *spherical harmonics*.

- What are the eigenvalues of $\widehat{L}^2$?
- We have

$$\widehat{L}^2 Y_{lm}(\theta, \phi) = l(l+1)\hbar^2\,Y_{lm}(\theta, \phi),$$

so the eigenvalues of $\widehat{L}^2$ are $l(l+1)\hbar^2$, where l is zero or a positive integer, the *orbital angular momentum quantum number*.

Thus $K = l(l+1)\hbar^2$, and the radial equation becomes

$$\left(-\frac{\hbar^2}{2\mu}\frac{1}{r^2}\frac{\mathrm{d}}{\mathrm{d}r}\left(r^2\frac{\mathrm{d}}{\mathrm{d}r}\right) + \frac{l(l+1)\hbar^2}{2\mu r^2} - \frac{e^2}{4\pi\varepsilon_0 r}\right)R(r) = E\,R(r). \quad (2.11)$$

It is helpful to rewrite this equation in a slightly different form. We make use of the following identity, which applies for any function $f(r)$:

$$\frac{1}{r^2}\frac{\mathrm{d}}{\mathrm{d}r}\left(r^2\frac{\mathrm{d}f(r)}{\mathrm{d}r}\right) \equiv \frac{1}{r}\frac{\mathrm{d}^2}{\mathrm{d}r^2}\left(rf(r)\right).$$

You can confirm this identity by expanding both sides, and showing that they are equal.

Using this identity in Equation 2.11, and multiplying through by r, we obtain

$$\left(-\frac{\hbar^2}{2\mu}\frac{\mathrm{d}^2}{\mathrm{d}r^2} + \frac{l(l+1)\hbar^2}{2\mu r^2} - \frac{e^2}{4\pi\varepsilon_0 r}\right)rR(r) = E\,rR(r),$$

which, on letting $u(r) = r\,R(r)$, becomes

$$\left(-\frac{\hbar^2}{2\mu}\frac{\mathrm{d}^2}{\mathrm{d}r^2} + \frac{l(l+1)\hbar^2}{2\mu r^2} - \frac{e^2}{4\pi\varepsilon_0 r}\right)u(r) = E\,u(r). \quad (2.12)$$

We call this the **reduced radial equation** for a hydrogen atom and call $u(r)$ the **reduced radial function**. Notice that Equation 2.12 is similar to a time-independent Schrödinger equation in one dimension. The first term in the round brackets is the kinetic energy operator for a particle with the reduced mass μ, while the second and third terms correspond to an effective potential energy function

$$V_{\text{eff}}(r) = \frac{l(l+1)\hbar^2}{2\mu r^2} - \frac{e^2}{4\pi\varepsilon_0 r}. \tag{2.13}$$

We are only interested in the region $r \geq 0$. Assuming that $R(r)$ does not diverge as $r \to 0$, we can set $u(0) = 0$. We shall also assume that $u(r)$ does not diverge as $r \to \infty$, otherwise we should not be able to normalize $R(r)$ (see Equation 2.27 below).

Before we solve the radial equation, let us consider the form of $V_{\text{eff}}(r)$. The second term on the right-hand side is simply the Coulomb potential energy function, which describes the attraction between the electron and the proton. The first term has the opposite sign, and so describes a repulsive interaction. This term was described as the *centrifugal barrier* in the previous chapter. Notice that it depends on the value of l. The form of the effective potential energy function of the hydrogen atom is illustrated for various values of l in Figure 2.3.

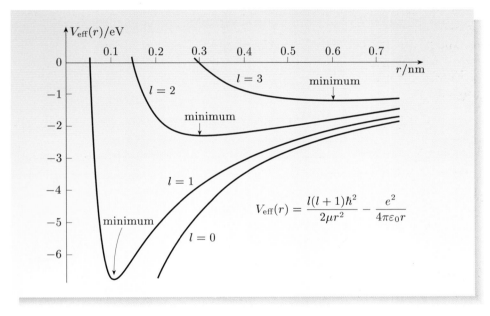

Figure 2.3 The effective potential energy $V_{\text{eff}}(r)$ for the hydrogen atom for various values of the orbital angular momentum quantum number l.

Note that for $l = 0$ there is no centrifugal barrier, and the Coulomb attraction provides an attractive well; nothing then prevents the electron and proton positions coinciding. For $l > 0$, however, centrifugal repulsion dominates over Coulomb attraction for very small values of r. It provides a barrier that becomes infinite as the electron–proton separation tends to zero, keeping the electron and proton separated. Note also that the curve for $l = 0$ has no minimum, but for $l > 0$ there is a minimum in the curve, and this moves to larger electron–proton distances as l increases.

Because Equation 2.12 has the form of a one-dimensional time-independent Schrödinger equation, its bound-state solutions will have discrete energy eigenvalues. For a given effective potential energy well (i.e. fixed l) we can assume that:

Chapter 2 The hydrogen atom

- the ground-state eigenfunction in the given well has no nodes;
- when arranged in order of increasing energy, each successive eigenfunction in the given well has one more node than its predecessor.

These properties will be useful in our study of the hydrogen atom.

2.3 Solutions of the radial equation

2.3.1 Limiting behaviour

Rather than attempt to solve for $R(r)$ or $u(r)$ straight away, we shall first consider the expected limiting behaviour at very small and very large values of r. To simplify the algebra, we define two dimensionless variables,

$$\rho = r/a_0 \quad \text{and} \quad \varepsilon = E/E_R,$$

See Equations 2.3 and 2.4.

where a_0 and E_R are the Bohr radius and the Rydberg energy introduced earlier.

Expressing Equation 2.12 in terms of these variables, and noting that

$$\frac{d^2}{dr^2} = \frac{1}{a_0^2}\frac{d^2}{d\rho^2}, \quad \frac{\hbar^2}{2\mu} = a_0^2 E_R \quad \text{and} \quad \frac{e^2}{4\pi\varepsilon_0} = 2a_0 E_R,$$

we obtain

$$-\frac{d^2 u(\rho)}{d\rho^2} + \frac{l(l+1)}{\rho^2}u(\rho) - \frac{2}{\rho}u(\rho) = \varepsilon\, u(\rho). \tag{2.14}$$

We are looking for solutions that represent bound states of the hydrogen atom. Because the energy zero has been taken to be at infinity, any bound state will have a negative energy, so ε is negative. We impose this condition by writing $\varepsilon = -\beta^2$, where β is real, and, without any loss of generality, take β to be positive. The reduced radial equation then becomes

$$-\frac{d^2 u(\rho)}{d\rho^2} + \frac{l(l+1)}{\rho^2}u(\rho) - \frac{2}{\rho}u(\rho) = -\beta^2 u(\rho). \tag{2.15}$$

Limiting behaviour at large r

As $r \to \infty$, and therefore $\rho \to \infty$, the two potential energy terms in Equation 2.15 tend to zero, and the equation becomes

$$\frac{d^2 u(\rho)}{d\rho^2} - \beta^2 u(\rho) = 0. \tag{2.16}$$

As you can easily verify, the general solution of this differential equation is

$$u(\rho) = Ce^{-\beta\rho} + De^{\beta\rho},$$

where C and D are arbitrary constants.

- Show that the function $u(\rho) = Ce^{-\beta\rho} + De^{\beta\rho}$ is the general solution of Equation 2.16.

○ Substituting $u(\rho) = Ce^{-\beta\rho} + De^{\beta\rho}$ into the left-hand side of Equation 2.16 gives

$$\frac{d^2 u(\rho)}{d\rho^2} - \beta^2 u(\rho) = \beta^2 Ce^{-\beta\rho} + \beta^2 De^{\beta\rho} - \beta^2 (Ce^{-\beta\rho} + De^{\beta\rho}) = 0,$$

so $u(\rho) = Ce^{-\beta\rho} + De^{\beta\rho}$ satisfies Equation 2.16 for any C and D. It is the general solution because it contains two arbitrary constants, as appropriate for a second-order differential equation.

In any bound state, we expect the probability of finding the electron and proton to be very far apart to be vanishingly small. This means that we must discard the term $De^{\beta\rho}$, which diverges exponentially as ρ tends to infinity. We do this by setting $D = 0$. We therefore conclude that:

As ρ tends to infinity, $u(\rho)$ behaves as $Ce^{-\beta\rho}$.

Limiting behaviour at small r

Let us suppose that $l \neq 0$. In this case, the centrifugal barrier dominates at low values of ρ, and we can approximate Equation 2.15 by

$$-\frac{d^2 u(\rho)}{d\rho^2} + \frac{l(l+1)}{\rho^2} u(\rho) = 0. \tag{2.17}$$

We shall investigate a trial solution of the form $u(\rho) = \rho^k$. Substituting this trial solution into Equation 2.17, it is easy to see that the condition for it to be a solution is $k(k-1) = l(l+1)$. This equation has the solutions $k = -l$ and $k = l + 1$, so the general solution for small ρ is of the form

$$u(\rho) = A\rho^{-l} + B\rho^{l+1}.$$

Exercise 2.2 Verify that, for $u(\rho) = \rho^k$ to be a solution of Equation 2.17, we must have $k = -l$ or $k = l + 1$. ■

We can reject the term $A\rho^{-l}$, which diverges as ρ tends to zero, leading to an infinite probability density. This is done by setting $A = 0$. We conclude that:

As ρ tends to zero, $u(\rho)$ behaves as $B\rho^{l+1}$.

We have not justified this limiting behaviour for $l = 0$, but it turns out to apply in this case too.

2.3.2 Nodeless solutions of the radial equation

Given that we expect $u(\rho)$ to behave like $e^{-\beta\rho}$ at very large values of ρ, and like ρ^{l+1} at very small values of ρ, we shall try a solution of the form

$$u(\rho) = C\rho^{l+1} e^{-\beta\rho},$$

where C is a constant. Note that this trial solution is the product of a positive function ρ^{l+1} that increases with ρ, and a positive function $e^{-\beta\rho}$ that decreases with ρ, so it has no nodes. It is therefore a candidate for the lowest energy reduced radial eigenfunction in the effective potential energy well defined by l.

Note that a zero at $\rho = 0$ is not counted as a node.

Exercise 2.3 Show that $u(\rho) = C\rho^{l+1}e^{-\beta\rho}$ is a solution of Equation 2.15 provided that $\beta = 1/(l+1)$. ∎

Since $\varepsilon = -\beta^2$, and Exercise 2.3 showed that $\beta = 1/(l+1)$, we have established that $\varepsilon = -1/(l+1)^2$ for our restricted class of trial solutions.

- What values can l take? What values can ε take?
- ○ The orbital angular momentum quantum number l can take the values $l = 0, 1, 2, 3, \ldots$. Thus $\varepsilon = -1/(l+1)^2$ can take the values $-1, -1/4, -1/9, -1/16, \ldots$. We can therefore write $\varepsilon = -1/n^2$, where $n = 1, 2, 3, \ldots$.

To summarize: for each value $l = 0, 1, 2, \ldots$, we have found solutions of the form

$$u(\rho) = C\rho^{l+1}e^{-\rho/n} \quad \text{where } n = l+1,$$

where C is a constant that is chosen to ensure normalization. (We write n rather than $l+1$ in the exponential factor to help comparison with expressions that will appear later.) Recalling that $\rho = r/a_0$ and that the radial function is $R(r) = u(r)/r$, we see that

$$R(r) = A\left(\frac{r}{a_0}\right)^l e^{-r/na_0}, \tag{2.18}$$

where A is another normalization constant. The first few radial functions of this form are shown in Figure 2.4.

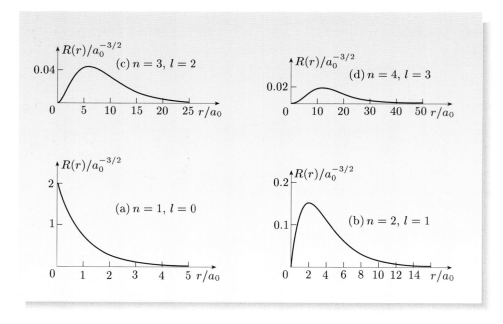

Figure 2.4 Nodeless radial functions of the form given in Equation 2.18. Note that the curves are plotted on different scales.

Since $l = n-1$, there is one solution for each positive integer n. The corresponding energy eigenvalues are

$$E_n = \varepsilon E_R = -\frac{E_R}{n^2} \quad \text{for } n = 1, 2, 3, \ldots. \tag{2.19}$$

These are the same as the energy levels given by the Bohr model, which we know are consistent with the measured spectrum of atomic hydrogen, so the result is most encouraging.

However, we should not celebrate too early. We have only found a special class of solutions for the radial wave equation. The solutions we have found are all nodeless. You know from Book 1 that the ground-state eigenfunction in a one-dimensional well is nodeless, and that there is generally a set of higher energy eigenfunctions, with the number of nodes increasing as the energy increases. For example, Figure 2.5 shows energy eigenfunctions of a harmonic oscillator. The eigenfunction for the ground state ($n = 0$, $E_0 = \frac{1}{2}\hbar\omega_0$) has no nodes, and the number of nodes increases by one for each successive energy level. We will therefore look for additional solutions of the time-independent Schrödinger equation for the hydrogen atom in which the radial function $R(r)$ has one or more nodes.

2.3.3 General solutions of the radial equation

For each value of $l = 0, 1, 2, \ldots$, we have discovered a nodeless solution $u(\rho) = C\rho^{l+1}e^{-\beta\rho}$, corresponding to the lowest energy level in the effective potential energy well labelled by l. We now need to generalize, and find all the possible bound-state solutions, including those with nodes.

A natural way of doing this is to consider a trial function of the form

$$u(\rho) = C(\rho)\,\rho^{l+1}e^{-\beta\rho}, \tag{2.20}$$

where $C(\rho)$ is no longer a constant, but is an arbitrary function of ρ. We shall suppose that $C(\rho)$ can be expressed as a power series, so that

$$C(\rho) = \sum_{k=0}^{\infty} c_k \rho^k \tag{2.21}$$

for some set of coefficients c_k.

Now, we can substitute Equation 2.20 (with $C(\rho)$ given by Equation 2.21) into the radial wave equation (Equation 2.14) and collect together all terms in a given power of ρ. It is not worth going through the lengthy details here (although of course it was worthwhile for Schrödinger!). Instead, we just quote the final result: the function in Equation 2.20 does satisfy the radial wave equation provided that successive coefficients in Equation 2.21 are related as follows:

$$c_{k+1} = \frac{2\beta(k+l+1) - 2}{(k+1)(k+2l+2)} c_k. \tag{2.22}$$

An equation like this is called a *recurrence relation* because it can be used recursively (that is, repeatedly) — first to find c_1 from c_0, then to find c_2 from c_1, and so on, so that all the coefficients in Equation 2.21 can be found in terms of a single constant c_0, whose value must be chosen to ensure normalization.

Now for the crucial point. Consider the behaviour of $C(\rho)$ at very large values of ρ; in this case, the main contribution to the sum on the right-hand side of Equation 2.21 comes from terms with large values of k, so we need to consider the above recurrence relation in the limit of large k. In this limit, we have

$$\frac{c_{k+1}}{c_k} \simeq \frac{2\beta}{k}.$$

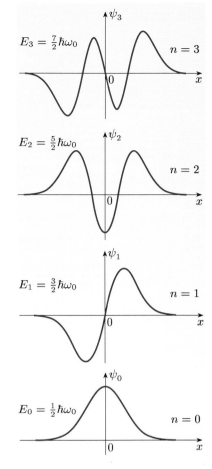

Figure 2.5 Energy eigenfunctions of a simple harmonic oscillator with classical angular frequency ω_0.

A similar ratio of successive coefficients is found in the Taylor series expansion of the exponential function

$$e^{2\beta\rho} = 1 + (2\beta\rho) + \cdots + \frac{(2\beta\rho)^{k-1}}{(k-1)!} + \frac{(2\beta\rho)^k}{k!} + \cdots,$$

and this implies that, in the limit of large ρ, $C(\rho)$ will behave very much like $e^{2\beta\rho}$. This looks like a disaster! We are looking for a function $u(\rho)$ that tends to zero as ρ tends to infinity but, combining the factor $e^{2\beta\rho}$ with the factor $e^{-\beta\rho}$ in Equation 2.20, we have found a function that diverges at infinity and is certainly not an acceptable eigenfunction.

Yet we can snatch victory from the jaws of defeat, because there is a loophole in the above argument. Suppose that the recurrence relation generates a coefficient $c_{k_{\max}+1}$ that is equal to zero. If this happens, all the higher coefficients will be equal to zero, and the function $C(\rho)$ will be a polynomial of order $k_{\max}$. Now, the exponential function $e^{-\beta\rho}$ in Equation 2.20 decreases sufficiently rapidly to kill off *any* polynomial function, so the disaster is averted: when $C(\rho)$ is a polynomial, $u(\rho)$ tends to zero as ρ tends to infinity.

Looking back to Equation 2.22, we can see that $c_{k_{\max}+1}$ will be the first coefficient to vanish if and only if

$$\frac{1}{\beta} = k_{\max} + l + 1. \qquad (2.23)$$

Here, $k_{\max}$ and l are both integers greater than or equal to 0, so $1/\beta$ must be an integer n that is greater than or equal to 1. This implies that the bound-state energies are given by

$$E_n = -\beta^2 E_R = -\frac{E_R}{n^2} \quad \text{for } n = 1, 2, 3, \ldots, \qquad (2.24)$$

exactly as predicted in the Bohr model, and supported by the observed spectra. We obtained this result earlier in the special case $l = n - 1$; we now see that it is true for all bound states in the hydrogen atom. Note that energy quantization arises directly from the need to prevent the reduced radial function $u(\rho)$ from diverging at infinity. The quantum number n that determines the energy levels is called the **principal quantum number** of the hydrogen atom.

The principal quantum number also determines which values of l correspond to a given energy level. Putting $1/\beta = n$ in Equation 2.23, we see that $n = k_{\max} + l + 1$. Since $k_{\max} \geq 0$, it follows that $n \geq l + 1$, and so

$$l \leq n - 1. \qquad (2.25)$$

This is an important restriction on the orbital angular momentum quantum number l. The lowest energy level, $n = 1$, can only have $l = 0$. The next energy level, $n = 2$, can have $l = 0$ or $l = 1$, and so on. The energy levels themselves depend only on the principal quantum number n and are independent of l, so the energy levels are degenerate.

The restriction on the allowed values of l, for a given value of n, can be understood from Figure 2.6, which reproduces the effective potential energy functions shown in Figure 2.3, but with the energy eigenvalues of the hydrogen atom, $E_n = -E_R/n^2$, shown as dashed red lines.

2.3 Solutions of the radial equation

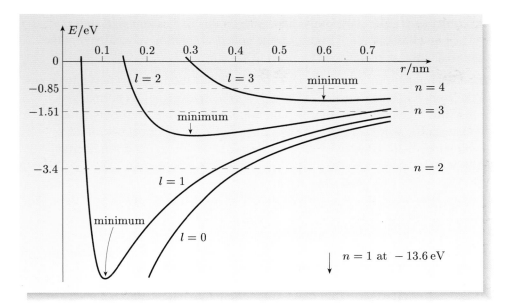

Figure 2.6 Effective potential energy wells for the hydrogen atom for various values of l. The dashed lines show the energy eigenvalues $E_n = -E_R/n^2$ for $n = 2, 3$ and 4; the energy eigenvalue for $n = 1$ is off the scale, below the bottom of the diagram.

For a particular value of n, there are solutions with various values of l, each corresponding to an energy eigenfunction in an effective potential energy well. It is important to realize that a given potential energy well has no solutions *below* the minimum energy of the well. If we take $n = 3$ as an example, there are solutions in the $l = 0$, $l = 1$ and $l = 2$ wells, but no solutions in wells with $l = 3$ or greater whose minima lie above the $n = 3$ energy level. The solution in the $l = 2$ well is the ground state in that well, and is one of the nodeless solutions considered earlier. The solution in the $l = 1$ well is the first excited state in that well, and will have one node, and the solution in the $l = 0$ well is the second excited state in that well, and will have two nodes.

The other striking point about Figure 2.6 is the way the dashed horizontal lines cut across different wells, corresponding to the fact that the energy eigenvalues for the different values of l are identical — they are determined by the value of n alone, and are independent of l. This result is only true in the Coulomb model of the hydrogen atom and similar systems. It is a consequence of the $1/r$-dependence of the Coulomb potential energy term in the Hamiltonian. For atoms with more than one electron, the potential energy function does not vary as $1/r$, and states with different values of l and the same value of n have different energies. This means that spectra for atoms with two or more electrons are much more complex than the spectrum of atomic hydrogen.

Finally, we consider the form of the radial functions. From the above discussion, we know that $u(\rho)$ takes the form

$$u(\rho) = \rho^{l+1} \times (\text{polynomial in } \rho) \times e^{-\rho/n}.$$

We introduced the reduced radial function $u(\rho)$ to simplify the mathematics, but we are really interested in the radial function $R(r)$. Using $\rho = r/a_0$ and $u(r) = r R(r)$, we have

$$R_{nl}(r) = \left(\frac{r}{a_0}\right)^l \times \left(\text{polynomial in } \frac{r}{a_0}\right) \times e^{-r/na_0}, \qquad (2.26)$$

where we have labelled the radial function R by the quantum numbers n and l on which it depends. The polynomial appearing in this expression is of order $n-l-1$ and has $n-l-1$ nodes, so the nodeless solutions considered in the previous subsection correspond to $l = n-1$.

The polynomial in Equation 2.26 contains a normalization factor, chosen to ensure that the complete eigenfunction, $\psi_{nlm}(\mathbf{r}) = R_{nl}(r)Y_{lm}(\theta,\phi)$, is normalized. Because this eigenfunction is three-dimensional, the normalization integral extends over the whole of three-dimensional space. In spherical coordinates, the volume element is given by $dV = r^2 \sin\theta\, dr\, d\theta\, d\phi$, so the normalization condition becomes

$$\langle \psi_{nlm} | \psi_{nlm} \rangle = \int_0^{2\pi} \int_0^{\pi} \int_0^{\infty} |R_{nl}(r)|^2 |Y_{lm}(\theta,\phi)|^2 r^2 \sin\theta\, dr\, d\theta\, d\phi = 1.$$

In Chapter 1 you saw that the spherical harmonics are normalized, so

$$\int_0^{2\pi} \int_0^{\pi} |Y_{lm}(\theta,\phi)|^2 \sin\theta\, d\theta\, d\phi = 1.$$

We therefore require that $R_{nl}(r)$ obeys the normalization condition

$$\int_0^{\infty} |R_{nl}(r)|^2 r^2\, dr = 1. \tag{2.27}$$

It is also possible to show that

$$\int_0^{\infty} R^*_{n_1 l}(r)\, R_{n_2 l}(r)\, r^2\, dr = \delta_{n_1 n_2}. \tag{2.28}$$

This helps to ensure that different energy eigenfunctions $\psi_{n_1 l_1 m_1}$ and $\psi_{n_2 l_2 m_2}$ are orthonormal when integrated over the whole of three-dimensional space.

Essential skill

Obtaining the normalization constant for an eigenfunction of the hydrogen atom

Worked Example 2.1

The ground-state radial function for a hydrogen atom is $R(r) = A e^{-r/a_0}$, where A is a normalization constant. Find a suitable value for A.

Solution

We require that

$$1 = \int_0^{\infty} |R(r)|^2 r^2\, dr = |A|^2 \int_0^{\infty} e^{-2r/a_0} r^2\, dr.$$

Using a standard integral given inside the back cover, we have

$$1 = |A|^2 \frac{2!}{(2/a_0)^3} = \frac{|A|^2 a_0^3}{4}.$$

Choosing A to be real and positive, we conclude that $A = 2/a_0^{3/2}$.

Exercise 2.4 The $n=2$, $l=1$ radial function is given by $R(r) = (Br/a_0)\, e^{-r/2a_0}$, where B is a normalization constant. Find a suitable value for B. ■

We shall not derive explicit formulae for the energy eigenfunctions, although this could clearly be done by using the recurrence relation (Equation 2.22) and

2.3 Solutions of the radial equation

imposing the normalization condition. For reference purposes, however, the first few normalized radial functions are listed in Table 2.1, and graphs of these functions are given in Figure 2.7.

Table 2.1 Normalized radial functions $R_{nl}(r)$ for the hydrogen atom.

n	l	$R_{nl}(r)$
1	0	$\left(\dfrac{1}{a_0}\right)^{3/2} 2\, e^{-r/a_0}$
2	0	$\left(\dfrac{1}{2a_0}\right)^{3/2} 2\left(1 - \dfrac{1}{2}\left(\dfrac{r}{a_0}\right)\right) e^{-r/2a_0}$
2	1	$\left(\dfrac{1}{2a_0}\right)^{3/2} \dfrac{1}{\sqrt{3}}\left(\dfrac{r}{a_0}\right) e^{-r/2a_0}$
3	0	$\left(\dfrac{1}{3a_0}\right)^{3/2} 2\left(1 - \dfrac{2}{3}\left(\dfrac{r}{a_0}\right) + \dfrac{2}{27}\left(\dfrac{r}{a_0}\right)^2\right) e^{-r/3a_0}$
3	1	$\left(\dfrac{1}{3a_0}\right)^{3/2} \dfrac{4\sqrt{2}}{9}\left(\dfrac{r}{a_0}\right)\left(1 - \dfrac{1}{6}\left(\dfrac{r}{a_0}\right)\right) e^{-r/3a_0}$
3	2	$\left(\dfrac{1}{3a_0}\right)^{3/2} \dfrac{2\sqrt{2}}{27\sqrt{5}}\left(\dfrac{r}{a_0}\right)^2 e^{-r/3a_0}$

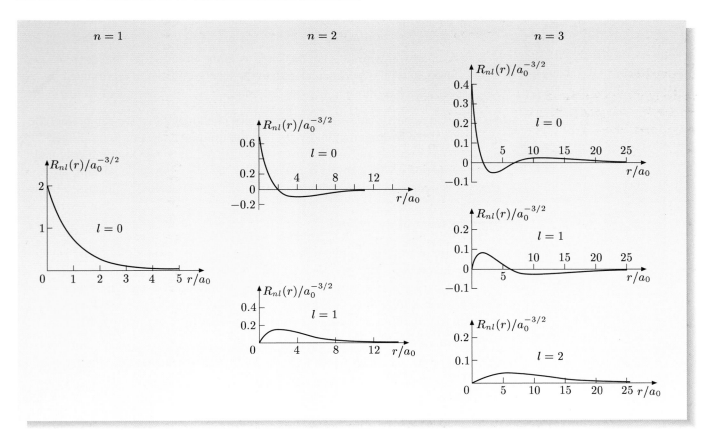

Figure 2.7 Radial functions $R_{nl}(r)$ for the hydrogen atom for $n = 1, 2$ and 3.

Exercise 2.5 From Figure 2.7, determine the number of nodes for each radial function shown, and verify that this number is equal to $n - l - 1$. ■

- How many solutions of the radial equation are there for $n = 5$? How many nodes does each of these solutions have?
- For $n = 5$, the possible values of l are $0, 1, 2, 3$ and 4, so there are five solutions.

  The number of nodes is $n - l - 1$, so for $l = 0$ there are four nodes, for $l = 1$ there are three nodes, for $l = 2$ there are two nodes, for $l = 3$ there is one node, and for $l = 4$ there are no nodes.

Let us summarize what has been achieved so far.

1. We have shown that, in the centre-of-mass frame, the time-independent Schrödinger equation for the Coulomb model of a hydrogen atom can be written as an equation for a single particle in a spherically-symmetric well.
2. As a consequence, the time-independent Schrödinger equation can be separated into two differential equations, one in θ and ϕ, and the other in r. The eigenfunctions $\psi_{nlm}(\mathbf{r})$ can be written as the products of the solutions of these equations: $\psi_{nlm}(\mathbf{r}) = R_{nl}(r)\, Y_{lm}(\theta, \phi)$.
3. The functions $Y_{lm}(\theta, \phi)$ are spherical harmonics.
4. We investigated limiting forms of the radial functions at small and large r.
5. The requirement that the radial functions should not diverge at infinity leads to energy quantization, and produces radial functions of the general form given in Equation 2.26.
6. The energy eigenvalues are the same as in the Bohr model, and are consistent with experimental data to the level of precision represented by spectra such as those in Figures 2.1 and 2.2.

So we seem to have achieved our original aim of showing that we can predict the spectrum of the hydrogen atom by solving the time-independent Schrödinger equation for a proton and an electron interacting purely via a Coulomb potential energy function. A further test is provided by the spectrum of deuterium, an isotope of hydrogen with a proton and a neutron in its nucleus, as discussed below.

The spectrum of deuterium

A naturally-occurring sample of hydrogen contains about 0.015% of the heavier isotope deuterium ($^{2}_{1}\text{H}$). At high resolution and sensitivity, spectral lines due to deuterium atoms can be observed with slightly different wavelengths from those produced by ordinary hydrogen atoms. Why should the spectral lines of deuterium have different wavelengths from those of ordinary hydrogen?

If you look at the formula for the energy levels (Equation 2.19), and combine it with Equations 2.3 and 2.4, you will see that the energy depends on the *reduced mass*, μ:

$$E_n = -\frac{E_\text{R}}{n^2} = -\left(\frac{e^2}{4\pi\varepsilon_0}\right)^2 \frac{\mu}{2\hbar^2} \frac{1}{n^2}.$$

So the energies of the photons that are emitted,

$$E_{\text{photon}} = -\left(\frac{1}{n_i^2} - \frac{1}{n_f^2}\right) E_R,$$

are proportional to μ, and since the wavelengths are given by $\lambda = hc/E_{\text{photon}}$, the wavelengths are proportional to $1/\mu$.

The reduced mass of a one-electron atom is given by $\mu = m_e M/(m_e + M)$, where M is the mass of the nucleus. So the ratio of the reduced masses of deuterium and hydrogen is

$$\frac{\mu_D}{\mu_H} = \frac{m_d(m_e + m_p)}{m_p(m_e + m_d)}.$$

When appropriate values of the masses m_e, m_p and m_d of the electron, proton and deuterium nucleus are substituted, we find $\mu_D/\mu_H = 1.000\,272$. This is very close to 1, but it is sufficient to separate the spectral lines for deuterium from the hydrogen lines, so that their wavelengths can be measured experimentally. Indeed, when experiments were performed to produce deuterium-enriched hydrogen, the presence of deuterium was monitored by observing one of the lines in the Balmer series. Because the reduced mass of deuterium is greater than that of hydrogen, the deuterium lines have slightly shorter wavelengths. The wavelength of the Balmer series line used to monitor deuterium enrichment was 486.1326 nm. The predicted wavelength of the corresponding deuterium line was therefore $(486.1326/1.000\,272)$ nm $= 486.0003$ nm. The experimentally measured wavelength was 486.0000 nm, which agreed well with the theoretical prediction.

Harold Urey was awarded the 1934 Nobel prize in chemistry for his discovery of deuterium using high-resolution spectroscopy.

2.4 Putting it together: the complete eigenfunctions

2.4.1 The complete solutions

The complete solutions of the time-independent Schrödinger equation (in the Coulomb model) are given by $R_{nl}(r)\,Y_{lm}(\theta, \phi)$, where $R_{nl}(r)$ is given by Equation 2.26, and $Y_{lm}(\theta, \phi)$ is a spherical harmonic. The complete eigenfunctions therefore depend on three quantum numbers, n, l and m, and they are simultaneously eigenfunctions of the Hamiltonian operator $\widehat{H}$ and the angular momentum operators $\widehat{L}^2$ and $\widehat{L}_z$, with eigenvalues $-E_R/n^2$, $l(l+1)\hbar^2$ and $m\hbar$, respectively.

This should not come as a surprise: in Section 2.5.3 of Book 2 we showed that $\widehat{H}$, $\widehat{L}^2$ and $\widehat{L}_z$ form a set of mutually commuting operators, so that each operator commutes with the other two. This means that it must be possible to find a set of functions that are simultaneous eigenfunctions of all three operators, and these eigenfunctions correspond to states in which the system simultaneously has definite values of energy, the square of the magnitude of the orbital angular momentum, and the z-component of the orbital angular momentum.

For each value of the principal quantum number n, there are n values of the orbital angular momentum quantum number l, ranging from $l = 0$ to $l = n - 1$. For each of these values of l, there are $(2l + 1)$ values of the magnetic angular momentum quantum number m, ranging from $m = -l$ to $m = +l$. So the number of different eigenfunctions associated with a given value n of the principal quantum number is

$$\sum_{l=0}^{n-1}(2l+1) = 2\sum_{l=0}^{n-1}l + \sum_{l=0}^{n-1}1 = 2\frac{n(n-1)}{2} + n = n^2. \qquad (2.29)$$

Thus for $n = 1$ there is only one eigenfunction, and for $n = 2$ there are four ($= 2^2$) eigenfunctions, and so on.

2.4.2 Spectroscopic notation

In the Coulomb model of the hydrogen atom, the energy levels depend only on n. However, we have just seen that there are n^2 eigenfunctions for each value of n. For each eigenfunction the electron can be spin-up ($m_s = +\frac{1}{2}$) or spin-down ($m_s = -\frac{1}{2}$) so the degeneracy of each energy level is $2n^2$.

Figure 2.8 shows the calculated energy levels for eigenfunctions with $n \leq 5$. In this figure, and elsewhere, we use **spectroscopic notation** to label the eigenfunctions (see Table 2.2). In spectroscopic notation, states and eigenfunctions with $l = 0$ are denoted as s, those with $l = 1$ as p, those with $l = 2$ as d, and those with $l = 3$ as f. For eigenfunctions with $l = 4$ and higher, the designation goes alphabetically from g. Particular eigenfunctions are then specified by attaching the value of n as a prefix and the value of m as a subscript. The ground state of hydrogen is thus labelled $1s_0$, since $n = 1$, $l = 0$ and $m = 0$. An eigenfunction with $n = 5$, $l = 3$ and $m = -3$ would be labelled $5f_{-3}$. The subscript m is often omitted, so that the ground-state eigenfunction of hydrogen is usually written simply as 1s.

Table 2.2 Spectroscopic notation for hydrogen atom eigenfunctions. These letters are derived from the names given to various series of spectral lines in sodium and other atoms: s, p, d, f stand for sharp, principal, diffuse and fundamental, but these names have little importance today.

l	Symbol
0	s
1	p
2	d
3	f
4	g
5	h
⋮	⋮

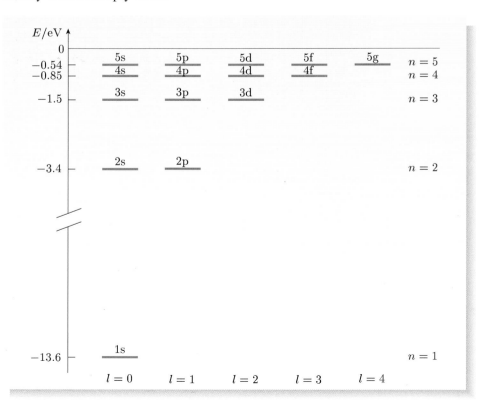

Figure 2.8 Energies of different states in the hydrogen atom. Ignoring spin, each horizontal red line corresponds to $2l + 1$ different eigenfunctions, e.g. 2p corresponds to $2p_{-1}$, $2p_0$ and $2p_{+1}$.

- What is the label for the eigenfunction with $n = 4, l = 3, m = +2$?
○ $4f_{+2}$.

2.4.3 Visualizing the complete eigenfunctions

Probability density

Suppose that we had a microscope sufficiently powerful to image the electron density around an atomic nucleus (this is becoming possible with scanning tunnelling microscopy), and that we obtained an image which was darker where there was a higher probability of finding an electron, and lighter where the probability was lower. This image would resemble a plot of the probability density $\psi^*\psi$. Figure 2.9 shows a representation of the probability densities corresponding to eigenfunctions of the hydrogen atom. The more dots there are in a region, the higher the probability density in that region. Note that each of these diagrams shows the probability density in the xz-plane. The shape of the probability density in three dimensions can be obtained by rotating these diagrams about the z-axis.

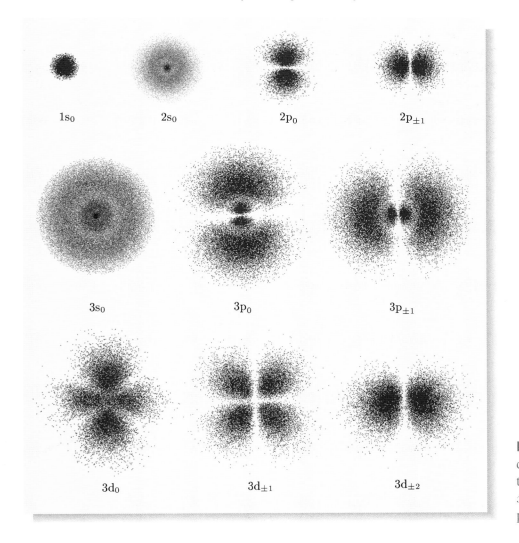

Figure 2.9 Probability densities for eigenfunctions of the hydrogen atom in the xz-plane, with the z-axis pointing up the page.

The DVD that accompanies this book includes software for visualizing hydrogen atom eigenfunctions.

The probability densities for states with $l = 0$ (the s states) are spherically symmetric since the eigenfunctions do not depend on θ or ϕ. For $l > 0$, the probability density is non-spherical, and its shape and orientation depend on the values of l and $|m|$. For example, compare the two different 2p probability densities shown in Figure 2.9.

An isolated hydrogen atom is in a spherically-symmetric environment, so you might wonder how it is possible to obtain the shapes given for the probability densities in Figure 2.9, which are clearly not spherically symmetric. However, it is important to realize that states with different angular momenta are degenerate with one another. This means that we have no reason to suppose that an atom in the $n = 2$ energy level is in any one of the 2s or 2p states — it could equally well be in a mixture of them. To observe probability densities like those shown for $2p_0$ (Figure 2.9), it would be necessary to place the atom in an environment where one direction was singled out by, for example applying an electric or magnetic field.

The probability density $\psi_{nlm}^{*}\psi_{nlm}$ must, of course, be a positive number, but the eigenfunction can be complex. On diagrams such as those in Figure 2.9, it is common to use colours to indicate whether the real (or imaginary) part of an eigenfunction is positive or negative in a particular region. Figure 2.10 shows probability densities for the solutions with $n = 1$, 2 and 3, with the regions where the real part of the eigenfunction is positive indicated by blue dots, and the regions where it is negative indicated by red dots.

Of course, we are free to multiply any eigenfunction by an arbitrary phase factor, so it would be equally valid to reverse the positive and negative regions. However, the important point is that the real (and imaginary) parts of the eigenfunction have positive and negative regions. This will be important in the discussion of molecules in Chapter 6. You will see that overlapping eigenfunctions from different atoms can interfere constructively or destructively, leading to high or low concentrations of electrons, and this is crucial for molecular bonding.

The radial probability density

Another way of describing the eigenfunctions is to specify the probability that the electron–proton separation lies within a small interval between r and $r + \delta r$. Since the radial function is one-dimensional and we are now dealing with the complete eigenfunction which is three-dimensional, we cannot simply take the square of the modulus of the radial function. Instead, we consider a thin spherical shell, of radius r and thickness δr, centred on the proton.

Now the probability that the electron is found within a small volume δV is simply the product of the probability density $\psi^{*}\psi$ and the volume element δV. In spherical coordinates, the volume element is $\delta V = r^2 \sin\theta \, \delta r \, \delta\theta \, \delta\phi$ (see Figure 1.3). To obtain the probability that the electron is found within the thin shell, we integrate the probability density over the volume of the shell. For a thin shell, we do not need to integrate over r.

2.4 Putting it together: the complete eigenfunctions

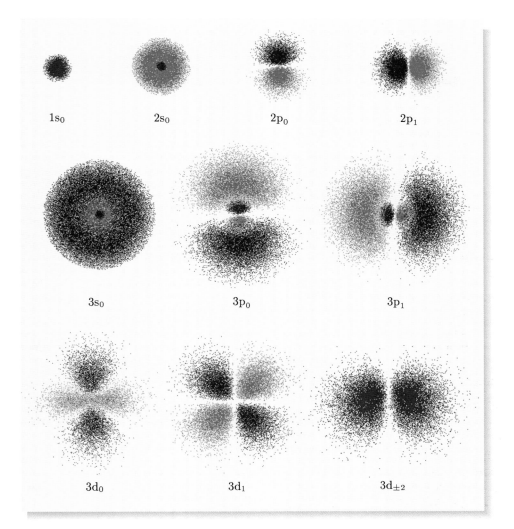

Figure 2.10 Probability densities for hydrogen atom eigenfunctions with $n = 1$, 2 and 3, and all possible values of l. All distributions are shown in the xz-plane, with the z-axis pointing up the page. Blue and red dots indicate regions where the real part of the eigenfunction is positive or negative, respectively.

These colours have *nothing* to do with the sign of charge, which is negative throughout as we are dealing with the probability density for an electron.

The probability is thus

$$\text{probability} = r^2 \, \delta r \int_0^{2\pi} \int_0^{\pi} \psi^* \psi \, \sin\theta \, \mathrm{d}\theta \, \mathrm{d}\phi.$$

Since $\psi_{nlm}(\mathbf{r}) = R_{nl}(r) \, Y_{lm}(\theta, \phi)$, we have

$$\text{probability} = R_{nl}^2(r) \, r^2 \, \delta r \int_0^{2\pi} \int_0^{\pi} Y_{lm}^* \, Y_{lm} \, \sin\theta \, \mathrm{d}\theta \, \mathrm{d}\phi.$$

The spherical harmonics Y_{lm} are normalized, so the double integral over angles is equal to 1. The probability that the electron is found within the shell is therefore $R_{nl}^2(r) \, r^2 \, \delta r$ — this is the probability that the electron–proton separation is between r and $r + \delta r$. The quantity $R_{nl}^2(r) \, r^2$ is known as the **radial probability density**, and this is the probability *per unit radial distance* of finding the electron at distance r from the proton. Figure 2.11 shows plots of the radial probability densities for $n = 1$, 2 and 3.

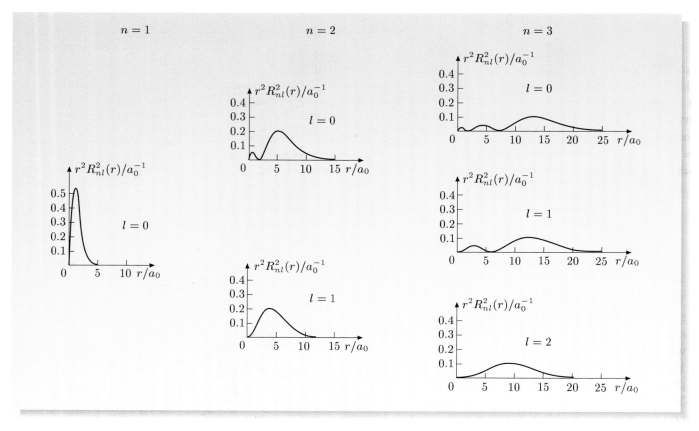

Figure 2.11 Radial probability densities for $n = 1$, 2 and 3. Note that the scale on the horizontal axis is in units of the Bohr radius a_0.

● Compare the three plots with $l = n - 1$ in Figures 2.7 and 2.11, and comment on the positions of the maxima of the curves.

○ In both figures the maxima for $l = n - 1$ occur at increasing values of r as n increases. However, the maxima in Figure 2.11 occur at larger distances than the corresponding maxima in Figure 2.7. This is a consequence of the r^2 factor in the radial probability density.

Interestingly, the maxima for the radial probability densities for states with $l = n - 1$ occur at $n^2 a_0$. These are the distances that Bohr predicted for the radii of the orbits of the electron around the proton. So for $l = n - 1$, the most probable distance is equal to the radius of the corresponding Bohr orbit.

This does not hold for other values of l. States with $l < n - 1$ have several maxima, and the most probable distance corresponds to the highest maximum. Note that the distance at which this highest maximum occurs is greater than the radius of the corresponding Bohr orbit.

● From Figure 2.11, what is the most probable electron–proton separation for the state with $n = 2, l = 0$, and for the state with $n = 2, l = 1$? Are these distances the same?

○ From Figure 2.11, the maximum for $n = 2, l = 0$ occurs at around $r = 5a_0$, whereas the highest maximum for $n = 2, l = 1$ occurs at around $r = 4a_0$. The most probable electron–proton separation is greater for $n = 2, l = 0$ than for $n = 2, l = 1$; in general the spatial extent of a hydrogen atom eigenfunction depends on both n and l.

You can explore the forms of the radial functions and the radial distributions further in a multimedia sequence on the DVD that accompanies this book.

2.4 Putting it together: the complete eigenfunctions

Nodal surfaces of the eigenfunctions

The complete eigenfunctions are zero when either $R_{nl}(r)$ is zero or $Y_{lm}(\theta, \phi)$ is zero. When $R_{nl}(r)$ is zero, the complete eigenfunction is zero on the surface of a sphere. Zeros in the spherical harmonics produce either planes or conical surfaces on which the eigenfunction is zero. In general, the eigenfunctions have **nodal surfaces** — surfaces where the eigenfunction vanishes — from both the radial function and the spherical harmonic (Figure 2.12). However, eigenfunctions with $l = n - 1$ have nodal surfaces only from the spherical harmonics, and those with $l = 0$ have spherical nodal surfaces only from the radial functions. The ground-state eigenfunction 1s is spherically symmetric with no nodal surfaces.

Worked Example 2.2

Where do the eigenfunctions for the $3p_0$ and $3d_0$ states vanish?

Solution

The eigenfunctions vanish where the values of the radial function or the spherical harmonic are zero. For $3p_0$, the radial function is zero where $r = 6a_0$ (see Table 2.1), so there will be a spherical surface with this radius where the eigenfunction vanishes. In addition, the spherical harmonic $Y_{1,0}(\theta, \phi)$ is proportional to $\cos\theta$ (see Table 1.1), so there will be a nodal plane for $\theta = 90°$. These surfaces are shown as a circle and lines on the $3p_0$ probability density in Figure 2.12. For $3d_0$, the radial function is non-zero for finite $r > 0$. There will therefore be nodal surfaces only where $Y_{lm}(\theta, \phi)$ is zero. The spherical harmonic $Y_{2,0}(\theta, \phi)$ is proportional to $3\cos^2\theta - 1$ (see Table 1.1), so it vanishes when $\cos\theta = 1/\sqrt{3}$, that is, $\theta = \cos^{-1}(1/\sqrt{3}) = 54.7°$ or $125.3°$. The black lines on the $3d_0$ probability density in Figure 2.12 represent a cross-section through this conical surface.

Essential skill

Predicting nodal surfaces of eigenfunctions for atomic hydrogen

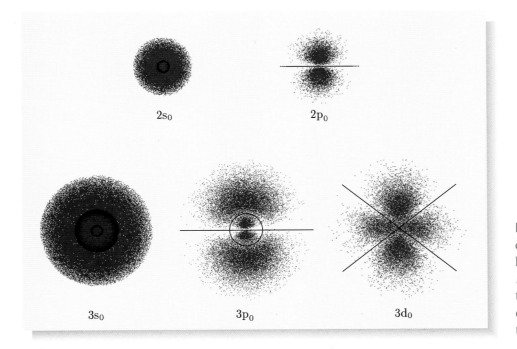

Figure 2.12 Probability density for five states of the hydrogen atom with $m = 0$. The z-axis points towards the top of the page. The black lines and circles indicate cross-sections through nodal surfaces.

Exercise 2.6 Where does the 3d$_1$ eigenfunction vanish?

> **Multimedia sequence: The hydrogen atom**
>
> **Radial functions**
>
> In this sequence, you can view plots of the radial functions and the radial probability densities for values of n up to 8.
>
> You can confirm that there are always $n - l - 1$ nodes, and observe how the locations of the maxima vary with l for a fixed value of n.
>
> **3D display**
>
> In this sequence you can see how the radial functions $R_{nl}(r)$ combine with the spherical harmonics $Y_{lm}(\theta, \phi)$.
>
> **Orbital Viewer**
>
> The Orbital Viewer program allows you to visualize probability densities for the hydrogen atom for values of n up to 30. The distributions can be rotated to give you an idea of their appearance in three dimensions. Files for $n = 1, 2, 3, 4$ have been provided in the format used in Figure 2.10. The program allows you to change the representation method.

2.5 Expectation values and uncertainties

Now that we know the energy eigenfunctions, we can calculate the expectation values and uncertainties of various quantities in different states of the hydrogen atom.

2.5.1 Expectation values

One way to quantify the size of a hydrogen atom is to specify $\langle r \rangle$, the expectation value of the electron–proton separation.

- How can $\langle r \rangle$ be calculated?

○ In quantum mechanics, the expectation value of any quantity is given by a sandwich integral. For the electron–proton separation r, the expectation value is

$$\langle r \rangle = \langle \psi | r | \psi \rangle = \int_0^{2\pi} \int_0^{\pi} \int_0^{\infty} \psi^* \, r \, \psi \, r^2 \sin\theta \, \mathrm{d}r \, \mathrm{d}\theta \, \mathrm{d}\phi.$$

Using $\psi_{nlm}(\mathbf{r}) = R_{nl}(r) Y_{lm}(\theta, \phi)$, and the fact that the spherical harmonics are normalized, the expectation value of r is given by

$$\langle r \rangle = \int_0^{\infty} r^3 R_{nl}^2(r) \, \mathrm{d}r \int_0^{2\pi} \int_0^{\pi} |Y_{lm}(\theta, \phi)|^2 \sin\theta \, \mathrm{d}\theta \, \mathrm{d}\phi = \int_0^{\infty} r^3 R_{nl}^2(r) \, \mathrm{d}r,$$

and more generally

$$\langle r^k \rangle = \int_0^{\infty} r^{k+2} R_{nl}^2(r) \, \mathrm{d}r. \tag{2.30}$$

2.5 Expectation values and uncertainties

Worked Example 2.3

Evaluate the expectation value of the electron–proton separation for the 1s ground state of the hydrogen atom.

Essential skill

Calculating an expectation value

Solution

For the 1s state, $R_{1,0}(r) = (2/a_0^{3/2}) e^{-r/a_0}$, and so

$$\langle r \rangle = \frac{4}{a_0^3} \int_0^\infty r^3 e^{-2r/a_0} \, dr.$$

Using the standard integral $\int_0^\infty x^n e^{-\alpha x} \, dx = n!/\alpha^{n+1}$ given inside the back cover, we obtain

$$\langle r \rangle = \frac{4}{a_0^3} \frac{3! \, a_0^4}{2^4} = \tfrac{3}{2} a_0.$$

So the expectation value of the electron–proton separation, $\langle r \rangle$, is $\tfrac{3}{2} a_0$.

Exercise 2.7 Calculate the expectation value of r for a hydrogen atom in a 3d state. ∎

You can see from Exercise 2.7 that the expectation value of r increases with n. It also decreases with l. The general expression for $\langle r \rangle$ is

$$\langle r \rangle = \frac{a_0}{2} \left[3n^2 - l(l+1) \right]. \tag{2.31}$$

For a nodeless state with $l = n - 1$, this reduces to

$$\langle r \rangle = \frac{a_0}{2} n(2n+1). \tag{2.32}$$

Note that $\langle r \rangle > n^2 a_0$, which is the most probable electron–proton separation for $l = n - 1$.

- Verify that Equation 2.32 predicts that the expectation value for a 3d state of the hydrogen atom is $21 a_0/2$, as found in Exercise 2.7.

○ For a 3d state, $n = 3$ and $l = 2$, so $l = n - 1$ and $\langle r \rangle$ is found by substituting the value $n = 3$ into Equation 2.32:

$$\langle r \rangle = \frac{a_0}{2}(3 \times 7) = \frac{21 a_0}{2}.$$

Exercise 2.8 Calculate the expectation value of the Coulomb potential energy between the electron and the proton in the ground state of a hydrogen atom. Express your answer in terms of the Rydberg energy.

Exercise 2.9 Distinguish between the following quantities in the ground state of a hydrogen atom: (a) the radial coordinate of a tiny volume element of fixed size, located so as to maximize the probability that the electron will be found within it; (b) the most likely electron–proton distance and (c) the expectation value of the electron–proton distance. ∎

2.5.2 Uncertainty in electron–proton separation

The uncertainty ΔA in any observable A is defined by $\Delta A = \sqrt{\langle A^2 \rangle - \langle A \rangle^2}$ (see Section 4.4 in Book 1).

The *uncertainty* in the electron–proton separation r is a measure of the spread of r and is given by $\Delta r = \sqrt{\langle r^2 \rangle - \langle r \rangle^2}$. We showed in Section 2.5.1 how to evaluate $\langle r \rangle$; to calculate Δr we need to find $\langle r^2 \rangle$. We shall do this for the ground state of the hydrogen atom, 1s.

Using the same method as used to obtain $\langle r \rangle$, the integral we need is

$$\langle r^2 \rangle = \int_0^\infty r^4 R_{1,0}^2(r)\, dr = \frac{4}{a_0^3} \int_0^\infty r^4\, e^{-2r/a_0}\, dr.$$

Using the standard formula for this form of integral from inside the back cover,

$$\langle r^2 \rangle = \frac{4}{a_0^3} \frac{4!\, a_0^5}{2^5} = 3a_0^2.$$

Substituting this result, and the value $\langle r \rangle = 3a_0/2$ calculated in Worked Example 2.3, into the expression for Δr, we obtain

$$\Delta r = \sqrt{\langle r^2 \rangle - \langle r \rangle^2} = \left(3a_0^2 - \frac{9a_0^2}{4} \right)^{1/2} = \frac{\sqrt{3}}{2} a_0.$$

So the uncertainty Δr is $\sqrt{3}a_0/2$.

The general formula for the uncertainty in r in a state characterized by quantum numbers l and n is

$$\Delta r = \frac{a_0}{2} \sqrt{n^4 + 2n^2 - l^2(l+1)^2}. \tag{2.33}$$

For a nodeless state with $l = n - 1$, this reduces to

$$\Delta r = \frac{a_0}{2} n\sqrt{2n+1}. \tag{2.34}$$

2.5.3 Rydberg states of the hydrogen atom

States of the hydrogen atom with very large n are known as **Rydberg states**. In these states, $\langle r \rangle$ can be much greater than the distance between neighbouring atoms, so that the eigenfunctions of the atoms overlap.

- Evaluate $\langle r \rangle$ for a hydrogen atom with $n = 100$, $l = 99$.
- From Equation 2.32,

$$\langle r \rangle = \frac{a_0}{2} (100 \times 201) = 1.005 \times 10^4 a_0.$$

Since $a_0 = 0.0529$ nm, this value of $\langle r \rangle$ is of the order of 5.3×10^{-7} m. However, at normal temperatures and pressures, hydrogen atoms are not found in such states. It would require nearly 13.6 eV to excite atoms from the ground state to states with $n \sim 100$, and if a Rydberg state did form, the atom would be readily ionized.

Exercise 2.10 Calculate the energy of a hydrogen atom in a state with $n = 100$. Hence deduce the energy required to ionize an atom that was in this state, and the energy required to excite an atom to this state from the ground state. ∎

The ionization energy of 1.36 meV calculated in Exercise 2.10 for a state with $n = 100$ is typical of the energy provided by a photon of infrared radiation. A typical thermal energy at room temperature (300 K) is 25 meV, and this energy would be more than sufficient to ionize the atom.

Rydberg states of atoms have been prepared by cooling gas samples to very low temperatures (of the order of 100 mK), and then exciting the atoms with a pulse of laser light. The very small amount of thermal energy present was sufficient to cause ionization of some excited atoms. The free electrons produced could easily collide with other atoms, causing them in turn to ionize. The result was the formation of a plasma, a state of matter consisting of free electrons and nuclei. Plasmas are usually formed using extremely high temperatures, so such a low temperature plasma is of great interest.

- ● Find the uncertainty in r, for a hydrogen atom with $n = 100$, $l = 99$.
- ○ Using Equation 2.34, $\Delta r = 50 a_0 \sqrt{201} \approx 700 a_0$.

So not only does a hydrogen atom with $n = 100$ have a very large value of $\langle r \rangle$, but it also has a large uncertainty, Δr. Note, however that although the uncertainty increases with n, the *fractional* uncertainty decreases. For states with $l = n - 1$,

$$\frac{\Delta r}{\langle r \rangle} = \frac{(a_0/2)\, n\, \sqrt{2n+1}}{(a_0/2)\, n\, (2n+1)} = \frac{1}{\sqrt{2n+1}},$$

which becomes small at very large values of n. For Rydberg states, the eigenfunctions can resemble well-defined Bohr orbits. Figure 2.13 shows the probability density for a state with $n = 28$, $l = 27$ and $m = 27$. This provides a good illustration of the correspondence principle mentioned in Chapter 5 of Book 1: the predictions of quantum physics often approach those of classical physics in the limit of high quantum numbers.

Figure 2.13 Probability density for a Rydberg state with $n = 28$, $l = 27$ and $m = 27$ shown for the $z = 0$ plane, with the z-axis pointing out of the page. The red/blue regions show where the real part of the eigenfunction is positive/negative.

2.5.4 Momentum distribution in the ground state

The probability density for an electron in an isolated hydrogen atom cannot be measured by any experiment yet devised. However, it is possible to measure the probability density of the electron's momentum. An experiment that did this was described by McCarthy and Weigold in 1981. In their experiment, free electrons of known momentum $\mathbf{p}_0$ collided with stationary hydrogen atoms. Although each hydrogen atom was stationary, the proton and electron comprising it were not.

At the time of impact, the electron in the atom had an initial momentum, which we denote by $\mathbf{p}'$, and the proton had initial momentum $-\mathbf{p}'$, equal in magnitude but opposite in direction to that of the electron. The change in the proton's momentum was negligible compared with its initial momentum. Thus, by conservation of momentum, $\mathbf{p}_0 + \mathbf{p}' = \mathbf{p}_A + \mathbf{p}_B$, where $\mathbf{p}_A$ is the momentum of the electron ejected from the hydrogen atom, and $\mathbf{p}_B$ is the momentum of the free electron after the collision. The initial momentum of the free electron and the final momenta of both electrons were measured, so it was possible to calculate the initial momentum of the electron in the hydrogen atom at the time of impact: $\mathbf{p}' = \mathbf{p}_A + \mathbf{p}_B - \mathbf{p}_0$. A classical picture of the momenta involved in this experiment is shown in Figure 2.14.

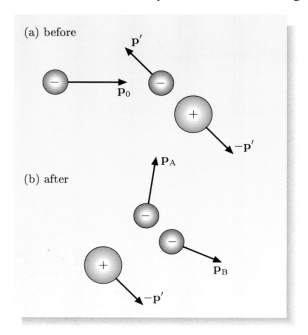

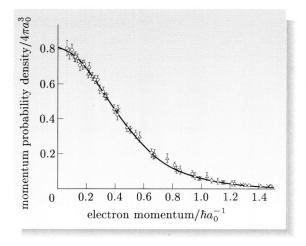

Figure 2.14 (a) An electron with momentum $\mathbf{p}_0$ approaches a stationary hydrogen atom that comprises an electron with momentum $\mathbf{p}'$ and a proton with momentum $-\mathbf{p}'$. (b) The free electron ejects the bound electron with momentum $\mathbf{p}_A$, and its own momentum is changed to $\mathbf{p}_B$. The proton's momentum is unchanged.

Figure 2.15 Momentum probability density for an electron in the ground state of a hydrogen atom. The crosses, circles and triangles refer to three different energies of the electrons fired at the atoms, and the continuous line is the momentum probability density predicted by quantum mechanics.

For a fixed value of the initial momentum of the free electron, $\mathbf{p}_0$, a count was made over a long period of time of the occurrence of particular pairs of momenta $(\mathbf{p}_A, \mathbf{p}_B)$, each pair corresponding to a particular value of $\mathbf{p}'$. Of course, it was not possible to distinguish which electron came from the atom. The results of the experiment are shown in Figure 2.15, along with a curve showing the prediction of quantum mechanics.

We now show how the momentum probability density is calculated. This is optional reading and will not be assessed.

For a given $\mathbf{p}_0$, the probability of detecting two electrons with momenta $\mathbf{p}_A$ and $\mathbf{p}_B$ is given by the probability of the collision occurring multiplied by the probability that the electron in the hydrogen atom is found to have a particular momentum $\mathbf{p}'$ at the time of the collision. The probability of the collision occurring can be calculated, but we shall not be concerned with that here.

Just as the distance of the electron from the proton is not a fixed quantity but has a distribution of values, so the momentum of the electron also has a probability distribution. We need to find the probability that the electron in the hydrogen atom has momentum $\mathbf{p}' = \hbar\mathbf{k}$.

For a one-dimensional wave packet, the probability of finding the momentum in a small interval $\hbar\,\delta k$, centred on $\hbar k$, is $|A(k,t)|^2\,\delta k$, where the momentum amplitude function $A(k,t)$ is the Fourier transform of the wave function $\Psi(x,t)$:

$$A(k,t) = \frac{1}{\sqrt{2\pi}} \int_{-\infty}^{\infty} \Psi(x,t)\,\mathrm{e}^{-\mathrm{i}kx}\,\mathrm{d}x. \tag{2.35}$$

See Chapter 6 of Book 1.

For a hydrogen atom, we need to extend this result to three dimensions, that is, we need a momentum amplitude function, $A(\mathbf{k},t)$, such that the probability of finding the momentum of the electron in a small region of extent $(\hbar\,\delta k_x)(\hbar\,\delta k_y)(\hbar\,\delta k_z)$, centred on $\hbar\mathbf{k}$, is given by $|A(\mathbf{k},t)|^2\,\delta k_x\,\delta k_y\,\delta k_z$. For any stationary state, $|A(\mathbf{k},t)|^2$ is independent of time, and the appropriate three-dimensional generalization of Equation 2.35 turns out to be

$$A(\mathbf{k}) = \frac{1}{(2\pi)^{3/2}} \int_{-\infty}^{\infty}\int_{-\infty}^{\infty}\int_{-\infty}^{\infty} \psi(\mathbf{r})\,\mathrm{e}^{-\mathrm{i}\mathbf{k}\cdot\mathbf{r}}\,\mathrm{d}x\,\mathrm{d}y\,\mathrm{d}z. \tag{2.36}$$

This is the **three-dimensional Fourier transform** of $\psi(\mathbf{r})$. Note that it involves the scalar product $\mathbf{k}\cdot\mathbf{r} = k_x x + k_y y + k_z z$.

To determine the momentum amplitude $A_{1,0,0}(\mathbf{k})$ for the ground state of the hydrogen atom, we take the three-dimensional Fourier transform of the eigenfunction $\psi_{1,0,0}(\mathbf{r})$. We will work in spherical coordinates, with the z-axis chosen to point in the direction of $\mathbf{k}$. Equation 2.36 then becomes

The subscripts $1,0,0$ denote the ground state of the hydrogen atom with $n=1$, $l=0$ and $m=0$.

$$A_{1,0,0}(\mathbf{k}) = \frac{1}{(2\pi)^{3/2}} \int_0^{2\pi}\int_0^{\pi}\int_0^{\infty} \psi_{1,0,0}(r)\,\mathrm{e}^{-\mathrm{i}kr\cos\theta}\,r^2\sin\theta\,\mathrm{d}r\,\mathrm{d}\theta\,\mathrm{d}\phi,$$

where the coordinate θ is the angle between the directions of $\mathbf{k}$ and $\mathbf{r}$. There are no ϕ-dependent terms in the integral, so the integration over ϕ gives 2π. To perform the integration over θ, we make the substitution $w = \cos\theta$. Then $\mathrm{d}w = -\sin\theta\,\mathrm{d}\theta$, and $\theta = 0$ corresponds to $w = 1$, while $\theta = \pi$ corresponds to $w = -1$, so we have

$$\int_0^{\pi} \mathrm{e}^{-\mathrm{i}kr\cos\theta}\sin\theta\,\mathrm{d}\theta = -\int_1^{-1} \mathrm{e}^{-\mathrm{i}krw}\,\mathrm{d}w = \frac{\mathrm{e}^{\mathrm{i}kr} - \mathrm{e}^{-\mathrm{i}kr}}{\mathrm{i}kr}.$$

The final step is the integration over r. Taking the ground-state energy eigenfunction from Tables 2.1 and 1.1, we have

$$A_{1,0,0}(\mathbf{k}) = \frac{2\pi}{(2\pi)^{3/2}} \frac{2}{a_0^{3/2}} \frac{1}{(4\pi)^{1/2}} \frac{1}{\mathrm{i}k} \int_0^{\infty} \mathrm{e}^{-r/a_0}\left(\mathrm{e}^{\mathrm{i}kr} - \mathrm{e}^{-\mathrm{i}kr}\right)r\,\mathrm{d}r.$$

Using a standard result given inside the back cover, the remaining integral is straightforward but tedious. We shall just quote the final answer:

$$A_{1,0,0}(\mathbf{k}) = \frac{2\sqrt{2}}{\pi}\,\frac{a_0^{3/2}}{\left(1+a_0^2 k^2\right)^2}.$$

This is the momentum amplitude, so the corresponding momentum probability density is

$$|A_{1,0,0}(\mathbf{k})|^2 = \frac{8 a_0^3}{\pi^2}\,\frac{1}{\left(1+a_0^2 k^2\right)^4}.$$

Figure 2.15 used this formula to plot the momentum probability density against the magnitude of electron's momentum in a hydrogen atom. The experimental data are adjusted to allow for the factor describing the probability of collision. As you can see, the calculated plot is a very good fit to the experimental data.

The optional reading ends here.

2.5.5 Postscript

In this chapter you have seen that the time-independent Schrödinger equation, applied to the Coulomb model of a hydrogen atom, accounts for the spectral lines shown in Figure 2.1. However, more precise measurements reveal that the longest-wavelength line shown in Figure 2.1 is composed of a number of lines, as shown in Figure 2.16. In Chapter 4, we will refine our simple model of the hydrogen atom to include a number of small effects, such as those due to electron spin. The refined model predicts that the energy levels depend on l as well as n, and accounts for the 'fine structure' shown in Figure 2.16. First, though, we need to develop some tools that will allow us to get approximate answers in situations where exact calculations become hard or impossible. The next chapter discusses two of these techniques: the variational method and perturbation theory.

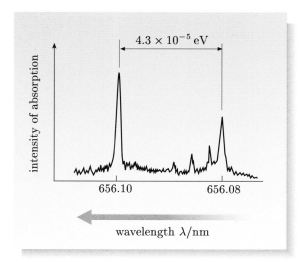

Figure 2.16 High-resolution absorption spectrum of stationary hydrogen atoms. This spectrum corresponds to the emission line at $\lambda = 656$ nm shown in Figure 2.1.

Summary of Chapter 2

Section 2.1 The Bohr model predicts energy levels for hydrogen that are consistent with the measured emission spectra, but the model is not satisfactory.

Section 2.2 The time-independent Schrödinger equation can be solved exactly for the Coulomb model of the hydrogen atom in which the electron and proton are treated as point-like particles interacting only via a Coulomb potential energy. Solutions are obtained using the method of separation of variables in spherical coordinates. The terms that depend on θ and ϕ are spherical harmonics, $Y_{lm}(\theta, \phi)$, while the r-dependence is governed by the radial equation

$$\left(-\frac{\hbar^2}{2\mu}\frac{1}{r^2}\frac{\mathrm{d}}{\mathrm{d}r}\left(r^2\frac{\mathrm{d}}{\mathrm{d}r}\right) + \frac{l(l+1)\hbar^2}{2\mu r^2} - \frac{e^2}{4\pi\varepsilon_0 r}\right)R(r) = E\,R(r).$$

Section 2.3 Acceptable solutions of the radial equation are of the form

$$R_{nl}(r) = \left(\frac{r}{a_0}\right)^l \times \left(\text{polynomial in } \frac{r}{a_0}\right) \times \mathrm{e}^{-r/na_0},$$

where the principal quantum number $n = 1, 2, 3, \ldots$ and the allowed values of l are $0, 1, \ldots, n-1$. The polynomial is of order $n-l-1$ and has $n-l-1$ nodes.

The corresponding energy eigenvalues are $E_n = -E_R/n^2$, where E_R is the Rydberg energy.

Section 2.4 The complete eigenfunction is the product of $R_{nl}(r)$ and a spherical harmonic, $Y_{lm}(\theta, \phi)$. Spectroscopic notation labels states with $l = 0, 1, 2, 3, 4, \ldots$ by s, p, d, f, g, ... The principal quantum number n is attached as a prefix, and the magnetic quantum number m is sometimes added as a subscript.

The radial probability density $R_{nl}^2(r)\, r^2$ is the probability per unit radial distance of finding an electron at distance r from the proton. The highest maximum in the radial probability density corresponds to the most probable electron–proton separation.

In general, the eigenfunctions have nodal surfaces from both the radial function and the spherical harmonic. Those due to the radial function are spherical surfaces. Those due to the spherical harmonic are either planar or conical surfaces.

Section 2.5 The eigenfunctions $\psi_{nlm}(r, \theta, \phi)$ can be used to calculate expectation values of quantities for the hydrogen atom. For large n, the expectation value $\langle r \rangle$ and the uncertainty Δr are both large, but the fractional uncertainty $\Delta r / \langle r \rangle$ decreases as n increases. The momentum probability density calculated for the ground-state eigenfunction is in very good agreement with experimental data.

Achievements from Chapter 2

After studying this chapter, you should be able to:

2.1 Explain the meanings of the newly defined (emboldened) terms and symbols, and use them appropriately.

2.2 Interpret the wavelengths of light emitted by atoms in terms of differences between energy levels.

2.3 Explain the significance of terms in the time-independent Schrödinger equation for a hydrogen atom, and in the radial and angular equations derived from it.

2.4 Recall that the general form of the radial function is $(r/a_0)^l \times$ (polynomial in r/a_0 of order $n - l - 1$) $\times\, e^{-r/na_0}$. Given a particular radial function, find appropriate normalization constants.

2.5 Explain how energy quantization arises for a hydrogen atom, and recall the restrictions on the quantum number l for a given n.

2.6 Write down and interpret spectroscopic notation for hydrogen atom eigenfunctions.

2.7 Interpret diagrams showing energy eigenfunctions, probability densities and radial probability densities for the hydrogen atom.

2.8 Calculate probability densities and the probability of finding the electron within a given region.

2.9 Identify nodal surfaces for hydrogen atom eigenfunctions.

2.10 Calculate expectation values of r^n and the uncertainty in r for hydrogen atom eigenfunctions.

Chapter 3 Time-independent approximation methods

Introduction

This chapter introduces approximation methods. These are crucially important in quantum mechanics. If we want to use quantum theory to understand the properties of real systems, such as atoms, molecules or solids, then we need practical methods of applying this theory.

Most of the examples that we have looked at so far in this course have dealt with simplified models. The reason for this is that Schrödinger's equation is almost impossible to solve analytically unless the potential energy function is very simple. For a particle of mass m with a potential energy function $V(\mathbf{r})$, the Hamiltonian operator is given by

$$\widehat{H} = -\frac{\hbar^2}{2m}\nabla^2 + V(\mathbf{r}).$$

When $V(\mathbf{r})$ has a simple form, such as one of the one-dimensional wells shown in Figure 3.1, or even the three-dimensional Coulomb potential energy $V(r) = -e^2/4\pi\varepsilon_0 r$ discussed in Chapter 2, we can find analytical solutions to Schrödinger's equation and hence determine both the stationary states for the system and the energies of those states. If the potential energy function is even slightly complicated, however, the task of finding an analytical solution becomes extremely difficult. Despite this apparent problem, quantum mechanics is used very successfully to study many physical phenomena, from the way electrons move through solids, to the nature of the light emitted by stars in distant galaxies. This success is accomplished by using mathematical methods that give *approximate* solutions to Schrödinger's equation for complicated physical systems.

In this chapter, we will introduce two of the most important approximation methods used in quantum mechanics. In Section 3.1 we discuss the *variational method*, and in Section 3.2 we discuss *perturbation theory*. The variational method will be used in Chapters 6 and 7 in studies of diatomic molecules and solids. Perturbation theory will be used in Chapters 4 and 5 to make more accurate predictions of the properties of hydrogen-like atoms than can be achieved with the Coulomb model of Chapter 2, and to obtain approximations valid for many-electron atoms. Both approximation methods are described as being time-independent because they give approximate energy eigenvalues and eigenfunctions that satisfy the time-independent Schrödinger equation. The last chapter in this book will show how approximations are made in time-dependent situations.

The importance of approximation methods should not be underestimated. Quantum mechanics textbooks, including these course books, are dominated by the few cases for which analytical solutions can be found, for the good reason that these examples provide much insight into the underlying physics. However, almost every application of quantum mechanics in the real world, from atomic physics to lasers, semiconductors and quantum computing, is based on solutions

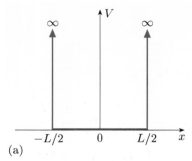

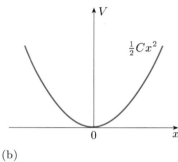

Figure 3.1 Simple one-dimensional potential energy wells: (a) infinite square well of width L, (b) harmonic well with force constant C. The solutions of the time-independent Schrödinger equation for these wells were discussed in Chapters 3 and 5 of Book 1.

obtained using approximation methods. Practical applications of quantum mechanics cannot proceed without them.

Although these methods are called approximation methods, you should not suppose that the solutions are merely rough estimates. The approximate solutions that are produced can be *extremely* accurate, in some cases surpassing the accuracy of experiments. The incredible success of quantum mechanics in predicting the behaviour of real systems is testament to the usefulness of these methods.

> There are not many exercises in this chapter, but some are quite lengthy. Do not miss them out as they cover essential skills.

3.1 The variational method

The variational method is perhaps the simplest of the approximation methods used in quantum mechanics. This method can be used in a variety of situations, but in this section we concentrate solely on finding approximate solutions to the time-independent Schrödinger equation. More specifically, we will be interested in estimating the energy of the ground state of a quantum system.

The variational method was devised by Lord Rayleigh in 1873 to study the vibrational modes of mechanical systems.

The basic principle of the **variational method** is straightforward: when we are unable to calculate the ground-state energy directly, we make an educated guess at the form of its eigenfunction and then find the expectation value of the energy in the state described by the guessed eigenfunction. We then vary the guessed eigenfunction in such a way as to minimize the expectation value. The minimum value provides an estimate of the ground-state energy of the system — the best estimate consistent with the assumptions made about the form of the eigenfunction.

Before starting, we should make our notation clear. For a specified Hamiltonian $\widehat{H}$, we write the unknown exact solutions for the eigenvalues as E_n and the unknown exact solutions for the eigenfunctions as ψ_n, where the subscript n denotes one particular state. These eigenfunctions and eigenvalues satisfy the time-independent Schrödinger equation

We shall often use the term **Hamiltonian** as a shorthand for *Hamiltonian operator*. This is standard practice in quantum mechanics.

$$\widehat{H}\psi_n = E_n \psi_n.$$

Now, we can obtain an expression for E_n in terms of the eigenfunctions by taking the inner product of both sides of this equation with the eigenfunction ψ_n:

$$\langle \psi_n | \widehat{H} | \psi_n \rangle = E_n \langle \psi_n | \psi_n \rangle,$$

and rearranging the result to give

$$E_n = \frac{\langle \psi_n | \widehat{H} | \psi_n \rangle}{\langle \psi_n | \psi_n \rangle}. \tag{3.1}$$

The right-hand side of this expression can be interpreted as the expectation value of the energy; this is equal to E_n because the energy has the definite value E_n in the state ψ_n.

As long as ψ_n is an eigenfunction of $\widehat{H}$, this expression for E_n is exact. However, we will be concerned with situations in which we do not know the eigenfunctions.

We then make an educated guess at the form of the eigenfunction, based on our knowledge of quantum mechanics and of the system under consideration. We shall call our guess the **trial function**, and denote it by $\phi_{n,t}(\mathbf{r})$, using ϕ rather than ψ to distinguish our guess from the exact solution, and including the subscript 't' for 'trial'. The expectation value of the energy in the state described by the trial function will be written as $E_{n,t}$, and this is given by a modified form of Equation 3.1:

$$E_{n,t} = \frac{\langle \phi_{n,t}| \widehat{H} |\phi_{n,t}\rangle}{\langle \phi_{n,t}|\phi_{n,t}\rangle}. \tag{3.2}$$

Notice that we have retained the term $\langle \phi_{n,t}|\phi_{n,t}\rangle$ in the denominator of Equation 3.2. If the trial function were normalized, we would have $\langle \phi_{n,t}|\phi_{n,t}\rangle = 1$, and there would be no need to include this denominator. However, it is often convenient to use trial functions without worrying about their normalization; by including the denominator we can estimate the ground-state energy using any trial function, whether normalized or not.

Clearly, if the trial function were exactly equal to the ground-state eigenfunction, so that $\phi_{n,t} = \psi_n$, then the expressions on the right-hand sides of Equations 3.1 and 3.2 would be identical, so we would have $E_{n,t} = E_n$ and the estimate for the eigenvalue would be exact too. However, this is unlikely to be the case, and the accuracy of our estimate $E_{n,t}$ depends on the extent to which the trial function is similar to the exact eigenfunction.

You may not have much confidence in your ability to pick an appropriate trial function that will give a reasonable estimate of the energy eigenvalue. Fortunately, the variational method provides a way to optimize the choice of trial function for the ground state, and in the next subsection we shall explain the theory underpinning this method.

3.1.1 The principle of the method

The variational method depends on the fact that, for any quantum system, the expectation value of the energy evaluated for *any* trial function is *always* greater than ...or equal to... the ground-state energy of the system. We will now show why this is so.

We assume that the exact energy eigenfunctions of the system, ψ_n, form a complete orthonormal set. This means that the trial function can be expanded as a linear combination of these eigenfunctions

$$\phi_{1,t}(\mathbf{r}) = \sum_n a_n \psi_n(\mathbf{r}), \tag{3.3}$$

This trial function carries the subscript '1' to indicate that it is intended to be an approximation for the ground-state eigenfunction, which we take to have quantum number $n = 1$.

where the unknown coefficients a_n are, in general, complex numbers. We can now substitute this expansion into Equation 3.2, with $n = 1$, to obtain

$$E_{1,t} = \frac{\langle \phi_{1,t}| \widehat{H} |\phi_{1,t}\rangle}{\langle \phi_{1,t}|\phi_{1,t}\rangle} = \frac{\left\langle \sum_m a_m\psi_m \middle| \widehat{H} \middle| \sum_n a_n\psi_n \right\rangle}{\left\langle \sum_m a_m\psi_m \middle| \sum_n a_n\psi_n \right\rangle}.$$

Now $\widehat{H}$ is a linear operator and the functions ψ_n are its eigenfunctions, so

$\widehat{H}\psi_n = E_n\psi_n$. We therefore have

$$E_{1,t} = \frac{\left\langle \sum_m a_m\psi_m \middle| \sum_n a_n E_n \psi_n \right\rangle}{\left\langle \sum_m a_m\psi_m \middle| \sum_n a_n\psi_n \right\rangle} = \frac{\sum_n \sum_m a_m^* a_n E_n \langle \psi_m|\psi_n\rangle}{\sum_n \sum_m a_m^* a_n \langle \psi_m|\psi_n\rangle}.$$

Moreover, the functions ψ_n are orthonormal, so $\langle \psi_m|\psi_n\rangle = \delta_{mn}$, giving

$$E_{1,t} = \frac{\sum_n \sum_m a_m^* a_n E_n \delta_{mn}}{\sum_n \sum_m a_m^* a_n \delta_{mn}} = \frac{\sum_n |a_n|^2 E_n}{\sum_n |a_n|^2}. \quad (3.4)$$

By definition, the ground-state energy eigenvalue E_1 is smaller than any other energy eigenvalue. We can therefore write $E_n \geq E_1$ for all n. Since each of the factors $|a_n|^2$ is positive, this means that

$$E_{1,t} = \frac{\sum_n |a_n|^2 E_n}{\sum_n |a_n|^2} \geq \frac{\sum_n |a_n|^2 E_1}{\sum_n |a_n|^2} = \frac{E_1 \sum_n |a_n|^2}{\sum_n |a_n|^2} = E_1, \quad (3.5)$$

where we have used the fact that E_1 is a constant to take it outside the sum in the penultimate step. We conclude that:

> The expectation value of the energy calculated using any trial function is always greater than or equal to the exact ground-state energy.

This is a very important result. It means that we can try various trial functions and select the one that gives the smallest possible energy expectation value; this will provide our best estimate of the ground-state energy.

For a real system, such as a complicated atom or a molecule composed of several atoms, we do not usually know the exact ground-state energy eigenfunction ψ_1. This is where the variational method is useful. The essence of the method is to choose a trial function that contains one or more adjustable parameters. For example, we might choose $\phi_{1,t}(r) = e^{-\alpha^2 x^2}$, which contains the adjustable parameter α. When we work out the expectation value of the energy using this trial function, we get an answer $E_{1,t} = E_{1,t}(\alpha)$ that depends on α. By varying α, we can minimize $E_{1,t}(\alpha)$, and the minimum energy generated is then our best estimate of the ground-state energy.

The accuracy of our estimate of the ground-state energy depends on our choice of trial function. If the trial function is very similar in form to the exact ground-state eigenfunction, then the variational method can produce a very good estimate for the ground-state energy. It is important to use all of the information that we have about the system when choosing a trial function. For example, we know that the ground state of a particle confined in a one-dimensional potential energy well has no nodes, so it is advisable to ensure that the trial function has no nodes. If we were to choose a function with one or more nodes, then the minimum energy would be very poor estimate of the ground-state energy.

It is often possible to improve the accuracy of the energy estimate by using more adjustable parameters in the trial function. This is analogous to improving the fit of a function, which includes adjustable parameters, to an arbitrary curve; the more free parameters there are, the closer the function can be made to fit the curve. If computers are used, there is no difficulty in using trial functions with many adjustable parameters. However, we will restrict our attention to trial functions with a single adjustable parameter.

3.1.2 Applying the technique

To illustrate the variational method, let us consider the example of a one-dimensional infinite square well. The exact ground-state energy and ground-state eigenfunction for this system were discussed in Chapter 3 of Book 1, but we will pretend that they are not known. We will use the variational method to estimate the ground-state energy eigenvalue, and then compare the estimate with the exact solution.

The potential energy function $V(x)$ for the infinite one-dimensional well shown in Figure 3.1a is

$$V(x) = \begin{cases} 0 & \text{for } -L/2 \leq x \leq L/2, \\ \infty & \text{elsewhere.} \end{cases}$$

As a first step, let us list some general properties that the ground-state eigenfunction in this well is expected to have.

- The potential energy well is symmetric, so the eigenfunctions must be either even or odd functions.
- The ground-state eigenfunction has no nodes inside the well, which means that it cannot be an odd function; it must therefore be even.
- The eigenfunctions are continuous and they must be equal to zero at the boundaries of the well, $x = -L/2$ and $x = L/2$.

Given these general conditions, it is sensible to choose a trial function that is even, nodeless, and is equal to zero at $x = -L/2$ and at $x = L/2$. One suitable choice is the simple quadratic function shown in Figure 3.2, which is represented by the equation

$$\phi_{1,t}(x) = \frac{L^2}{4} - x^2.$$

This function is not normalized, but this is not a problem because Equation 3.2 does not require the trial function to be normalized.

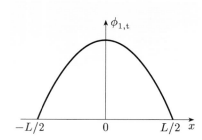

Figure 3.2 The trial function $\phi_{1,t} = \frac{L^2}{4} - x^2$.

The next step is to find the expectation value of the energy for this trial function. To do this, we evaluate the right-hand side of Equation 3.2, obtaining

$$E_{1,t} = \frac{\langle \phi_{1,t}| \widehat{H} |\phi_{1,t}\rangle}{\langle \phi_{1,t}|\phi_{1,t}\rangle}$$

$$= \frac{\int_{-L/2}^{L/2} \left(\frac{L^2}{4} - x^2\right) \left(-\frac{\hbar^2}{2m}\frac{d^2}{dx^2}\right) \left(\frac{L^2}{4} - x^2\right) dx}{\int_{-L/2}^{L/2} \left(\frac{L^2}{4} - x^2\right)^2 dx}. \qquad (3.6)$$

This expression can be evaluated to give

$$E_{1,\text{t}} = \frac{5\hbar^2}{mL^2},$$

and we know that this value must be greater than or equal to the ground-state energy.

Exercise 3.1 By evaluating the integrals in Equation 3.6, verify that $E_{1,\text{t}} = 5\hbar^2/mL^2$. ∎

How does the value $E_{1,\text{t}} = 5\hbar^2/mL^2$ compare with the exact value of the ground-state energy? In Book 1, you saw that the energy eigenvalues for this potential energy well are

$$E_n = \frac{n^2\pi^2\hbar^2}{2mL^2},$$

so the exact ground-state energy is $E_1 = \hbar^2\pi^2/2mL^2$. The value that we found using our trial function differs from this by a factor of $10/\pi^2$, so it is higher by a factor of 1.013. As anticipated, our estimate is greater than the exact value, but the estimate is very close, so we must have chosen a very reasonable trial function.

The exact ground-state eigenfunction is

$$\psi_1(x) = \sqrt{\frac{2}{L}}\cos\left(\frac{\pi x}{L}\right),$$

and this is compared with our trial function in Figure 3.3. To make the comparison a fair one, we have multiplied the trial function by a suitable normalization constant $\sqrt{30}/L^{5/2}$, but even so, you can see that the two functions are not the same, differing by up to 4 per cent in places. Our estimate for the ground-state energy is even better — within about 1 per cent of the exact value. This illustrates an important point about the variational method: the accuracy of the estimated energy is usually better than that of the trial eigenfunction.

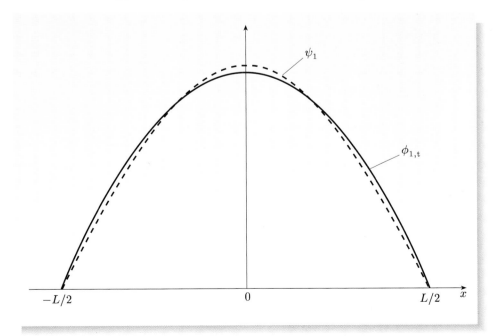

Figure 3.3 The normalized trial function $\phi_{1,\text{t}}$ for the ground state of a particle in a one-dimensional infinite square well, compared with the exact eigenfunction, ψ_1.

Chapter 3 Time-independent approximation methods

This example of how the ground-state energy can be estimated illustrates most of the procedures used in the variational method, but it does not include any optimization of the trial function to improve the estimate. The trial function $\phi_{1,t} = L^2/4 - x^2$ does not include any parameters that can be varied to minimize the energy. The following worked example shows that the variational method becomes much more powerful when the trial function includes an adjustable parameter.

Essential skill

Using the variational method

Here, we use the subscript 0 to denote the ground state, because the quantum number for the lowest energy state of the harmonic oscillator is $n = 0$.

Worked Example 3.1

Use the variational method to estimate the energy of the ground state of a harmonic oscillator for which the potential energy function is $V(x) = \tfrac{1}{2}Cx^2$ (Figure 3.1b). Take as a trial function $\phi_{0,t}(x) = e^{-b^2 x^2}$, where b is a real positive parameter. The standard integrals inside the back cover should be useful.

Solution

As always in the variational method, we begin by using Equation 3.2 to find the expectation value of the energy in the state described by the trial function. The Hamiltonian operator for a harmonic oscillator is

$$\widehat{H} = -\frac{\hbar^2}{2m}\frac{d^2}{dx^2} + \tfrac{1}{2}Cx^2$$

(see Section 5.2 in Book 1), so using the suggested trial function, the expectation value of the energy is

$$E_{0,t} = \frac{\displaystyle\int_{-\infty}^{\infty} e^{-b^2 x^2}\left(-\frac{\hbar^2}{2m}\frac{d^2}{dx^2} + \tfrac{1}{2}Cx^2\right) e^{-b^2 x^2}\, dx}{\displaystyle\int_{-\infty}^{\infty} e^{-2b^2 x^2}\, dx}.$$

In the numerator, we perform the differentiation

$$\frac{d^2}{dx^2}(e^{-b^2 x^2}) = \frac{d}{dx}(-2b^2 x\, e^{-b^2 x^2}) = (-2b^2 + 4b^4 x^2)\, e^{-b^2 x^2},$$

so that the entire numerator is

$$-\frac{\hbar^2}{2m}(-2b^2)\int_{-\infty}^{\infty} e^{-2b^2 x^2}\, dx + \left(-\frac{\hbar^2}{2m}(4b^4) + \tfrac{1}{2}C\right)\int_{-\infty}^{\infty} x^2 e^{-2b^2 x^2}\, dx.$$

Substituting this result back into the expression for the energy and simplifying gives

$$E_{0,t} = \frac{\hbar^2 b^2}{m} + \frac{\left(-\dfrac{2\hbar^2 b^4}{m} + \tfrac{1}{2}C\right)\displaystyle\int_{-\infty}^{\infty} x^2 e^{-2b^2 x^2}\, dx}{\displaystyle\int_{-\infty}^{\infty} e^{-2b^2 x^2}\, dx}. \qquad (3.7)$$

The two integrals can be evaluated using two of the standard integrals given inside the back cover:

$$\int_{-\infty}^{\infty} e^{-y^2}\, dy = \sqrt{\pi} \qquad \text{and} \qquad \int_{-\infty}^{\infty} y^2 e^{-y^2}\, dy = \frac{\sqrt{\pi}}{2}.$$

If we make the substitution $y = \sqrt{2}bx$, then the integral in the denominator becomes

$$\int_{-\infty}^{\infty} e^{-2b^2x^2}\,dx = \frac{1}{\sqrt{2}\,b}\int_{-\infty}^{\infty} e^{-y^2}\,dy = \frac{\sqrt{\pi}}{\sqrt{2}\,b},$$

and the integral in the numerator becomes

$$\int_{-\infty}^{\infty} x^2 e^{-2b^2x^2}\,dx = \left(\frac{1}{\sqrt{2}\,b}\right)^3 \int_{-\infty}^{\infty} y^2 e^{-y^2}\,dy = \frac{\sqrt{\pi}}{2}\left(\frac{1}{\sqrt{2}\,b}\right)^3.$$

Substituting these results for the integrals back into Equation 3.7 gives

$$E_{0,\mathrm{t}} = \frac{\hbar^2 b^2}{m} + \left(-\frac{2\hbar^2 b^4}{m} + \tfrac{1}{2}C\right)\frac{1}{4b^2} = \frac{\hbar^2 b^2}{2m} + \frac{C}{8b^2}. \tag{3.8}$$

This gives a value for the expectation value of the energy in a state described by our trial function; the answer depends on the adjustable parameter b.

We now minimize this energy with respect to b. To do this, we examine the first and second derivatives. First, we differentiate $E_{0,\mathrm{t}}(b)$ with respect to b, and equate the derivative to zero. This gives

$$\frac{dE_{0,\mathrm{t}}}{db} = \frac{\hbar^2 b}{m} - \frac{C}{4b^3} = 0,$$

which has the real positive solution

$$b = \left(\frac{Cm}{4\hbar^2}\right)^{1/4}.$$

This indicates that the function $E_{0,\mathrm{t}}(b)$ has a single turning point for positive b. To check that this is a minimum, we can examine the sign of the second derivative:

$$\frac{d^2 E_{0,\mathrm{t}}}{db^2} = \frac{\hbar^2}{m} + \frac{3C}{4b^4}.$$

Since C is positive, the second derivative is always positive, and so the turning point must be a minimum. As this is the *only* minimum, it must be the absolute minimum value that can be achieved with any function of the form $e^{-b^2 x^2}$. Hence substituting $b = (Cm/4\hbar^2)^{1/4}$ into Equation 3.8 gives the best approximation of the ground-state energy. We obtain

$$E_{0,\mathrm{t,min}} = \frac{\hbar^2}{2m}\sqrt{\frac{Cm}{4\hbar^2}} + \frac{C}{8}\sqrt{\frac{4\hbar^2}{Cm}} = \frac{\hbar}{2}\sqrt{\frac{C}{m}}.$$

The exact value for the ground-state energy derived in Book 1, Section 5.2 is $E_0 = \tfrac{1}{2}\hbar\omega_0 = \tfrac{1}{2}\hbar\sqrt{C/m}$, so the estimate obtained from the variational method is equal to the exact value of the ground-state energy in this case. This is rather unusual, and arises from the fact that by varying b, the trial function can be made identical to the exact eigenfunction for the ground state. If a slightly different trial function had been chosen, the estimated energy would have been greater than the true ground-state energy.

3.1.3 A step-by-step recipe for the ground-state energy

Worked Example 3.1 shows that the variational method can be applied in a systematic way to give a good estimate of the ground-state energy. The key to the method is choosing a suitable trial function, $\phi_{1,t}$, and then substituting it into the following expression:

Variational estimate of the ground-state energy

$$\text{ground-state energy} \simeq \text{minimum value of } \frac{\langle \phi_{1,t} | \widehat{H} | \phi_{1,t} \rangle}{\langle \phi_{1,t} | \phi_{1,t} \rangle} \qquad (3.9)$$

The minimum value is found by adjusting any free parameters in the trial function.

Application of the technique can be summarized in the following steps.

1. Gather relevant information about the ground-state eigenfunction, such as the number of nodes, symmetry and any boundary conditions.
2. Choose a trial function $\phi_{1,t}$, preferably with one or more adjustable parameters, that satisfies the conditions identified in step 1.
3. Find the expectation value of the energy for $\phi_{1,t}$ by evaluating the right-hand side of Equation 3.2.
4. Use the methods of calculus to minimize the expectation value of the energy with respect to any adjustable parameters of the trial function.

Essential skill
Using the variational method

Exercise 3.2 Use the variational method to estimate an upper bound for the ground-state energy of a one-dimensional oscillator in which a particle of mass m has the potential energy function $V(x) = \frac{1}{2}Dx^4$. Note that you can shorten the solution to this problem by considering how the terms involving the potential energy function $V(x) = \frac{1}{2}Cx^2$ in Worked Example 3.1 change when the potential energy is modified to $V(x) = \frac{1}{2}Dx^4$. ■

3.1.4 The variational method and excited states

The variational method can be used very successfully to estimate ground-state energies. An important factor in this success is Equation 3.5, which tells us that anything we do to reduce the value of $\langle \phi_{1,t} | \widehat{H} | \phi_{1,t} \rangle / \langle \phi_{1,t} | \phi_{1,t} \rangle$ will improve our estimate of the ground-state energy. By using functions with several adjustable parameters, and throwing enough computing power at the problem, we can get very accurate answers indeed.

We cannot apply exactly the same reasoning to excited states. However, an advanced mathematical technique can be used to establish the following result, which we just quote:

> If we search among all possible functions ϕ, then any function that produces a stationary value of $\langle\phi|\widehat{H}|\phi\rangle/\langle\phi|\phi\rangle$ is an eigenfunction of $\widehat{H}$, and the stationary value is the corresponding eigenvalue.
>
> Conversely, every eigenfunction of $\widehat{H}$ produces a stationary value in $\langle\phi|\widehat{H}|\phi\rangle/\langle\phi|\phi\rangle$, and this stationary value is the corresponding energy eigenvalue.

A stationary value is either a maximum, a minimum or a point of inflection; in all cases, $\langle\phi|\widehat{H}|\phi\rangle/\langle\phi|\phi\rangle$ is insensitive to small changes in the function ϕ. The precise details involve an advanced mathematical technique called the *calculus of variations*.

In other words, the problem of solving the time-independent Schrödinger equation can be recast as that of finding the functions that produce the stationary values of $\langle\phi|\widehat{H}|\phi\rangle/\langle\phi|\phi\rangle$.

There is, of course, a snag. In principle, we have to search among all possible functions — scarcely a feasible task! However, if we have good reasons for thinking that a given excited-state energy eigenfunction can be well-approximated by a function $\phi(x, \lambda)$ of a given form that depends an adjustable parameter λ, we can generally optimize the approximation by requiring that

$$\frac{d}{d\lambda}\frac{\langle\phi|\widehat{H}|\phi\rangle}{\langle\phi|\phi\rangle} = 0.$$

The method is less secure than when used on the ground state because we have no guarantee that the stationary value produced by our approximate function will be greater than the exact eigenvalue. With a bad choice, it may not even be close. Nevertheless, using physical insight, it is often possible to use this variational method to obtain sensible approximations for excited states, and you will see examples of this in Chapter 6 on diatomic molecules.

3.2 Perturbation methods

The second approximation technique that we shall discuss in this chapter is the **perturbation method**, which uses an alternative approach: we approximate a complicated Hamiltonian by a simpler Hamiltonian for which we can find exact solutions for the eigenfunctions and eigenvalues. We then use these exact solutions, together with the part of the complicated Hamiltonian that was initially omitted, as the basis of further, better, approximations.

The first subsection explains the basic principles of perturbation theory, while the second explains the concepts of approximation order and approximation accuracy. The third subsection shows how we use perturbation theory to find first-order approximate solutions, and the final subsection gives a brief outline of higher-order corrections.

3.2.1 Basic perturbation theory

Perturbation theory is used in many fields of physics, many of which pre-date quantum mechanics. For example, the orbit of the Moon around the Earth depends on many factors: it depends mostly on the gravitational attraction between the Earth and the Moon, but it also depends on effects such as tides, the non-spherical shape of the Earth, and the gravitational attraction of other planets and the Sun. However, if you wanted to calculate the orbit of the Moon, a good start would be to ignore all effects other than the gravitational attraction of the Earth. This simplification would reduce the problem to one with an exact solution, and this solution could then be used as the starting point for a series of successively better estimates. Newton used a version of perturbation theory in this way.

Perturbation theory in quantum mechanics shares the same basic approach. When we consider a complicated system for which we cannot calculate the exact form of the energy eigenfunctions and eigenvalues, we simplify the Hamiltonian to one for which we *can* do this. The exact solutions to the simplified problem are then used as the basis of more accurate estimates.

Before developing the basics of perturbation theory, we will define the notation that will be used. It is especially important to be clear about this because the notation in perturbation theory can become rather complicated. As in the variational method, we will be concerned with the time-independent Schrödinger equation. The solutions of this equation are the exact energy eigenfunctions and eigenvalues of the system, which we denote by ψ_n and E_n. Thus

$$\widehat{H}\psi_n = E_n\psi_n,$$

where $\widehat{H}$ is the exact Hamiltonian of the system. We shall assume that the ψ_n form a complete orthonormal set.

Suppose that the exact time-independent Schrödinger equation cannot be solved, but that we can make the equation soluble either by ignoring one or more terms in the Hamiltonian or by replacing the Hamiltonian by a simpler, more tractable one. (Think again about the example of calculating the Moon's orbit, in which many small effects are ignored to leave only the dominant effect, resulting in a problem that *can* be solved.) We identify the simplified Hamiltonian — the one for which we can calculate an exact solution — by a superscript 0 enclosed in parentheses, that is, $\widehat{H}^{(0)}$. The eigenvalues and eigenfunctions for this Hamiltonian will also be given a $^{(0)}$ superscript, so the simplified Schrödinger equation is written as

$$\widehat{H}^{(0)}\psi_n^{(0)} = E_n^{(0)}\psi_n^{(0)}.$$

The simplified Hamiltonian $\widehat{H}^{(0)}$ is often called the **unperturbed Hamiltonian**, while the full Hamiltonian $\widehat{H}$ is called the **perturbed Hamiltonian**. The key to perturbation theory is being able to find a simplified Hamiltonian that has known solutions $\psi_n^{(0)}$ and $E_n^{(0)}$, and that is not very different from the full Hamiltonian of interest.

The relationship between the energies of the simplified system and the full system is shown schematically in Figure 3.4. Each state of the simplified system corresponds to a state of the full system.

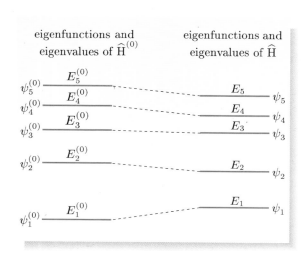

Figure 3.4 The one-to-one correspondence between the states of the simplified or unperturbed Hamiltonian $\widehat{H}^{(0)}$ and those of the full or perturbed Hamiltonian $\widehat{H}$.

There is nothing especially new about the idea of approximating an equation in order to solve it. What distinguishes perturbation theory is that we can use the terms that we initially discarded from the Hamiltonian to improve on the initial approximation. The discarded terms represent the difference between the two Hamiltonians $\widehat{H}$ and $\widehat{H}^{(0)}$. We call this difference the **perturbation**, and denote it by $\delta\widehat{H}$, that is,

$$\delta\widehat{H} = \widehat{H} - \widehat{H}^{(0)}.$$

This expression can be rearranged to give

$$\widehat{H} = \widehat{H}^{(0)} + \delta\widehat{H}. \tag{3.10}$$

So the full Hamiltonian is the sum of the unperturbed Hamiltonian and the perturbation.

To illustrate the ideas and terminology introduced so far, we shall briefly discuss two cases where simplified Hamiltonians can be readily identified.

A hydrogen atom with spin–orbit interaction

The first example is the hydrogen atom, with the spin–orbit interaction included. You saw in the previous chapter that the Hamiltonian operator for the Coulomb model of a hydrogen atom can be written as

$$\widehat{H} = -\frac{\hbar^2}{2\mu}\nabla^2 - \frac{e^2}{4\pi\varepsilon_0 r},$$

where μ is the reduced mass and $V(r) = -e^2/4\pi\varepsilon_0 r$ is the Coulomb potential energy. In the next chapter, we shall consider the effect of the interaction between the electron's spin angular momentum and its orbital angular momentum. Including this interaction adds an extra term to the Hamiltonian:

$$\widehat{H} = -\frac{\hbar^2}{2\mu}\nabla^2 - \frac{e^2}{4\pi\varepsilon_0 r} + V_{\text{so}}(r)\,\widehat{\mathbf{L}}\cdot\widehat{\mathbf{S}}, \tag{3.11}$$

where $\widehat{\mathbf{L}}$ is the orbital angular momentum operator and $\widehat{\mathbf{S}}$ is the spin angular momentum operator. Although this extra term is numerically small compared to the other terms in the Hamiltonian, it makes finding an exact solution for the energy eigenvalues and eigenfunctions of the Hamiltonian extremely difficult. In this case, it is convenient to use perturbation methods. We split the Hamiltonian in Equation 3.11 into an unperturbed part $\widehat{H}^{(0)}$ and a small perturbation $\delta\widehat{H}$,

$$\widehat{H} = \widehat{H}^{(0)} + \delta\widehat{H},$$

where the unperturbed part, for which we found exact solutions in the previous chapter, is

$$\widehat{H}^{(0)} = -\frac{\hbar^2}{2\mu}\nabla^2 - \frac{e^2}{4\pi\varepsilon_0 r},$$

and the small perturbation is given by

$$\delta\widehat{H} = V_{\text{so}}(r)\,\widehat{\mathbf{L}}\cdot\widehat{\mathbf{S}}.$$

A vibrating diatomic molecule — an anharmonic oscillator

The second example is a diatomic molecule, such as hydrogen chloride HCl. The Hamiltonian operator for the vibrational motion of such a molecule is

$$\widehat{H} = -\frac{\hbar^2}{2\mu}\frac{d^2}{dx^2} + V(x), \qquad (3.12)$$

where μ is the reduced mass of the molecule, $V(x)$ has the form of an asymmetric potential energy well, like the one shown in Figure 3.5, and x is the separation of the two atoms measured from their equilibrium separation. This system is called an **anharmonic oscillator**, since the potential energy function does not have the symmetrical parabolic shape that is associated with a harmonic oscillator, but rises more steeply for decreased separations than for increased separations.

Note that this reduced mass involves the masses of two nuclei; it is different from the reduced mass of a hydrogen atom, which involves the electron mass and the proton mass.

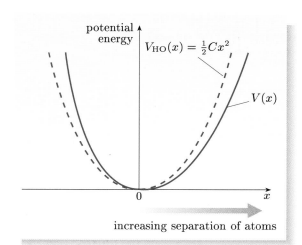

Figure 3.5 The potential energy function $V(x)$ for an anharmonic oscillator (solid line) compared with the potential energy function $V_{HO}(x) = \frac{1}{2}Cx^2$ for a harmonic oscillator (dashed line). The coordinate x is measured from the equilibrium separation $x = 0$ of the atoms.

The time-independent Schrödinger equation for an anharmonic oscillator may be difficult or even impossible to solve, depending on the exact form of the potential energy function $V(x)$. However, if the difference between $V(x)$ and the potential energy function for a harmonic oscillator, $V_{HO}(x) = \frac{1}{2}Cx^2$, is small, then we can use a perturbation approach to tackle this problem.

In this case, we express the total Hamiltonian, given by Equation 3.12, in the form $\widehat{H} = \widehat{H}^{(0)} + \delta\widehat{H}$, where the unperturbed Hamiltonian $\widehat{H}^{(0)}$ is for a harmonic oscillator,

$$\widehat{H}^{(0)} = -\frac{\hbar^2}{2\mu}\frac{d^2}{dx^2} + \tfrac{1}{2}Cx^2,$$

and $\delta\widehat{H}$ is a small perturbation, given by

$$\delta\widehat{H} = \widehat{H} - \widehat{H}^{(0)} = V(x) - \tfrac{1}{2}Cx^2.$$

The eigenvalues and eigenfunctions for the unperturbed Hamiltonian are those for a harmonic oscillator, given in Book 1, Chapter 5. The eigenvalues are

$$E_n^{(0)} = \left(n + \tfrac{1}{2}\right)\hbar\omega_0 \quad \text{for } n = 0, 1, 2, \ldots, \qquad (3.13)$$

where $\omega_0 = \sqrt{C/\mu}$ is the classical angular frequency of oscillation.

The aim of perturbation theory is to determine the corrections that need to be made to $E_n^{(0)}$ and to the corresponding eigenfunctions $\psi_n^{(0)}(x)$ to give better approximations for the eigenvalues and eigenfunctions of the perturbed Hamiltonian $\widehat{H}$.

Exercise 3.3 A particle is confined within a one-dimensional infinite square well of width L. There is a small crater inside the well, of width $w \ll L$, centred on the point $x = x_0$. The potential energy function is shown in Figure 3.6, and is given by

$$V(x) = \begin{cases} \infty & \text{for } x < -L/2 \text{ and } x > L/2, \\ v(x) & \text{for } |x - x_0| < w/2, \\ 0 & \text{elsewhere.} \end{cases}$$

Write down (a) the Hamiltonian for this problem, (b) a Hamiltonian for a related problem for which you know the solutions for the energy eigenvalues and eigenfunctions, and (c) the perturbation that represents the difference between the Hamiltonians in (a) and (b). ■

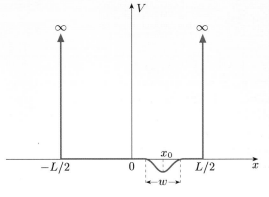

Figure 3.6 A small crater in a one-dimensional infinite square well.

3.2.2 Taylor expansions and orders of approximation

An important part of applying approximation methods is understanding the accuracy of the approximation. In perturbation theory, the accuracy of the calculation is closely related to an important concept called the **order of approximation**.

This concept might already be familiar to you in the slightly simpler context of Taylor expansions of functions. Taylor expansions are used to approximate the value of a function at a point in terms of information about the function at a nearby point. More specifically, if $f(x)$ is a differentiable function whose value and derivatives are known at a point $x = x_0$, then its value at a nearby point, $x_0 + \delta x$, is given by

$$f(x_0 + \delta x) = f(x_0) + \left[\frac{df}{dx}\right]_{x_0} \delta x + \frac{1}{2!}\left[\frac{d^2 f}{dx^2}\right]_{x_0} (\delta x)^2 + \cdots$$

$$+ \frac{1}{n!}\left[\frac{d^n f}{dx^n}\right]_{x_0} (\delta x)^n + \cdots, \quad (3.14)$$

where $[d^n f/dx^n]_{x_0}$ is the value of the nth derivative of f at $x = x_0$. Equation 3.14 is the *Taylor expansion* or *Taylor series* of the function $f(x)$ around the point $x = x_0$. If δx is small, this series usually converges, and if we could calculate an infinite number of terms, the expression would be exact. In real life, however, we generally cut off the calculation after a finite number of terms and hence obtain an approximate value for $f(x_0 + \delta x)$. The approximation improves as we include more and more terms.

The simplest and most crude approximation to $f(x_0 + \delta x)$ is the first term in Equation 3.14, namely

$$f(x_0 + \delta x) \simeq f(x_0).$$

This is called the **zeroth-order approximation**. We obtain a better approximation by including the next term in the expansion, hence approximating the function value as

$$f(x_0 + \delta x) \simeq f(x_0) + \left[\frac{\mathrm{d}f}{\mathrm{d}x}\right]_{x_0} \delta x.$$

This is the **first-order approximation**, and the term $[\mathrm{d}f/\mathrm{d}x]_{x_0} \delta x$ is called the *first-order correction*. Extrapolating from this, the term $[\mathrm{d}^n f/\mathrm{d}x^n]_{x_0} (\delta x)^n/n!$ is the *nth-order correction* and the sum on the right-hand side of Equation 3.14 up to this term is the n**th-order approximation** to the function value.

Taylor expansions of functions are concerned with the small differences between the values of a function at closely-spaced points. Perturbation theory in quantum mechanics is concerned with small differences between the eigenvalues and eigenfunctions for two closely-related Hamiltonians. This is a different and more complicated problem than the Taylor approximation of a function, but there are several common ideas. You have seen that we can express the full Hamiltonian as the sum of an unperturbed Hamiltonian and a perturbation: $\widehat{H} = \widehat{H}^{(0)} + \delta\widehat{H}$, and we will show that each eigenvalue of $\widehat{H}$ can be expressed as a series of terms involving the perturbation $\delta\widehat{H}$ and the eigenvalues and eigenfunctions of the *unperturbed* Hamiltonian $\widehat{H}^{(0)}$. If the perturbation is small, the first few terms in this series will provide a good approximation for the exact eigenvalue. Another common idea is that we can find increasingly accurate approximations by including additional correction terms. The same terminology is used as for Taylor expansions: we can calculate nth-order corrections and nth-order approximations to the eigenvalues and eigenfunctions.

3.2.3 First-order approximation for the energy eigenvalues

In this section we derive an equation for the first-order approximation to the energy eigenvalues of a quantum system.

We start by expressing the Hamiltonian operator for the given system in the form

$$\widehat{H} = \widehat{H}^{(0)} + \widehat{H}^{(1)},$$

where $\widehat{H}^{(0)}$ is a suitably chosen unperturbed Hamiltonian, and we now denote the perturbation $\delta\widehat{H}$ by $\widehat{H}^{(1)}$. The full Hamiltonian has (unknown) eigenvalues and eigenfunctions given by

$$\widehat{H}\psi_n = E_n\psi_n, \tag{3.15}$$

while the unperturbed Hamiltonian has known eigenvalues and eigenfunctions given by

$$\widehat{H}^{(0)}\psi_n^{(0)} = E_n^{(0)}\psi_n^{(0)}. \tag{3.16}$$

We can then formally write the exact eigenfunctions ψ_n and the exact eigenvalues E_n as follows:

$$\psi_n = \psi_n^{(0)} + \psi_n^{(1)} + \psi_n^{(2)} + \cdots, \tag{3.17}$$

$$E_n = E_n^{(0)} + E_n^{(1)} + E_n^{(2)} + \cdots, \tag{3.18}$$

> The superscript (1) will help us to remember that the perturbation $\delta\widehat{H}$ is assumed to be small compared to $\widehat{H}^{(0)}$.

where the superscripts in round brackets indicate the order of the correction. The lowest (zeroth) order of approximation corresponds to using the unperturbed Hamiltonian, eigenfunctions and eigenvalues. We label the first-order corrections by the superscript (1), and give subsequent refinements higher indices. We do not need to say what $\psi_n^{(1)}$ or $E_n^{(1)}$ are yet, but need only know that, when the perturbation is small, these terms are of first order in smallness (being proportional to $\widehat{H}^{(1)}$), while terms labelled (2) are of second order in smallness (being proportional to the *square* of $\widehat{H}^{(1)}$), and so on.

Using these expansions, we now derive an expression for the first-order approximation to the energy eigenvalue E_n. We start with the exact time-independent Schrödinger equation, and take the inner product of both sides with the nth eigenfunction:

$$\langle \psi_n | \widehat{H} | \psi_n \rangle = E_n \langle \psi_n | \psi_n \rangle.$$

The eigenfunction is assumed to be normalized, so $\langle \psi_n | \psi_n \rangle = 1$ and the exact energy eigenvalue is

$$E_n = \langle \psi_n | \widehat{H} | \psi_n \rangle.$$

Of course, this is not much use to us yet because we do not know the exact energy eigenfunction, but we can proceed by substituting the expansion for ψ_n given in Equation 3.17. Using Dirac notation, we have

$$|\psi_n\rangle = |\psi_n^{(0)}\rangle + |\psi_n^{(1)}\rangle + \cdots,$$

and

$$\langle \psi_n | = \langle \psi_n^{(0)} | + \langle \psi_n^{(1)} | + \cdots,$$

so

$$\langle \psi_n | \widehat{H} | \psi_n \rangle = \left(\langle \psi_n^{(0)} | + \langle \psi_n^{(1)} | + \cdots \right) \left(\widehat{H}^{(0)} + \widehat{H}^{(1)} \right) \left(|\psi_n^{(0)}\rangle + |\psi_n^{(1)}\rangle + \cdots \right).$$

Expanding the round brackets then gives

$$\begin{aligned}\langle \psi_n | \widehat{H} | \psi_n \rangle &= \langle \psi_n^{(0)} | \widehat{H}^{(0)} | \psi_n^{(0)} \rangle \\ &+ \langle \psi_n^{(0)} | \widehat{H}^{(1)} | \psi_n^{(0)} \rangle + \langle \psi_n^{(1)} | \widehat{H}^{(0)} | \psi_n^{(0)} \rangle + \langle \psi_n^{(0)} | \widehat{H}^{(0)} | \psi_n^{(1)} \rangle \\ &+ \langle \psi_n^{(0)} | \widehat{H}^{(1)} | \psi_n^{(1)} \rangle + \langle \psi_n^{(1)} | \widehat{H}^{(1)} | \psi_n^{(0)} \rangle + \cdots, \end{aligned} \quad (3.19)$$

where we have grouped terms of different orders on different lines. The zeroth-order approximation is given by the first line, which involves only the unperturbed quantities. The first-order correction is given by the second line. These terms are of first order in smallness because they each involve a single first-order quantity (either $\widehat{H}^{(1)}$, $\langle \psi_n^{(1)} |$ or $|\psi_n^{(1)}\rangle$). In general, the order of a term can be seen by taking the sum of the indices in round brackets. The terms on the last line are therefore of second order in smallness; there are some additional terms of the same order,

$$\langle \psi_n^{(2)} | \widehat{H}^{(0)} | \psi_n^{(0)} \rangle \quad \text{and} \quad \langle \psi_n^{(0)} | \widehat{H}^{(0)} | \psi_n^{(2)} \rangle,$$

but we have not bothered to write these down. Remember that $\langle \psi_n | \widehat{H} | \psi_n \rangle$ is the exact energy eigenvalue E_n, which we are aiming to approximate in a series of successively smaller terms. Equation 3.19 gives us the expansion we are looking for.

Using Equation 3.16 and taking the unperturbed energy eigenfunctions to be normalized, the zeroth-order approximation is

$$\langle \psi_n^{(0)} | \widehat{H}^{(0)} | \psi_n^{(0)} \rangle = E_n^{(0)} \langle \psi_n^{(0)} | \psi_n^{(0)} \rangle = E_n^{(0)}.$$

As expected, the zeroth-order approximation to the energy is given by the unperturbed energy eigenvalue.

The first-order correction is given by the terms in the second line of Equation 3.19:

$$E_n^{(1)} = \langle \psi_n^{(0)} | \widehat{H}^{(1)} | \psi_n^{(0)} \rangle + \langle \psi_n^{(1)} | \widehat{H}^{(0)} | \psi_n^{(0)} \rangle + \langle \psi_n^{(0)} | \widehat{H}^{(0)} | \psi_n^{(1)} \rangle. \quad (3.20)$$

Some simplifications can be made. Using the unperturbed time-independent Schrödinger equation (Equation 3.16), we see that

$$\langle \psi_n^{(1)} | \widehat{H}^{(0)} | \psi_n^{(0)} \rangle = E_n^{(0)} \langle \psi_n^{(1)} | \psi_n^{(0)} \rangle.$$

Also, because the unperturbed Hamiltonian operator is Hermitian, the last term on the right-hand side of Equation 3.20 can be written as

$$\langle \psi_n^{(0)} | \widehat{H}^{(0)} | \psi_n^{(1)} \rangle \equiv \langle \psi_n^{(0)} | \widehat{H}^{(0)} \psi_n^{(1)} \rangle = \langle \widehat{H}^{(0)} \psi_n^{(0)} | \psi_n^{(1)} \rangle.$$

Using the unperturbed Schrödinger equation again, together with the fact that the energy eigenvalues are real, we see that

$$\langle \psi_n^{(0)} | \widehat{H}^{(0)} | \psi_n^{(1)} \rangle = E_n^{(0)} \langle \psi_n^{(0)} | \psi_n^{(1)} \rangle.$$

Inserting these results into Equation 3.20 gives

$$E_n^{(1)} = \langle \psi_n^{(0)} | \widehat{H}^{(1)} | \psi_n^{(0)} \rangle + E_n^{(0)} \left(\langle \psi_n^{(1)} | \psi_n^{(0)} \rangle + \langle \psi_n^{(0)} | \psi_n^{(1)} \rangle \right). \quad (3.21)$$

Finally, let us investigate the term $\langle \psi_n^{(1)} | \psi_n^{(0)} \rangle + \langle \psi_n^{(0)} | \psi_n^{(1)} \rangle$. We can do this by writing down the normalization condition for the exact eigenfunction and expanding it in terms of the unperturbed eigenfunctions. We have

$$1 = \langle \psi_n | \psi_n \rangle = \left(\langle \psi_n^{(0)} | + \langle \psi_n^{(1)} | + \cdots \right) \left(| \psi_n^{(0)} \rangle + | \psi_n^{(1)} \rangle + \cdots \right)$$
$$= \langle \psi_n^{(0)} | \psi_n^{(0)} \rangle$$
$$+ \langle \psi_n^{(1)} | \psi_n^{(0)} \rangle + \langle \psi_n^{(0)} | \psi_n^{(1)} \rangle$$
$$+ \langle \psi_n^{(1)} | \psi_n^{(1)} \rangle + \cdots,$$

where, in the final expansion, terms of different orders appear on different lines. Now, the unperturbed eigenfunctions are normalized too, so we have $\langle \psi_n^{(0)} | \psi_n^{(0)} \rangle = 1$ and therefore, to first order,

$$\langle \psi_n^{(1)} | \psi_n^{(0)} \rangle + \langle \psi_n^{(0)} | \psi_n^{(1)} \rangle = 0.$$

Using this result in Equation 3.21 and reverting to the notation

$$\delta \widehat{H} = \widehat{H}^{(1)}$$

for the perturbation, we conclude that the first-order correction to the nth energy eigenvalue is

$$E_n^{(1)} = \langle \psi_n^{(0)} | \delta \widehat{H} | \psi_n^{(0)} \rangle. \quad (3.22)$$

So, recalling that $E_n = E_n^{(0)} + E_n^{(1)} + E_n^{(2)} + \cdots$, and truncating this expansion at the first-order term, we obtain the following important result:

> **First-order approximation for the nth energy eigenvalue**
>
> In first-order perturbation theory, the nth energy eigenvalue is approximated by
>
> $$E_n \simeq E_n^{(0)} + \langle \psi_n^{(0)} | \delta\widehat{H} | \psi_n^{(0)} \rangle, \tag{3.23}$$
>
> where $E_n^{(0)}$ is the nth unperturbed energy eigenvalue, $\delta\widehat{H}$ is the perturbation, and $\psi_n^{(0)}$ is the nth unperturbed energy eigenfunction. All the quantities on the right-hand side of the equation are known, or can be found by solving the unperturbed time-independent Schrödinger equation.

This result is very plausible. Without any approximations, we can say that the exact eigenvalue, E_n, is equal to the expectation value of the exact Hamiltonian, $\widehat{H}$, taken with respect to the exact eigenfunction, ψ_n. That is,

$$E_n = \langle \psi_n | \widehat{H} | \psi_n \rangle.$$

In the absence of precise information about the exact eigenfunctions, we make do with the eigenfunctions of the *unperturbed* Hamiltonian to obtain the first-order approximation:

$$\begin{aligned} E_n &\simeq \langle \psi_n^{(0)} | \widehat{H} | \psi_n^{(0)} \rangle \\ &= \langle \psi_n^{(0)} | (\widehat{H}^{(0)} + \delta\widehat{H}) | \psi_n^{(0)} \rangle \\ &= E_n^{(0)} + \langle \psi_n^{(0)} | \delta\widehat{H} | \psi_n^{(0)} \rangle, \end{aligned} \tag{3.24}$$

which is just the result in Equation 3.23.

It is worth noting that the first-order correction $E_n^{(1)}$ can be positive or negative, so the first-order approximation for the energy can be higher or lower than the energy for the unperturbed system. Contrast this with the estimates for the ground-state energy obtained using the variational method, where the values obtained must be greater than or equal to the exact ground-state energy.

Although the theory behind the first-order perturbation result is quite different from that behind the variational method, you should note the similarity between Equation 3.24 and Equation 3.2, which we used to approximate the ground-state energy by the variational method. In both cases, the exact energy eigenvalue is approximated by the expectation value of the Hamiltonian of the system with respect to an approximate function. In the variational method, we use a trial function — one that we essentially guess, and then tune a bit to get the best possible estimate. In the perturbation method, we use an eigenfunction of a Hamiltonian that we know is close to that of $\widehat{H}$. In both cases, the accuracy of the solution that we obtain depends on how close the approximate function is to the real eigenfunction.

Essential skill

Using first-order perturbation theory

Worked Example 3.2

A particle of mass m is in the one-dimensional infinite well shown in Figure 3.7.

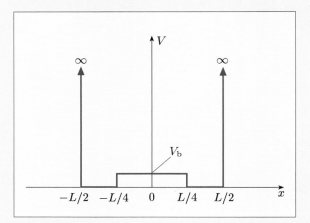

Figure 3.7 A one-dimensional infinite potential energy well with a small bump.

This well is described by the potential energy function

$$V(x) = \begin{cases} 0 & \text{for } -L/2 < x < -L/4, \\ V_b & \text{for } -L/4 \leq x \leq L/4, \\ 0 & \text{for } L/4 < x < L/2, \\ \infty & \text{elsewhere.} \end{cases}$$

Use perturbation theory to find an expression for the frequency of light emitted when the particle makes a transition from the first excited state to the ground state of this well.

Solution

Let the energy of the ground state be E_1 and the energy of the first excited state be E_2. Then the frequency of the emitted light is $(E_2 - E_1)/h$, so we need to estimate these two energy levels. We shall use first-order perturbation theory, taking the unperturbed Hamiltonian to be that for a particle of mass m in the corresponding infinite square well (without the bump), and the perturbation to be

$$\delta \widehat{H} = \begin{cases} V_b & \text{for } -L/4 \leq x \leq L/4, \\ 0 & \text{elsewhere.} \end{cases}$$

From Book 1, Chapter 3, the unperturbed energy eigenfunction and eigenvalue for the ground state are

$$\psi_1^{(0)}(x) = \sqrt{\frac{2}{L}} \cos\left(\frac{\pi x}{L}\right) \quad \text{and} \quad E_1^{(0)} = \frac{\pi^2 \hbar^2}{2mL^2}.$$

Using these results in Equation 3.23, we obtain the first-order approximation to the ground-state eigenvalue:

$$E_1 \simeq E_1^{(0)} + \int_{-L/4}^{L/4} \sqrt{\frac{2}{L}} \cos\left(\frac{\pi x}{L}\right) V_{\text{b}} \sqrt{\frac{2}{L}} \cos\left(\frac{\pi x}{L}\right) \mathrm{d}x$$

$$= E_1^{(0)} + \frac{2V_{\text{b}}}{L} \int_{-L/4}^{L/4} \cos^2\left(\frac{\pi x}{L}\right) \mathrm{d}x.$$

Changing the variable of integration to $y = \pi x/L$ and noting that $\mathrm{d}x = (L/\pi)\,\mathrm{d}y$ and that $x = \pm L/4$ corresponds to $y = \pm \pi/4$, gives

$$E_1 \simeq E_1^{(0)} + \frac{2V_{\text{b}}}{L} \frac{L}{\pi} \int_{-\pi/4}^{\pi/4} \cos^2 y \, \mathrm{d}y.$$

Using a standard integral given inside the back cover, we then obtain

$$E_1 \simeq E_1^{(0)} + \frac{2V_{\text{b}}}{\pi} \left[\frac{y}{2} + \frac{1}{4}\sin(2y)\right]_{-\pi/4}^{\pi/4}$$

$$= E_1^{(0)} + \frac{2V_{\text{b}}}{\pi} \left[\frac{\pi}{4} + \frac{1}{2}\right]$$

$$= E_1^{(0)} + V_{\text{b}} \left[\frac{1}{2} + \frac{1}{\pi}\right].$$

A similar calculation can be carried out for the first excited state. In this case, the unperturbed energy eigenfunction and eigenvalue are

$$\psi_2^{(0)}(x) = \sqrt{\frac{2}{L}} \sin\left(\frac{2\pi x}{L}\right) \quad \text{and} \quad E_2^{(0)} = \frac{4\pi^2 \hbar^2}{2mL^2},$$

and the first-order approximation to the excited-state eigenvalue is

$$E_2 \simeq E_2^{(0)} + \frac{2V_{\text{b}}}{L} \int_{-L/4}^{L/4} \sin^2\left(\frac{2\pi x}{L}\right) \mathrm{d}x.$$

Changing the variable of integration to $y = 2\pi x/L$ and noting that $\mathrm{d}x = (L/2\pi)\,\mathrm{d}y$ and that $x = \pm L/4$ corresponds to $y = \pm \pi/2$, gives

$$E_2 \simeq E_2^{(0)} + \frac{2V_{\text{b}}}{L} \frac{L}{2\pi} \int_{-\pi/2}^{\pi/2} \sin^2 y \, \mathrm{d}y.$$

Using a standard integral given inside the back cover, we then obtain

$$E_2 \simeq E_2^{(0)} + \frac{V_{\text{b}}}{\pi} \left[\frac{y}{2} - \frac{1}{4}\sin(2y)\right]_{-\pi/2}^{\pi/2}$$

$$= E_2^{(0)} + \frac{V_{\text{b}}}{2}.$$

In first-order perturbation theory, our estimate for the frequency of the emitted light is

$$f = \frac{E_2 - E_1}{h} \simeq \frac{1}{h}\left[\left(E_2^{(0)} + \frac{V_{\text{b}}}{2}\right) - \left(E_1^{(0)} + V_{\text{b}}\left[\frac{1}{2} + \frac{1}{\pi}\right]\right)\right]$$

$$= \frac{1}{h}\left[E_2^{(0)} - E_1^{(0)} - \frac{V_{\text{b}}}{\pi}\right]$$

$$= \frac{1}{h}\left[\frac{3\pi^2 \hbar^2}{2mL^2} - \frac{V_{\text{b}}}{\pi}\right].$$

Chapter 3 Time-independent approximation methods

> This frequency is *smaller* than for the unperturbed infinite well because the bump raises the energy of the ground state more than it raises the energy of the first excited state. This makes good sense because the unperturbed eigenfunctions suggest that the particle is more likely to be found near the centre of the well, where the perturbing bump is, in the ground state than in the first excited state.

Essential skill
Using first-order perturbation theory

Exercise 3.4 Use first-order perturbation theory to derive an expression for the energy of the first excited state of an anharmonic oscillator, with mass m, that has the potential energy function $V(x) = \frac{1}{2}Cx^2 + \frac{1}{2}Dx^4$. You may use the following results for the first excited state of a harmonic oscillator (quantum number $n = 1$):

$$E_1^{(0)} = \tfrac{3}{2}\hbar\left(\frac{C}{m}\right)^{1/2}, \quad \psi_1^{(0)} = \left(\frac{1}{2\sqrt{\pi}a}\right)^{1/2}\frac{2x}{a}\,\mathrm{e}^{-x^2/2a^2},$$

where $a = \hbar^{1/2}/(mC)^{1/4}$ is the characteristic length parameter of the oscillator. You should also assume that $Da^2/C \ll 1$.

Essential skill
Using first-order perturbation theory

Exercise 3.5 The Coulomb model of a hydrogen atom discussed in Chapter 2 assumes that the proton is a point charge, so the potential energy function in the Hamiltonian has the Coulomb form, $V(r) = -e^2/4\pi\varepsilon_0 r$. A better model replaces the point charge by a sphere of radius R, with the proton charge e spread uniformly throughout its volume. The potential energy function then turns out to be

$$V(r) = \begin{cases} -\dfrac{e^2}{4\pi\varepsilon_0}\left(\dfrac{3}{2R} - \dfrac{r^2}{2R^3}\right) & \text{for } r \leq R, \\ -\dfrac{e^2}{4\pi\varepsilon_0 r} & \text{for } r \geq R. \end{cases}$$

In this exercise we shall be concerned with the ground state of the atom.

(a) Write down the Hamiltonian operator for the hydrogen atom, assuming that the proton charge is uniformly spread through a sphere of radius R. Also write down an unperturbed Hamiltonian for the hydrogen atom for which the energy eigenvalues are known, and a perturbation to this Hamiltonian that corresponds to the effect of the finite size of the proton.

(b) Use first-order perturbation theory to show that this perturbation shifts the ground-state energy of the atom by a small amount given by the integral

$$-\frac{e^2}{\pi\varepsilon_0 a_0^3}\int_0^R \left(\frac{3r^2}{2R} - \frac{r^4}{2R^3} - r\right)\mathrm{e}^{-2r/a_0}\,\mathrm{d}r.$$

(c) Provided that $R \ll a_0$, it is a good approximation to replace e^{-2r/a_0} by 1 inside the above integral. Use this approximation to obtain a formula for the first-order energy shift of the ground state, expressed in terms of R/a_0. ∎

3.2.4 Higher-order perturbation theory

We have restricted our attention to using perturbation methods for determining the first-order approximation for the energy eigenvalues. The reason for this is that the mathematics involved in finding higher-order approximations is lengthy and time-consuming. However, it is worth indicating the approach used if more accurate approximations are required.

One way of improving our estimate of the energy is to go back to Equation 3.19 and to obtain the second-order relationship, ensuring that our approximation for the energy eigenfunction remains normalized to second order. We will not go into the details here, but merely state the final result. The *second-order correction* to the nth energy eigenvalue turns out to be

$$E_n^{(2)} = \sum_{p \neq n} \frac{|\langle \psi_p^{(0)}| \delta\widehat{H} |\psi_n^{(0)}\rangle|^2}{E_n^{(0)} - E_p^{(0)}}, \tag{3.25}$$

where the sum is over all states. The **second-order approximation** for the nth energy eigenvalue is therefore

$$E_n \approx E_n^{(0)} + \langle \psi_n^{(0)}| \delta\widehat{H} |\psi_n^{(0)}\rangle + \sum_{p \neq n} \frac{|\langle \psi_p^{(0)}| \delta\widehat{H} |\psi_n^{(0)}\rangle|^2}{E_n^{(0)} - E_p^{(0)}}. \tag{3.26}$$

This expression is much more difficult to use than the first-order correction. It involves a sum of terms, each of which involves evaluating a matrix element, usually by integration. Note that the expression appears to break down whenever the unperturbed eigenfunctions $\psi_n^{(0)}$ and $\psi_p^{(0)}$ are degenerate, since we then get an infinite term in the sum. However, it is always possible to choose the eigenfunctions such that $\langle \psi_p^{(0)}| \delta\widehat{H} |\psi_n^{(0)}\rangle = 0$ whenever $E_n^{(0)} = E_p^{(0)}$, and this avoids the infinite terms.

> Any quantity of the form $\langle f|\widehat{O}|g\rangle$ is called a **matrix element**.

The process of refining the eigenfunctions and eigenvalues could be continued. The second-order approximation for the energy could be used to obtain an improved approximation for the eigenfunction, which could then be used to improve the energy eigenvalue. You might think that it would always be advisable to use a large number of terms in the expansion, in order to get very good approximations. However, when using approximation methods to determine properties of real systems, we need to strike a balance between the accuracy of the approximation and the difficulty of performing the calculations. As Equation 3.26 indicates, calculating higher terms in perturbation expansions can quickly become rather complicated. Hence perturbation methods often calculate only the first- or second-order corrections. However, if $\delta\widehat{H}$ is small, even low-order approximations can provide very good solutions. This is an important point because although it is true that higher-order approximations will give more accurate solutions, in many circumstances low-order approximations can still be highly accurate — or, at least, sufficiently accurate for the problem under consideration.

Summary of Chapter 3

Section 3.1 The variational method is especially useful for finding an approximate value of the ground-state energy of a quantum system. When we do not know the energy eigenvalue or eigenfunction for the ground state, we use information that we have about the system to make an educated guess at the form of the eigenfunction. The guessed function — a trial function $\phi_{1,t}(\mathbf{r})$ — may depend on one or more adjustable parameters. For a system characterized by a Hamiltonian $\widehat{H}$, the exact ground-state energy E_1 satisfies the inequality

$$E_1 \leq E_{1,t} = \frac{\langle \phi_{1,t} | \widehat{H} | \phi_{1,t} \rangle}{\langle \phi_{1,t} | \phi_{1,t} \rangle}.$$

The expression for $E_{1,t}$ is minimized with respect to the adjustable parameter(s) in the trial function, and the minimum value gives the optimum estimate of the ground-state energy. Under favourable circumstances, variational methods can also be used to find approximate energy eigenfunctions and eigenvalues for excited states.

Section 3.2 The perturbation method is a technique for obtaining approximate values for the energy eigenvalues and eigenfunctions of a quantum system. When we cannot solve the time-independent Schrödinger equation for a system with a complicated Hamiltonian, we simplify the Hamiltonian into a form for which we can find exact solutions, and use these solutions for the energy eigenvalues and eigenfunctions as the basis for obtaining more accurate solutions for the real Hamiltonian.

For a system characterized by the Hamiltonian $\widehat{H} = \widehat{H}^{(0)} + \delta\widehat{H}$, where $\widehat{H}^{(0)}$ is the unperturbed Hamiltonian with eigenvalues $E_n^{(0)}$ and eigenfunctions $\psi_n^{(0)}$, and $\delta\widehat{H}$ is the perturbation, the first-order approximation for the energy eigenvalues E_n is

$$E_n \simeq E_n^{(0)} + \langle \psi_n^{(0)} | \delta\widehat{H} | \psi_n^{(0)} \rangle.$$

It is also possible to obtain higher-order approximations.

Achievements from Chapter 3

After studying this chapter, you should be able to:

3.1 Explain the meanings of the newly defined (emboldened) terms and symbols, and use them appropriately.

3.2 Explain the need for approximation methods in quantum mechanics.

3.3 Explain the basic principles of the variational method.

3.4 Use the variational method to obtain an estimate for the energy eigenvalue for the ground state of a system.

3.5 Explain the basic principles of perturbation methods.

3.6 Explain the relationship between the order of a perturbation approximation and the accuracy of the result.

3.7 Use perturbation methods to calculate the first-order approximation to the energy eigenvalues in simple cases.

Chapter 4 Hydrogen-like systems

Introduction

When Schrödinger used his equation and the Coulomb model to derive the energy levels of hydrogen atoms, and hence the Balmer series of spectral lines, it was a huge triumph. The patchwork of semi-quantum theories going back to the Bohr model was replaced by a coherent theory, capable of explaining behaviour of matter in ways that were previously unimagined.

And yet ... Schrödinger's derivation was just the first step, even for hydrogen. As noted in Chapter 2, the Balmer series is not a complete account of the visible spectrum for hydrogen atoms. Close examination shows that what appear as single spectral lines in a low-resolution spectroscope are actually sets of closely-spaced lines; the spectrum has *fine structure*. In fact, the old *ad hoc* semi-quantum model of Bohr, as developed relativistically by Sommerfeld and others, had explanations for some of these lines, and this success suggested that an explanation would be found by putting quantum theory and relativity together. This was something that Schrödinger did not achieve. One aim of this chapter is to explain this fine structure. Just as the gross (Balmer) structure led to one momentous conclusion, that Schrödinger's equation was a profound advance, so the fine structure led to equally momentous discoveries, not least the prediction of antiparticles.

Before we delve into the fine structure of hydrogen, we shall apply the hydrogen-atom formalism to a number of hydrogen-like systems. The reason for doing this is partly to reinforce your understanding of the basic hydrogen atom solutions, but also to show the power of these solutions to account for behaviour of great historical and practical importance. The historical example we discuss is the helium ion spectrum, and the practical examples include muonic atoms and X-ray spectra.

Section 4.1 reminds you of a few key features of the hydrogen atom solutions that you met in Chapter 2. We shall then show that two equations derived from the hydrogen atom formalism allow us to understand a surprising range of phenomena. For example, they lead to techniques that allow us to measure not just the sizes of atomic nuclei, but also the different distributions of protons and neutrons within the nuclei.

In Section 4.2, we examine the fine structure of the energy levels of the hydrogen atom. Accounting for this structure required a variety of very significant developments of quantum theory, and calculating the splitting of the energy levels provides examples of perturbation theory in action. However, a deeper understanding of the details of the hydrogen spectrum had to await Dirac's relativistic quantum mechanics. In Section 4.3, we describe the results of this work and some of the remarkable and unexpected spin-offs. Notable among these are the discovery of antiparticles and the birth of quantum field theory.

Chapter 4 Hydrogen-like systems

4.1 Hydrogen-like atoms

The energy levels of a hydrogen atom depend upon the charges of the electron and of the nucleus, and upon the mass of the electron and, to a lesser extent, the mass of the nucleus. In Chapter 2, you saw how deuterium revealed its existence through the small effect of the mass of the nucleus upon the reduced mass of the atom. There are other systems which have essentially the same energy eigenfunctions as hydrogen, apart from scaling due to changes in the masses and charges of the two interacting particles. These systems are of considerable historical and practical importance, and also provide a new window into the hydrogen atom solutions of Chapter 2. Before we discuss the helium ion, muonic atoms and other hydrogen-like systems, we remind you of some key properties of the hydrogen atom solutions.

4.1.1 Key results for the hydrogen atom

In Chapter 2, we saw that a state of the hydrogen atom with principal quantum number n has energy eigenvalue

$$E_n = -\frac{E_R}{n^2}, \tag{4.1}$$

where E_R is the *Rydberg energy* given by

$$E_R = \left(\frac{e^2}{4\pi\varepsilon_0}\right)^2 \frac{\mu_H}{2\hbar^2} = 13.6\,\text{eV}. \tag{4.2}$$

Here e is the magnitude of the charge of the electron or proton, and μ_H is the reduced mass of the hydrogen atom:

$$\mu_H = \frac{m_e m_p}{m_e + m_p}. \tag{4.3}$$

> In general, the reduced mass of a pair of particles with masses m_1 and m_2 is $m_1 m_2/(m_1 + m_2)$. In this case, m_e is the mass of an electron and m_p is the mass of a proton.

We use the subscript H for the reduced mass of a hydrogen atom because we will be considering reduced masses for a number of different systems in this chapter. The postscript at the end of Chapter 2 noted that each of the energy levels E_n is not a single energy level, but a set of closely-spaced energy levels. These closely-spaced energy levels lead to a spectrum in which each line of the Balmer series is, under closer inspection, a set of closely-spaced lines. This is known as the **fine structure** of the spectrum, and is one of the main topics of this chapter.

The radial functions $R_{nl}(r)$ in the hydrogen atom eigenfunctions are all dependent on the ratio r/a_0, where the *Bohr radius* a_0 is given by

$$a_0 = \frac{4\pi\varepsilon_0}{e^2}\frac{\hbar^2}{\mu_H} = 5.29 \times 10^{-11}\,\text{m}. \tag{4.4}$$

The Bohr radius therefore sets the scale for the radial extent of the hydrogen atom wave functions as can be seen from Equation 2.26. In Section 2.5.1, for example, you saw that the values of $\langle r \rangle$ for 1s and 3d states are $\frac{3}{2}a_0$ and $\frac{21}{2}a_0$, respectively. If a_0 were to be reduced by some factor, the expectation value of the radial coordinate, $\langle r \rangle$, would be reduced by the same factor.

Unlike the radial-dependence $R_{nl}(r)$, the angular-dependence of the hydrogen atom energy eigenfunctions is determined by the spherical harmonics $Y_{lm}(\theta, \phi)$, which do not in any way depend on physical properties such as charge or reduced mass.

4.1.2 The helium ion and similar systems

The expressions for the Rydberg energy E_R and the Bohr radius a_0 both contain physical constants e^2 and μ_H that depend on the properties of the electron and the nucleus. You have already seen in Chapter 2 that Equation 4.2 can be used to calculate the stationary-state energies of the deuterium atom simply by changing the reduced mass to account for the greater nuclear mass. In Chapter 2, μ_D/μ_H was calculated to be $1.000\,272$, and using this value leads to correct predictions for the wavelengths in the Balmer series for deuterium. Similarly, applying this modification to Equation 4.4 allows us to calculate the radial extent of the states of deuterium.

Exercise 4.1 Calculate the ratio of the values of $\langle r \rangle$ for the 1s states of the deuterium atom and the hydrogen atom. ■

From Exercise 4.1, we deduce that the deuterium atom is very slightly smaller than the hydrogen atom; we shall find a much more dramatic — and useful — case of atomic shrinkage shortly.

The example of the deuterium spectrum suggests that Equations 4.2–4.4 can be generalized to other systems. Having solved Schrödinger's equation for hydrogen, we have effectively solved it for all 'hydrogen-like' atoms. By **hydrogen-like atom**, I mean a single negatively-charged particle interacting with a positively-charged nucleus through the Coulomb attraction. This definition covers many systems besides hydrogen, deuterium and tritium. All hydrogen-like atoms have a series of energy levels and a radial size determined by Equations 4.1–4.4, except for two modifications.

Tritium is the heaviest hydrogen isotope, having two neutrons as well a proton in the nucleus. It is radioactive.

1. The factor e^2, the product of the magnitudes of the electron and proton charges, becomes Ze^2 for a particle of charge $-e$ interacting with a nucleus of charge Ze, where Z is the atomic number (the number of protons in the nucleus).
2. The reduced mass for hydrogen, μ_H, is replaced by the appropriate reduced mass of the system, μ.

In short, there is a wide range of hydrogen-like atoms for which we can calculate the energy levels and radial extent of the energy eigenfunctions by using the following prescription:

Properties of hydrogen-like atoms

If we denote the reduced mass for ordinary hydrogen by μ_H, then the energy eigenvalues for a hydrogen-like system with reduced mass μ and nuclear charge Ze are given by

$$E_n = -\frac{E_R^{\text{scaled}}}{n^2}, \qquad (4.5)$$

which is Equation 4.1 with the replacement

$$E_R \Longrightarrow E_R^{\text{scaled}} = Z^2 \frac{\mu}{\mu_H} E_R. \qquad (4.6)$$

The radial properties can be calculated by making the following replacement for a_0 in expressions for quantities such as $\langle r \rangle$ that quantify the spatial extent of the eigenfunctions:

$$a_0 \Longrightarrow a_0^{\text{scaled}} = \frac{1}{Z} \frac{\mu_H}{\mu} a_0. \quad (4.7)$$

We shall call E_R^{scaled} and a_0^{scaled} the **scaled Rydberg energy** and the **scaled Bohr radius** respectively.

For example, consider a singly-ionized helium ion, He$^+$, a system consisting of a helium nucleus and a single electron. The nucleus has two protons and two neutrons, and is therefore about four times the mass of a proton. We shall assume throughout that protons and neutrons both have a mass 1840 times that of an electron, m_e. This leads to a ratio μ_{He^+}/μ_H of the reduced masses of He$^+$ and H given by

$$\frac{\mu_{\text{He}^+}}{\mu_H} = \left(\frac{m_e \times 4 \times 1840 m_e}{m_e + 4 \times 1840 m_e}\right) \bigg/ \left(\frac{m_e \times 1840 m_e}{m_e + 1840 m_e}\right)$$

$$= \frac{7360}{7361} \times \frac{1841}{1840} = 1.000\,41. \quad (4.8)$$

More precisely, a neutron is about 1.5 parts in a thousand heavier than a proton, but taking them both to have 1840 times the electron mass m_e still gives μ_D/μ_H or μ_{He^+}/μ_H correct to 7 significant figures.

Essential skill

Extending the hydrogen atom formalism to hydrogen-like systems

Worked Example 4.1

Calculate the energies and the expectation values of the radial coordinate, $\langle r \rangle$, for 1s, 2p and 3d states of the singly-ionized helium ion, He$^+$. (For 1s, 2p and 3d states of the hydrogen atom, $\langle r \rangle$ has values of $\frac{3}{2}a_0$, $5a_0$ and $\frac{21}{2}a_0$, respectively, where $a_0 = 5.29 \times 10^{-11}$ m.)

Solution

The ratio of the reduced masses, μ_{He^+}/μ_H, is $1.000\,41$, as we have just seen. Inserting this value in Equation 4.6, the energies depend on the principal quantum number n through

$$E_n = -\frac{13.6 \times 4 \times 1.000\,41}{n^2} \text{ eV} = -\frac{54.4}{n^2} \text{ eV}$$

to 3 significant figures (the value 13.6 eV for E_R is no more precise than that). The energies of states with $n = 1, 2, 3$ are, respectively, -54.4 eV, -13.6 eV and -6.04 eV.

Using Equation 4.7, the scaled Bohr radius for He$^+$ is

$$a_0^{\text{scaled}} = \frac{a_0}{Z \times 1.000\,41} = \frac{a_0}{2.000\,82},$$

which we can take as $\frac{1}{2}a_0$. Since the values of $\langle r \rangle$ for 1s, 2p and 3d states for hydrogen are $\frac{3}{2}a_0$, $5a_0$ and $\frac{21}{2}a_0$, respectively, where $a_0 = 5.29 \times 10^{-11}$ m, we find that the values of $\langle r \rangle$ for the same three states in He$^+$ are, respectively, $0.75a_0$, $2.5a_0$ and $5.25a_0$, i.e. 3.97×10^{-11} m, 1.32×10^{-10} m and 2.78×10^{-10} m.

- Describe briefly the main differences between (i) the energies of the stationary states, and (ii) the radial extents of stationary states (as measured by $\langle r \rangle$), for singly-ionized helium and for hydrogen.

○ (i) The energies of the stationary states of He$^+$ are about a factor of four larger in magnitude than the energies of the corresponding states of H.

(ii) The radial sizes of the stationary states of He$^+$ are about a factor of two smaller than those of the corresponding states in hydrogen. More precisely, the energies are increased and the radial sizes are reduced, by a further factor of 1.000 41 in each case. But these additional changes are smaller than the precision to which quantities are quoted.

Exercise 4.2 Calculate the energies in eV and the expectation values $\langle r \rangle$ for 1s, 2p and 3d states of the doubly-ionized lithium ion, Li^{2+}. Assume that a lithium nucleus has three protons and four neutrons. ∎

The historical importance of the He$^+$ spectrum

The helium ion spectrum played an important part in the pre-history of quantum theory. In 1896, Pickering in the USA found spectral lines in starlight that had a pattern similar to lines of the Balmer series, and the same lines were found in the laboratory by Fowler in London. Bohr realized that these newly-discovered spectral lines could be fitted by multiplying by four the energies of the stationary states of hydrogen in his semi-quantum model. His interpretation of this was that the states responsible for the newly-discovered spectrum were those of the helium ion. This was important supporting evidence for the Bohr model, which was viewed by many scientists as having many *ad hoc*, not to say disconcerting, features, such as the mysterious jumps in energy. Until then, the Bohr model could explain the hydrogen spectrum but not much else. However, the factor of four did not, in fact, give a perfect fit to the very precise spectroscopy of the time. Fowler pointed out that a factor of 4.0016 was required to get the best fit. Bohr responded by introducing, for the first time, reduced masses into his calculations of the hydrogen spectrum and of the helium ion spectrum, which led to a value of 4.001 63 for the factor by which the hydrogen energies must be multiplied. To quote Pais: 'Up to that time no one had ever produced anything like it in the realm of spectroscopy, agreement between theory and experiment to five significant figures.' This was in 1913, 12 years before the arrival of the full quantum theory.

This account of the Bohr, Pickering, Fowler story is based on the biography of Bohr, *Niels Bohr's times* by Abraham Pais.

4.1.3 Muonic atoms and their uses

We now come to a rather different hydrogen-like system, one that has been of considerable scientific importance over the years: a muonic atom. A **muon**, which we denote μ^-, is a particle rather like an electron in many respects, having the same electric charge and spin, and not being subject to nuclear-type forces. However, its mass is about 207 times greater than that of an electron, and it is unstable, decaying with a half-life of about 2.2×10^{-6} seconds into an electron and a pair of neutrinos. That might seem a short time, but we shall see that it leaves plenty of time for muons to do some interesting and useful things within atoms.

Muons also have positively-charged antiparticles, μ^+, which do not play a role at this point.

Chapter 4 Hydrogen-like systems

A muon can be captured by an atom, forming what is known as a **muonic atom**. The simplest muonic atom consists of a proton with a negatively-charged muon bound to it. The energy levels and the radial extent of a muonic atom can be calculated following the methods used for the helium ion, except in this case it is the change of reduced mass that is important rather than the changes of nuclear charge. Denoting the electron mass by m_e, and taking the mass of the muon to be $m_\mu = 207 m_e$ and the mass of the proton to be $m_p = 1840 m_e$, we find that the reduced mass of a muonic hydrogen atom is

Custom compels us to use the symbol μ both for reduced mass and for the muon particle; however, the context should make the meaning clear.

$$\mu_\mu = \frac{207 m_e \times 1840 m_e}{207 m_e + 1840 m_e} = \frac{207 \times 1840}{207 + 1840} m_e = 186 m_e. \qquad (4.9)$$

Using $\mu_H = (1840/1841) m_e = 0.99946 m_e$, we can say that the ratio of the reduced mass for muonic hydrogen to the reduced mass for ordinary hydrogen, μ_μ / μ_H, is 186 to 3 significant figures. In the cases of deuterium and the helium ion, the increased mass of the nucleus led to a small increase of the reduced mass of the system. In the case of muonic hydrogen, the large increase in the mass of the lighter particle leads to a very substantial increase in the reduced mass, and this has large effects, as the next exercise shows.

Exercise 4.3

(a) Calculate the energy of the photon emitted when the first excited state of muonic hydrogen decays to the ground state.

(b) Calculate $\langle r \rangle$ for the ground state of muonic hydrogen. ∎

- ● Explain in one brief sentence how the energies of the states of muonic hydrogen differ from those of ordinary hydrogen. Also, explain in one brief sentence how the spatial extents of the states for muonic hydrogen differ from the spatial extents of the corresponding states for ordinary hydrogen.

- ○ The energies of the states of muonic hydrogen are 186 times greater in magnitude than the energies of the corresponding states of hydrogen. The states of muonic hydrogen have spatial extents which are 186 times smaller than for the corresponding states of hydrogen.

Muons can be captured by any atoms, and muonic atoms have provided a great deal of important information about atomic nuclei, as well as about fundamental processes. Exercise 4.3 revealed key features that are important for muonic atom experiments in general, and these are summarized below.

X-ray photons have energies in the range 0.1 keV to 1000 keV.

Energies of the states and transitions between states. The photons emitted by muonic atoms are *not* in the eV range, but hundreds of eV, or even keV; these are *X-ray photons*. In practical terms, this means that instead of using optical spectroscopes, detection methods use solid-state detectors similar to those used for measuring the energy of gamma rays.

Radial extent of the eigenfunctions. The radial function $R_{nl}(r)$ depends on the scaled Bohr radius; this is why $\langle r \rangle$ is a factor of 186 smaller for muonic hydrogen than for ordinary hydrogen. One important consequence of this is that the finite size of the nucleus must be taken into account in a model of the muonic atom. For the ground state of ordinary hydrogen, with $n = 1$ and $l = 0$, the most probable place to find the electron is at the centre of the proton, since this is where $|\psi|^2$ has its largest value. However, the probability of finding the electron within the tiny volume of the proton, roughly 10^{-15} m in radius, is very small indeed. This makes the Coulomb model of a hydrogen atom, in which the proton is treated as a point charge, a good approximation. However, because of its highly contracted radial eigenfunction, a muon in muonic hydrogen has a much greater probability of being found within the volume of the proton. The Coulomb potential energy, proportional to $1/r$, is therefore not valid over a significant region of the eigenfunction. If one knows the distribution of charge within the proton, the correct form of $V(r)$ can be calculated using standard electromagnetic theory. The deviation from a pure Coulomb interaction can then be corrected for using perturbation theory (as in Exercise 3.5).

Measuring atomic nuclei with muons

The detailed pattern of the energy levels of a muonic atom can be used to determine the size of an atomic nucleus, and indeed it provides significant information about the way the protons are distributed within the nucleus. We now shift our attention to atoms with many electrons and with a large value of atomic number Z — anything up to 92 for uranium, or beyond. When a muon is captured by such an atom, it is not inhibited from occupying a state with the same set of quantum numbers as a state that is occupied by an electron. The Pauli exclusion principle does not apply in this case because muons and electrons are not identical particles. The muon rapidly loses energy, briefly stopping off in states with successively smaller values of the principal quantum number, n, and finally occupying a state with $n = 1$ and a very small value of $\langle r \rangle$.

The Pauli exclusion principle was introduced in Chapter 4 of Book 2. Its relevance to the structure of atoms is fully discussed in Chapter 5 of this book.

- Which *two* factors conspire to make $\langle r \rangle$ particularly small for states of muons in heavy atoms?

○ Referring to Equation 4.7, the quantity corresponding to a_0, which determines the spatial extent of the muon state, has a factor of Z in the denominator. Also, μ for muonic atoms is very much larger than for electronic atoms. Both of these factors reduce the spatial extent of the muonic eigenfunction.

The answer to the above question makes the implicit assumption that the state of a muon within an atom that already contains many electrons can be described as being hydrogen-like. To justify this, we use a fact from the classical theory of electromagnetism: the electric field at a distance r from the centre of a spherically-symmetric charge distribution is the same as that due to a point charge Q located at the centre of the distribution, where Q is the total amount of charge that is *closer to the centre* than r. Assuming that the electrons in the atom are distributed in a spherically-symmetric way around the nucleus, and bearing in mind that the muon is much closer to the nucleus than nearly all of the electron charge distribution, this means that the muon will scarcely feel any force due to the electrons; it only feels the Coulomb force towards the centre of the nucleus. This is reasonable because, very close to the nucleus, we would expect the forces due to different parts of the electron charge distribution to cancel out.

This is a consequence of a result called Gauss's law.

To take a definite example, let us consider a muon, in a state with quantum numbers $n = 1, l = 0$, in a tin atom with atomic number $Z = 50$ and mass number $A = 120$. In this case, the muon energy eigenfunction is roughly 207 times smaller in extent, as measured by $\langle r \rangle$, than an electron energy eigenfunction with the same quantum numbers.

The **mass number** is the sum of the number of protons and the number of neutrons in the nucleus.

Exercise 4.4 Explain why the radial functions for muonic tin are 207 times smaller than the corresponding electronic radial functions, whereas the radial functions for muonic hydrogen are 186 times smaller than the corresponding electronic radial functions. ■

So here is the picture of a muonic atom when the muon is in a state with a small value of its principal quantum number n: the muon and the nucleus are, in effect, a hydrogen-like system because the radial extent of the muon's eigenfunction is so small that essentially all the charge associated with the electrons is located outside the region roughly defined by the muon's $\langle r \rangle$. The muon and the nucleus do not constitute a hydrogen-like system if the muon is in a state with very large n, like the Rydberg states you met in Chapter 2. However, muons in such states would soon find themselves in the low-n states, as we shall see.

We now calculate $\langle r \rangle$ for a muon with quantum numbers $n = 1$ and $l = 0$, bound to a tin atom with $Z = 50$. We follow the argument that we have used several times before, noting that for hydrogen, the value of $\langle r \rangle$ for an electron with these quantum numbers is $\frac{3}{2} \times 5.29 \times 10^{-11}$ m, much greater than the radius of the tin nucleus, which is approximately 6×10^{-15} m. For a muon bound to a tin nucleus in an orbital with $n = 1$,

$$\langle r \rangle = \frac{1}{50 \times 207} \times \tfrac{3}{2} \times 5.29 \times 10^{-11} \text{ m} = 7.67 \times 10^{-15} \text{ m}, \quad (4.10)$$

so the spatial extent of this muon eigenfunction is comparable to the size of the nucleus.

Strictly speaking, the calculation we have just done is only a rough approximation. This is because Equation 4.4 is derived in Chapter 2 with the implicit assumption that the nucleus is a point charge. However, the tin nucleus is *not* very small in comparison to the spatial extent of muon states with low n. In such a case, the energy eigenfunctions and energy eigenvalues for a muon bound to a nucleus cannot be found using the simple Coulomb model that postulates a point charge, since the problem involves an extended charge. Finding the energy eigenfunctions and eigenvalues for a specific distribution of nuclear charge requires the use of perturbation theory and numerical methods on a computer. The good news is that this provides an opportunity to turn the whole process around — the size of a nucleus can be deduced from the energy levels of muonic atoms.

Here is an outline of how muons are used to measure the size of atomic nuclei. Muons are produced copiously by particle accelerators, and can be captured by atoms containing the particular nucleus to be studied. The muons are stopped within a target containing the appropriate atoms, and are initially captured into orbitals with very large values of n. They then rapidly lose energy within the atoms as they make a sequence of transitions from higher-energy states to lower-energy states, sometimes ejecting electrons from that atom, but in the final stages emitting photons. The time taken for a muon to end up in the $n = 1$ ground state is less than 10^{-14} seconds, about 10^8 times smaller than the muon half-life.

The energy ranges of X-ray photons and γ-ray photons overlap.

The photons that are emitted in the last few transitions are customarily referred to as X-ray photons because of the way they are produced, but in fact their energies are well within the range of energies of γ-rays produced by nuclei. In Exercise 4.3, you found that the photon emitted when a muon bound to a proton makes a transition from an $n = 2$ state to the $n = 1$ state is about 2 keV. For tin, with $Z = 50$, this energy will be multiplied by a factor of $50^2 \times 207/186$, which would make it about 5 MeV.

The scheme for measuring nuclear sizes with muons is as follows. The charge density of the nucleus is postulated to have the general form shown in Figure 4.1, which is represented by the equation

$$\rho(r) = \frac{\rho_0}{1 + \exp[(r-c)/a]}, \tag{4.11}$$

where $c \gg a$ and ρ_0 is essentially the charge density at the centre of the nucleus, $r = 0$. The parameter c is the radius at which the charge density falls to $\rho_0/2$, and the parameter a determines the distance over which the charge density drops off at the surface of the nucleus. The *skin thickness* τ is the difference between the radii at which $\rho(r) = 0.1\rho_0$ and $\rho(r) = 0.9\rho_0$, and this is given by $\tau = 4a \ln 3$; a larger value of τ means that the charge density falls off more slowly. The total nuclear charge is Ze, so

$$4\pi \int_0^\infty \rho(r) r^2 \, dr = Ze, \tag{4.12}$$

and ρ_0 must be adjusted so that this holds true.

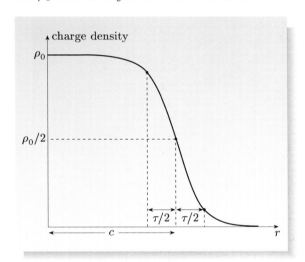

Figure 4.1 The postulated nuclear charge density, showing the central density ρ_0, the half-density radius c, and the skin thickness τ.

The form of $\rho(r)$ in Equation 4.11 embodies the general property of atomic nuclei that they have a region of approximately constant charge density at the nuclear centre, and this charge density is much the same for all but the lightest nuclei. Given this expression for the charge density $\rho(r)$, the electrostatic potential energy function $V_\mu(r)$ of the muon can be calculated using classical electrostatics. Inserting $V_\mu(r)$ calculated in this way into the time-independent Schrödinger equation, we can obtain the energy eigenvalues for the muon. These energy eigenvalues can be recalculated using different functions $V_\mu(r)$ that are determined using a range of values of the parameters c and a. These parameters are varied until the energy eigenvalues of the muon states are consistent with the energy differences revealed by the measured X-ray energies. The parameters that give the best fit to the experimental data can be used to find a *mean radius*, a number characterizing the nuclear size. This quantity can be measured to four significant figures. Such measurements provide an important experimental foundation for our understanding of the properties of atomic nuclei.

Because the muon moves very rapidly, accurate calculations include relativistic corrections, or replace the time-independent Schrödinger equation by its relativistic cousin, the Dirac equation (see Section 4.3.1). Because of these complications, and the detailed form of the potential energy function, the solutions are obtained numerically, rather than as analytic functions.

Short-lived radioactive nuclei are of great scientific interest, not least for the important role they play in stellar events, such as supernovas, and in the production of the chemical elements. Such nuclei cannot be fixed in targets, so it is difficult to use the muon technique we have described, or rival techniques based on high-energy electron diffraction. However, beams of short-lived radioactive nuclei can be produced, and, as this chapter is being written, there is an active research program aimed at measuring the properties of short-lived nuclei by producing beams of their muonic atoms. An Anglo–Japanese collaboration is currently exploring this possibility (see Figure 4.2).

Figure 4.2 Equipment used to produce a beam of short-lived radioactive muonic ions. The large dark green structure, A, is concrete shielding, within which muons are created in a process involving much radiation. The muons emerge through a vacuum beam line B, and are absorbed by a solid hydrogen film, at C, kept at a temperature of $3\,\text{K}$ by a cryostat, the shiny stainless steel vertical cylinder D. At the same time, a beam of ions passes through another vacuum beam line, E and F, leading from the ion source in the pale blue cube G, via a dark blue bending magnet H, to the solid hydrogen film within C, where the muons have been captured. The ions become neutral atoms here, and pick up muons that have been absorbed in the solid hydrogen. (Research collaboration between RIKEN in Japan and Rutherford Appleton Laboratory in the UK, where there is a good source of muons.)

4.1.4 Not just muons...

Muons and electrons are not the only negatively-charged particles that can form a hydrogen-like system with a positively-charged nucleus. The same thing can be done with **antiprotons**, which have the same mass as protons but the opposite charge.

An antiproton interacts strongly with neutrons and so, if captured by an atom to form an antiprotonic atom, provides the opportunity for measuring the distribution of neutrons within nuclei. This is of interest because there are reasons for believing that the distribution of neutrons differs somewhat from the distribution of protons, especially for heavy nuclei like $^{208}_{82}$Pb.

Exercise 4.5 For a given value of n, how much smaller (as measured by $\langle r \rangle$) is the energy eigenfunction of an antiproton than the corresponding eigenfunction of a muon? Assume that the particles are bound to a heavy nucleus, such as $^{208}_{82}$Pb. ∎

Now, antiproton eigenfunctions with low values of n would be located deep within the nucleus, as Exercise 4.5 demonstrated. However, antiprotons quickly annihilate with protons in the nucleus and never reach states with very low n. But the many transitions between states with n in the range 10–20 can be exploited to provide evidence concerning the neutron distribution. Evidence was presented in 2007, on the basis of experiments with antiprotons, that the neutrons in $^{208}_{82}$Pb extend about 0.14×10^{-15} m further than the protons.

Finally, we note that many other negatively-charged particles, such as **pions**, can be captured by atoms and fall into states with low n. These are again hydrogen-like systems in which the negatively-charged particle is in a state that has a large overlap with the nucleus. Such systems have also been exploited as sources of information about nuclei.

Pions are particles that are about 30% heavier than muons. Unlike muons but like antiprotons, pions interact strongly with nuclei. They are produced copiously in collisions between matter and high-energy protons from particle accelerators.

4.1.5 Deep within atoms: X-ray spectra

For our final example of a hydrogen-like atom, we turn to the production of X-rays. When electrons with energies of tens of thousands of electronvolts strike a block of metal, such as molybdenum or tungsten, X-rays are emitted. When the intensity of the X-rays is measured as a function of the wavelength, we find an *X-ray spectrum*, rather like that shown for molybdenum in Figure 4.3. There is a smooth curve, which drops sharply to zero at a wavelength λ of about 0.037 nm, with two sharp peaks below and just above 0.07 nm.

The smooth curve and the peaks arise from X-rays generated in quite different ways. The X-rays that depend smoothly on λ are known as **bremsstrahlung**, from the German for 'braking radiation'; this radiation is generated by the rapid slowing down of the electrons as they strike the metal. However, the X-rays corresponding to the sharp peaks at 0.065 nm and 0.071 nm are of more interest here. These peaks are spectral lines like those in the Balmer series, but with photons of much higher energies. Like the lines in the Balmer series, they correspond to photons of definite energy emitted when the molybdenum atom makes a transition from one stationary state to another of lower energy.

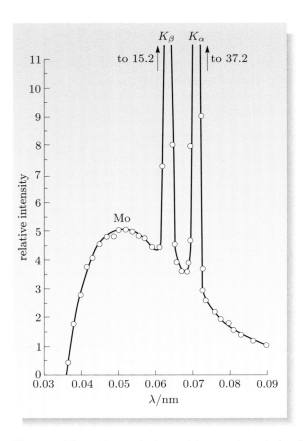

Figure 4.3 The variation with wavelength λ of the intensity of the X-rays emitted when 35 keV electrons strike a block of molybdenum, $Z = 42$. The K_α peak corresponds to a transition from an $n = 2$ state to an $n = 1$ state, while the K_β peak corresponds to a transition from an $n = 3$ state to an $n = 1$ state.

How are such high-energy photons generated within atoms? The photons in the optical spectrum that are emitted when a material is heated in a flame, for example, have energies of a few eV, whereas X-ray photons have something like a thousand times that energy. These high-energy photons arise from quantum jumps by deep-lying electrons within atoms. Here, 'deep-lying' means that these electrons have their eigenfunctions deep within the atom, with small values of $\langle r \rangle$, and small values of the principal quantum number n. The full meaning of 'deep-lying' will be clearer after the discussion of many-electron atoms in the next chapter, but the general idea is this. Each electron in an atom has a set of quantum numbers, n, l, m, m_s. Because electrons are identical, the Pauli exclusion principle comes into play, and only one electron can have a particular set of quantum numbers. The electrons with the smallest value of n are the most tightly bound, meaning that their (negative) energy has the greatest magnitude. This is some tens of keV for an element of medium atomic number, like molybdenum, and becomes even larger for very heavy elements like uranium.

Now, when a lump of molybdenum is struck by a high-energy electron, there is a finite chance that one of the low-lying electrons will be knocked out of the atom. When this happens, an electron with higher energy makes a transition to the vacated state, and as it does so, an X-ray photon carries off the energy.

For essentially the same reasons as discussed on page 95, an electron with $n = 1$ within a heavy atom hardly feels any effect of the charge of the many less-deeply-lying electrons. Apart from the fact that there is another $n = 1$ electron, it would, together with the nucleus, constitute a hydrogen-like system, and its energy would be

$$E_n = -\frac{13.6 Z^2}{n^2} \text{ eV}, \tag{4.13}$$

where we have left n in the expression, although we have $n = 1$ in mind. It is the Z^2 factor in this expression, which is 1764 for molybdenum ($Z = 42$), that leads to X-ray photons having keV rather than eV energies. It turns out that the effect of the second $n = 1$ electron can be taken into account reasonably well by replacing Z by $Z - 1$ in Equation 4.13. In around 1913, Moseley systematically measured the X-ray wavelengths for many elements, and found that the energies of the photons that correspond to peaks like the one marked K_α in Figure 4.3 could be fitted by the expression

$$E_{K_\alpha} = \left[\frac{1}{1^2} - \frac{1}{2^2}\right](Z - 1)^2 \, 13.6 \text{ eV} = \tfrac{3}{4} \times (Z - 1)^2 \, 13.6 \text{ eV}. \tag{4.14}$$

This is equivalent to the energy released when a hydrogen-like atom, in which an electron is bound to a nucleus of charge $(Z - 1)e$, jumps from an $n = 2$ state to the $n = 1$ state. The fact that $Z - 1$ rather than Z occurs in Equation 4.13 takes account of the presence of the second $n = 1$ electron, a so-called *screening effect*.

The other emission peak in Figure 4.3, labelled K_β, corresponds to the higher-energy photon that is emitted when an electron jumps from a state with $n = 3$ to a state with $n = 1$. It should be said that going beyond the simple picture that gives Equation 4.14 to an exact account of the X-ray spectra of atoms is a very non-trivial task.

The energy-dependence discovered by Moseley, and given in Equation 4.14, was of immense scientific importance. It was this that established the significance of the atomic number Z as the fundamental ordering principle of the chemical elements, rather than the **relative atomic mass** as had been thought. This sorted out some annoying anomalies that had appeared in early versions of the Periodic Table: for example, chemists had concluded already that argon, with a relative atomic mass of 39.95, should come *before* potassium, which has a lower atomic mass, 39.10. Moseley's X-ray results showed that argon did indeed have a lower atomic number, $Z = 18$, than potassium, $Z = 19$. The anomalous order of the atomic masses is due to the fact that the predominant potassium isotope is $^{39}_{19}\text{K}$ (with 20 neutrons), whereas the predominant argon isotope is $^{40}_{18}\text{A}$ (with 22 neutrons).

The relative atomic mass of an atom is the ratio of the mass of the atom to $1/12$ of the mass of the ^{12}C atom. The relative atomic mass as measured by chemists is an average of the relative atomic masses of the isotopes, weighted by their relative abundances.

Exercise 4.6 Use Equation 4.14 to calculate the wavelength of the K_α X-ray lines in copper, Cu ($Z = 29$) and tungsten, W ($Z = 74$). Comment on the difference in wavelengths of these two X-ray lines. ∎

4.2 The hydrogen spectrum 'under the microscope'

Rather than explore further varieties of hydrogen-like systems, we shall now look more closely at the archetypal hydrogen-like system, the hydrogen atom itself.

4.2.1 Fine structure of the hydrogen spectrum

A high-resolution spectrometer reveals that the lines of the Lyman, Balmer and other series in the hydrogen spectrum are *not* single lines, but multiple lines. This is because there are a number of very closely-spaced energy levels for each value of the principal quantum number n, as shown in the second column of Figure 4.4, where the departures from the energy levels given by Equation 4.2 are magnified by a factor of 137^2.

The reason for choosing a factor of 137^2 will emerge shortly.

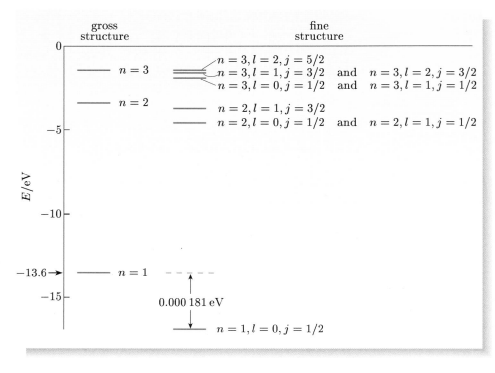

Figure 4.4 On the left are the three lowest energy levels of a hydrogen atom as presented in Chapter 2. High-resolution spectroscopy reveals that each of these energy levels is actually a number of very closely-spaced levels, and on the right we show this fine structure expanded by a factor of 137^2.

It is apparent from this figure that the level with principal quantum number n is split into n components. For example, the level with $n = 3$ is split into three. The lowest of these energy levels corresponds to two sets of quantum numbers: $(l = 0, \ j = \frac{1}{2})$ and $(l = 1, \ j = \frac{1}{2})$.

- Allowing for the fact that states with a given value of j may have different values of m_j, how many different states in total correspond to the lowest energy level with $n = 3$?

○ For each pair of quantum numbers (l, j), there are $2j + 1$ different values of m_j. Since the $n = 3$ level has two l values corresponding to $j = \frac{1}{2}$, the total number of states in this energy level is $2(2 \times \frac{1}{2} + 1) = 4$.

The next higher energy for $n = 3$ corresponds to $j = \frac{3}{2}$ and to two values of l (1 and 2); finally the, highest energy corresponds to $(l = 2, j = \frac{5}{2})$.

A remarkable feature of the energy level splitting shown in Figure 4.4 is that states with the same j but different l apparently have the same energy. A successful theory for the fine structure must explain this surprising feature.

We shall now give an explanation of the origin of the fine structure of hydrogen. The arguments we give here are based on the time-independent Schrödinger equation and are correct as far as they go, but Section 4.3 will briefly mention an alternative approach based on Dirac's relativistic quantum mechanics.

As we pointed out earlier, all of the energy shifts shown in Figure 4.4 have been magnified by a factor of 137^2. The changes in energy are clearly very small compared to the original energies. This suggests that they can be accounted for using first-order perturbation theory, as described in Chapter 3. In other words, the energy shift for a state described by the unperturbed eigenfunction ψ_{nlj} is given to a first-order approximation by

$$E^{(1)}_{nlj} \simeq \langle \psi_{nlj} | \delta\widehat{H} | \psi_{nlj} \rangle,$$

where $\delta\widehat{H}$ is the perturbation to the Coulomb model Hamiltonian. It turns out that there are three main contributions to the perturbation $\delta\widehat{H}$:

1. Relativistic effects modify the form of the kinetic energy term.

2. Spin–orbit effects must be taken into account for states with $l \neq 0$.

3. An additional term, called the Darwin term, arises only if $l = 0$.

You met spin–orbit interactions in Chapter 1 and saw how they produce a splitting of the energy levels of sodium atoms, leading to two closely-spaced yellow spectral lines. In hydrogen, the other two effects listed above are of a similar magnitude to the spin–orbit effect, and all three are grouped together as fine-structure perturbations. Even smaller perturbations exist (such as that due to the finite size of the proton); these are called hyperfine-structure effects, and will be discussed later.

We shall now go through the three fine-structure effects in turn.

1. The relativistic correction to the kinetic energy

The kinetic energy term in the Hamiltonian operator in Schrödinger's equation is derived from the classical non-relativistic expression for the kinetic energy of a particle, $p^2/2m$. However, this expression is no longer correct for speeds approaching the speed of light. An idea of the importance of relativistic corrections can be obtained by considering the ground state of hydrogen. You saw in Chapter 2 (Exercise 2.8) that the expectation value of the Coulomb potential energy in the ground state of a hydrogen atom is $\langle V \rangle = -2E_R$. Since the total energy in the ground state is $E_1 = -E_R$, it follows that the expectation value of the kinetic energy in the ground state is

$$\langle E_{\text{kin}} \rangle = E_1 - \langle V \rangle = -E_R + 2E_R = E_R.$$

To get some measure of the importance of relativistic effects, we interpret $\langle E_{\text{kin}} \rangle$ as a classical kinetic energy $\frac{1}{2}m_e v^2$. Doing this, we get

$$v = \sqrt{\frac{2E_R}{m_e}} \tag{4.15}$$

so

$$\frac{v}{c} = \sqrt{\frac{2E_R}{m_e c^2}}, \tag{4.16}$$

where c is the speed of light. In special relativity, the quantity $m_e c^2$ is called the *rest energy* of the electron; it is the energy that a stationary electron has by virtue of possessing mass (an example of Einstein's famous equation $E = mc^2$).

In this book, 'relativity' refers to special relativity, not general relativity (the relativistic theory of gravity).

For the present discussion, we ignore the difference between the electron mass m_e and its reduced mass μ in hydrogen. We also use the non-relativistic expression $\frac{1}{2}mv^2$ for kinetic energy, anticipating the fact that relativistic effects will be small.

Using Equation 4.2, we obtain

$$\frac{v}{c} = \frac{e^2}{4\pi\varepsilon_0 \hbar c} \simeq \frac{1}{137}. \tag{4.17}$$

Relativistic corrections are usually of order v^2/c^2, which is smaller than 10^{-4} in our case. So, relativistic effects are expected to be small in hydrogen, but given the precision of spectroscopic measurements, they are not completely negligible.

The above argument is not rigorous, but it provides a fair guide, and the dimensionless quantity $e^2/4\pi\varepsilon_0 \hbar c$ turns out to be important in all discussions of relativistic corrections in atoms. The so-called **fine structure constant** α is defined as

$$\alpha \equiv \frac{e^2}{4\pi\varepsilon_0 \hbar c}. \tag{4.18}$$

The value of α is now known to great precision,

$$\alpha = 1/(137.035\,999\,679 \pm 0.000\,000\,094), \tag{4.19}$$

i.e. to 0.7 parts per billion. Note that the factor 137^2 by which the fine structure deviations have been scaled in Figure 4.4, in order that they be visible, is $1/\alpha^2$.

We now estimate the contribution to $\delta\widehat{H}$ from the relativistic correction to the kinetic energy. In what follows we shall gloss over the small difference between m_e and the reduced mass μ in order to concentrate on the essential point. According to special relativity, the total energy of a particle of mass m having momentum of magnitude p is $E_{\rm rel} = [(mc^2)^2 + p^2c^2]^{1/2}$. Subtracting the rest energy mc^2 of the particle, we get the relativistic expression for the kinetic energy:

$$E_{\rm kin} = [(mc^2)^2 + p^2c^2]^{1/2} - mc^2. \tag{4.20}$$

Using a Taylor (or binomial) expansion, we obtain

$$\begin{aligned} E_{\rm kin} &= mc^2\left(1 + \frac{p^2}{m^2c^2}\right)^{1/2} - mc^2 \\ &= mc^2\left(1 + \frac{1}{2}\frac{p^2}{m^2c^2} - \frac{1}{8}\frac{p^4}{m^4c^4} + \cdots\right) - mc^2 \\ &= \frac{p^2}{2m} - \frac{1}{8}\frac{p^4}{m^3c^2} + \cdots. \end{aligned} \tag{4.21}$$

The first term is the standard non-relativistic expression for the kinetic energy. We get the usual non-relativistic kinetic energy term in the Hamiltonian $\widehat{H}$ by noting that

$$p^2 = p_x^2 + p_y^2 + p_z^2$$

and by making the replacements

$$p_x \Longrightarrow \widehat{p}_x = -i\hbar\frac{\partial}{\partial x} \quad \text{etc.,}$$

leading to

$$\frac{p^2}{2m} = -\frac{\hbar^2}{2m}\nabla^2.$$

The second term in Equation 4.21 can be rewritten using the same substitution, and this represents a relativistic perturbation to the kinetic energy term in the Hamiltonian:

$$\delta \widehat{H}_{\rm rel} = -\frac{1}{8}\frac{\hbar^4(\nabla^2)^2}{m^3 c^2}. \qquad (4.22)$$

This perturbation might surprise you; most of the examples of perturbations discussed in Chapter 3 were modifications to the potential energy rather than the kinetic energy. Nevertheless, sandwiching $\delta\widehat{H}_{\rm rel}$ between the bra and ket representing the unperturbed eigenfunction for each state will give the first-order correction to the energy resulting from this relativistic correction to the kinetic energy.

We shall see what contribution the relativistic kinetic energy correction makes when we put all the contributions together, but first we look into another contribution to the fine structure.

2. The spin–orbit correction

In Chapter 1, you saw that the spin–orbit interaction produces a term in the Hamiltonian of the form $V_{\rm so}(r)\,\widehat{\mathbf{L}}\cdot\widehat{\mathbf{S}}$. This term ensures that states with the same values of n and $l(\neq 0)$, but with different values of j (namely $j = l + \frac{1}{2}$ and $j = l - \frac{1}{2}$) have different energies.

The spin–orbit interaction has its origins in relativistic quantum theory, but it can be made plausible by the following semi-classical argument. In classical terms, an electron described by an eigenfunction with $l > 0$, can be thought of as being in a circular orbit around the nucleus. In the rest frame of the electron, however, we would put this the other way round: the electron sees an orbiting, positively-charged nucleus. The orbiting nucleus constitutes an electric current $I \propto v/r \propto |\mathbf{L}|/r^2$, where v is the speed, r is the radius, and $\mathbf{L}$ is the angular momentum of the electron in the rest frame of the nucleus. The current due to the nucleus behaves like a current loop, producing a magnetic field of magnitude $B \propto I/r$ at the position of the electron. The direction of this field is the same as the direction of $\mathbf{L}$. The electron behaves like a tiny magnet with a magnetic dipole moment $\boldsymbol{\mu}$ that is proportional to its spin, $\mathbf{S}$. Since, the potential energy of a magnetic dipole $\boldsymbol{\mu}$ in a magnetic field $\mathbf{B}$ is $-\boldsymbol{\mu}\cdot\mathbf{B}$, the spin–orbit interaction is proportional to $\mathbf{L}\cdot\mathbf{S}$. When all the details are included, including a correction for the fact that the rest frame of an orbiting electron is not an inertial frame, the following expression for the spin–orbit interaction is obtained:

$$\begin{aligned}\delta\widehat{H}_{\rm so} = V_{\rm so}(r)\,\widehat{\mathbf{L}}\cdot\widehat{\mathbf{S}} &= \frac{e^2}{8\pi\varepsilon_0 m_{\rm e}^2 c^2}\frac{1}{r^3}\,\widehat{\mathbf{L}}\cdot\widehat{\mathbf{S}} \\ &= \frac{\alpha\hbar}{2m_{\rm e}^2 c}\frac{1}{r^3}\,\widehat{\mathbf{L}}\cdot\widehat{\mathbf{S}}. \end{aligned} \qquad (4.23)$$

Equation 4.23 defines $V_{\rm so}(r)$ and makes its dependence on the fine structure constant α explicit.

To evaluate the energy shift caused by the spin–orbit perturbation, we recall from Chapter 1 that $\widehat{\mathbf{L}}\cdot\widehat{\mathbf{S}} = \frac{1}{2}[\widehat{J}^2 - \widehat{L}^2 - \widehat{S}^2]$. When $\widehat{\mathbf{L}}\cdot\widehat{\mathbf{S}}$ operates on a state that is an eigenfunction of all three of $\widehat{J}^2$, $\widehat{L}^2$ and $\widehat{S}^2$, we need only insert the eigenvalues of these three operators in order to determine the eigenvalue of $\widehat{\mathbf{L}}\cdot\widehat{\mathbf{S}}$. For the states of an electron in hydrogen, we know that the eigenvalue of $\widehat{S}^2$ is what it always is for a spin-$\frac{1}{2}$ particle, $\frac{1}{2}(\frac{1}{2}+1)\hbar^2 = \frac{3}{4}\hbar^2$. Similarly, for a state labelled with quantum numbers l and j, the eigenvalues of $\widehat{L}^2$ and $\widehat{J}^2$ are, respectively, $l(l+1)\hbar^2$ and $j(j+1)\hbar^2$.

Exercise 4.7 Show that the eigenvalues of $\widehat{\mathbf{L}} \cdot \widehat{\mathbf{S}}$ for the two states with a given non-zero value of l, but different values of j, namely $j = l - \frac{1}{2}$ and $j = l + \frac{1}{2}$, are

$$-(l+1)\frac{\hbar^2}{2} \quad \text{and} \quad l\frac{\hbar^2}{2}.$$

What is the eigenvalue of $\widehat{\mathbf{L}} \cdot \widehat{\mathbf{S}}$ when $l = 0$ and what is the significance of this result? ∎

3. The Darwin term

There is no spin–orbit interaction for states with $l = 0$, but there is a term that applies *only* for $l = 0$. The characteristic feature of eigenfunctions for $l = 0$ states is that the radial factor $R_{nl}(r)$ is not zero at $r = 0$. In fact, for the hydrogen atom ground state (the $n = 1$, $l = 0$ state), the most likely point in the atom for the electron to be found is at the centre of the proton. It is the fact that $l = 0$ eigenfunctions are finite at the origin that brings into play a new term called the **Darwin term**. This term has its origin in relativistic quantum mechanics and takes the form

$$\delta\widehat{H}_{\text{Darwin}} = \frac{\hbar^2}{8m_e^2 c^2}\nabla^2 V(\mathbf{r}), \tag{4.24}$$

where $V(\mathbf{r})$ is the Coulomb potential energy function due to the charge on the proton. The term $\nabla^2 V(\mathbf{r})$ is zero outside the tiny volume of the proton.

The Darwin term is due to Charles Darwin's grandson.

Poisson's law of electrostatics tells us that $\nabla^2 V(\mathbf{r}) \propto \rho(\mathbf{r})$, where $\rho(\mathbf{r})$ is the *proton's* charge density at $\mathbf{r}$; this is zero outside the proton.

Putting it all together

Adding together the three fine-structure perturbation terms gives a total perturbation

$$\delta\widehat{H} = \delta\widehat{H}_{\text{rel}} + \delta\widehat{H}_{\text{so}} + \delta\widehat{H}_{\text{Darwin}}. \tag{4.25}$$

Then using first-order perturbation theory in the unperturbed state $|n,l,j\rangle = \psi_{nlj}(r,\theta,\phi)$, the first-order correction to the energy eigenvalue is

$$\delta E_{nlj} = \langle n,l,j | \delta\widehat{H} | n,l,j \rangle$$

$$= \int_0^{2\pi}\int_0^{\pi}\int_0^{\infty} \psi_{nlj}^*(r,\theta,\phi)\, \delta\widehat{H}\, \psi_{nlj}(r,\theta,\phi)\, r^2 \sin\theta\, \mathrm{d}r\, \mathrm{d}\theta\, \mathrm{d}\phi. \tag{4.26}$$

We shall simply quote the result of evaluating δE_{nlj} and adding it to the unperturbed energy eigenvalue $-E_R/n^2$:

$$E_{nlj}^{\text{total}} = -\frac{E_R}{n^2}\left[1 + \frac{\alpha^2}{n}\left(\frac{1}{j+\frac{1}{2}} - \frac{3}{4n}\right)\right]. \tag{4.27}$$

The factor of α^2 in Equation 4.27 indicates that the first-order correction to the energy is very small, and this gives us confidence that first-order perturbation theory is adequate for calculating the fine structure of hydrogen.

Note that the energy eigenvalue depends on n and j and does *not* depend on the quantum number l. In the following exercise you can verify that Figure 4.4 does indeed faithfully capture the qualitative features of Equation 4.27.

Exercise 4.8 Show that Equation 4.27 is consistent with the following features of Figure 4.4.

(a) The energy of the ground state is lowered.

(b) The energy levels for the different values of j that are allowed for a particular value of n are split by an amount that decreases as n increases.

(c) For $n = 3$, the splitting between the $j = \frac{1}{2}$ and $j = \frac{3}{2}$ levels is greater than the splitting between the $j = \frac{3}{2}$ and $j = \frac{5}{2}$ levels.

(d) The energy shifts shown in the figure are all exaggerated by a factor of 137^2.

(e) The energies of the states do not explicitly depend on l. ■

4.2.2 Hyperfine structure of the hydrogen spectrum

The refinements that we have been discussing are not exhaustive — the energy levels do, in fact, have even finer structure. Some of this so-called **hyperfine structure** arises from the fact that the proton is not a structureless point particle (Exercise 3.5 illustrated this effect). We concentrate here on another source of the hyperfine structure, due to magnetic interactions between the electron and the proton, which is of great interest to astronomers and cosmologists.

The proton at the centre of the hydrogen atom is, like the electron, a spin-$\frac{1}{2}$ particle that possesses a magnetic dipole moment. The energy of the pair of tiny magnets that constitutes a hydrogen atom depends, as for any pair of magnets, upon their relative orientation. Quantum mechanics limits the possible ways in which the two magnets can be oriented relative to each other. If $\widehat{\mathbf{S}}$ represents the spin operator of the electron, and $\widehat{\mathbf{S}}_\text{p}$ represents the spin operator of the proton, then we can define the total spin operator of the atom as $\widehat{\mathbf{F}} = \widehat{\mathbf{S}} + \widehat{\mathbf{S}}_\text{p}$, in much the same way that we defined the total angular momentum operator of a particle as $\widehat{\mathbf{J}} = \widehat{\mathbf{L}} + \widehat{\mathbf{S}}$. Just as $\widehat{J}^2$ has eigenvalues $j(j+1)\hbar^2$, where $j = l + \frac{1}{2}$ or $j = l - \frac{1}{2}$, so $\widehat{F}^2$ has eigenvalues $F(F+1)\hbar^2$, where the total spin quantum number F has two possible values: $\frac{1}{2} + \frac{1}{2} = 1$ or $\frac{1}{2} - \frac{1}{2} = 0$.

Let us see what this means for the electron and the proton in the state with $n = 1$, $l = 0$. The total spin quantum number of this system can have one of two values: either $F = 0$ (in which case we can think of the spins of the proton and electron as being aligned in opposite directions) or $F = 1$ (in which case their spins are in the same direction). But the electron is a little magnet (as you saw in connection with the Stern–Gerlach experiment) with magnetic dipole moment $\boldsymbol{\mu}$ equal to its spin gyromagnetic ratio (which is $-e/m_\text{e}$) times $\mathbf{S}$. The proton, having a much larger mass, has a much smaller magnetic dipole moment $\boldsymbol{\mu}_\text{p}$; in fact, it is a few multiples of e/m_p times $\mathbf{S}_\text{p}$. The state in which the dipole moments are parallel is lowest in energy, and because the electron and proton have opposite electric charges, this is the state with opposite spins, i.e. the state with $F = 0$. The difference in energy between the state with $F = 0$ and the state with $F = 1$ is about 5.9×10^{-6} eV. In other words, the $n = 1$ state in Figure 4.4 should really be represented by two lines, but the separation between them would have to be magnified by almost a million in order to be visible in the figure; hence the term *hyperfine structure*.

This tiny hyperfine splitting was predicted by van de Hulst in 1945, and first reported by Ewen and Purcell of Harvard in 1951 (Figure 4.5). It has, for many years, been of great interest to astronomers and cosmologists. A hydrogen atom excited into the $F = 1$ state (perhaps by colliding with another hydrogen atom) has a half-life in that state of about 10 million years, because the normal ways in which this excited state can decay by emitting a photon are strongly suppressed. The small energy difference $\Delta E = 5.9 \times 10^{-6}$ eV corresponds to a photon of frequency $f = \Delta E/h = 1.42$ GHz, which is in the UHF radio wave band and has a wavelength of about 21 cm. This is the famous **twenty-one centimetre radiation**. Isolated hydrogen atoms exist throughout our galaxy and every other galaxy, and the 21 centimetre radiation they emit is intensely studied by radio-astronomers and cosmologists because it conveys important information about the distribution of cold hydrogen atoms. For example, the Doppler shift of the 21 centimetre radiation has been used to measure the speed of the arms in our spiral galaxy.

Figure 4.5 Harold Ewen working on the horn antenna that helped detect cosmic 21 cm radiation for the first time. This was part of his PhD work with Edward Mills Purcell at Harvard University.

4.3 Putting quantum mechanics and relativity together

You have seen that the properties of hydrogen and hydrogen-like systems confirm the general correctness of quantum mechanics and are a key to understanding systems ranging from atomic nuclei to galaxies. But the study of hydrogen gave birth to even more profound revelations, which we can do no more than briefly survey.

4.3.1 The Dirac equation

In Section 4.2 we added relativistic corrections to the time-independent Schrödinger equation for hydrogen. However, it was recognized by the quantum pioneers that what was really needed was a fully relativistic wave equation. Schrödinger's equation is not satisfactory because it involves second-order derivatives of the spatial coordinates and first-order derivatives of the time coordinates. In special relativity, spatial and time coordinates are on an equal footing, getting 'mixed up' by Lorentz transformations.

In 1928 Dirac put forward a relativistic quantum-mechanical wave equation that, like the familiar Schrödinger equation, had a first-order derivative with respect to time, but, unlike the Schrödinger equation, had first-order derivatives with respect to the spatial coordinates. To produce such an equation, Dirac had to make some breathtakingly bold moves. He essentially had to factorize the Hamiltonian operator to obtain factors that depended only on the first-order spatial derivatives, rather than the second-order derivatives familiar in Schrödinger's equation. When he did this he obtained a wave equation that involved a 4×4 matrix operator, whose matrix elements were themselves partial derivative operators; this matrix operator then acted on a 4×1 spinor wave function whose components depend on space and time coordinates. Dirac may not have started with this structure in mind, but he found that it gave the simplest wave equation consistent with the requirements of special relativity.

We are familiar with the idea that a spin-$\frac{1}{2}$ particle is represented by a 2×1 spinor, and you have seen that the components of this function can be functions of position (see Section 1.4, for example). Leaving aside the third and fourth components of Dirac's spinor wave function for the moment, it turns out that the first and second components of this wave function can be interpreted as the components of the spinor of a spin-$\frac{1}{2}$ particle. This is a great gift from Dirac's formalism; the two-component spinors that describe spin-$\frac{1}{2}$ particles arise naturally from Dirac's marriage of quantum mechanics and special relativity. It also means that Dirac's equation describes only spin-$\frac{1}{2}$ particles — but that still leaves plenty of scope for applications because so much of physics focuses on the behaviour of electrons. However, explaining the origin of spin and spinors for electrons is not the end of the bounty from Dirac's equation.

1. Dirac's equation predicted the magnetic dipole moment of the electron to be

$$\boldsymbol{\mu} = -\frac{e}{m_e}\mathbf{S}, \tag{4.28}$$

   in close agreement with the measured value (see later for small corrections).

2. Dirac's equation provides a complete account of the hydrogen atom energy levels given by Equation 4.27, including the fine structure. No perturbation theory is required, and the mysterious Darwin contribution is included implicitly along with the relativistic kinetic energy and spin–orbit corrections. The spin–orbit correction, in particular, now appears naturally as a consequence of using a consistent relativistic wave equation.

3. Two components of the four-component spinor are required for describing spin, but what about the other two? These turned out, in some ways, to be the greatest gift of all, deserving a separate subsection, but in one word: antiparticles.

Components three and four: antiparticles

The third and fourth components of the Dirac four-component spinor wave function caused deep puzzlement at first. They appeared to represent states of electrons with negative energies, but, if this were so, why don't all electrons radiate, losing energy as they fall through an endless progression of negative-energy states? At one stage, Dirac suggested that all the negative energy states in the vacuum were filled, so that the Pauli exclusion principle would forbid electrons of positive energy from falling down into these states. Eventually, he proposed that the third and fourth components of the spinor wave function correspond to **antiparticles** of electrons — particles with the same mass and spin as an electron but with a positive charge, e. These antiparticles were discovered in 1932 by Carl Anderson, and became known as **positrons**.

Apart from positrons, you have already met antiprotons, which have the same mass as protons but a negative charge.

It is now understood that a consequence of combining quantum mechanics and special relativity is that *all* particles with mass have corresponding antiparticles of the same mass; no exception to this has been found.

When an electron encounters a positron, the pair of particles may form a bound state called **positronium**. The electron and positron attract one another via the Coulomb force, so positronium is a *hydrogen-like system* which can be analyzed using the methods developed in the first half of this chapter. Since the electron and positron have the same mass, the reduced mass of positronium is

$$\mu = \frac{m_e m_e}{m_e + m_e} = \tfrac{1}{2} m_e. \tag{4.29}$$

The ground state of positronium exists in two forms: a singlet state (opposite spins) and a triplet state (parallel spins). The singlet state is slightly lower in energy than the triplet state, but neither state is stable because the electron and positron can annihilate one another, leaving no particle possessing mass, but only photons. The singlet state emits two photons and has a half-life of 1.25×10^{-10} s; the triplet state emits three photons and has a half-life of 1.4×10^{-7} s.

Exercise 4.9 What is the minimum energy that would have to be supplied to the ground state of positronium in order to produce a free electron and positron? ∎

The annihilation of positrons by electrons is the key to an important medical technique called *positron emission tomography* (PET). Traces of chemicals containing positron-emitting nuclei are ingested and reach various parts of the body. The positrons emitted by the nuclei are rapidly annihilated by electrons, or form positronium and are annihilated shortly after. Ignoring the three-photon emission process, two photons are produced by each annihilation, each with energy $m_e c^2 = 511$ keV. Assuming that the positron and electron have no centre-of-mass motion when they annihilate, the photons will have opposite momenta and therefore travel in opposite directions. By detecting a pair of coincident 511 keV photons, we can find the line along which the annihilation took place, and by detecting many such coincidences we can find the places in the body where annihilations are frequent; this is where the chemicals have accumulated. This technique is particularly useful in determining whether tumours have spread or responded to treatment.

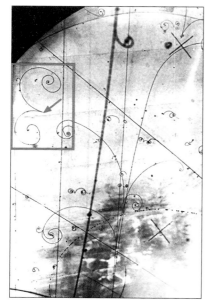

Figure 4.6 Electromagnetic radiation becomes matter: at the point indicated by the arrow, a γ-ray photon (that has left no track) was transformed into an electron–positron pair within the electromagnetic field of a nucleus. The two tracks that start from this point, which look like a backwards '3', were produced by two oppositely charged particles curving in opposite directions in an applied magnetic field.

The reverse process to annihilation — known as **pair production** — can also occur. A photon with energy $hf > 2m_e c^2$ can create an electron–positron pair, as shown in Figure 4.6. The energy criterion simply says that the photon energy must be greater than the energy equivalent, according to special relativity, of the combined rest mass of an electron and a positron. In Figure 4.6, a photon was transformed into an electron and a positron when it interacted with an atom and exchanged some momentum. Note that a single photon in free space, no matter how high its frequency, cannot transform into an electron–positron pair because energy and momentum cannot simultaneously be conserved in the process. It is necessary for the photon to interact with some other particle that can carry away the excess momentum.

4.3.2 The next step: quantum fields

Dirac could not have anticipated where his attempts to create a consistent relativistic quantum mechanics, and to apply it to the hydrogen atom, would subsequently lead. But the hydrogen fine structure, antiparticles, the spin and magnetic dipole moment of electrons, are far from the whole story, and we round out this chapter with a brief survey of other developments. The key idea is already manifest in Figure 4.6: the number of particles in the Universe is not constant. Particles can be created and destroyed; any change in the number of particles consistent with the various fundamental conservation rules is allowed, and everything that is allowed to happen will sometimes happen. For example, when high-energy particles collide, the more energy they have, the greater the number of particles that are created, and this is an everyday observation in large particle accelerators.

A description of these phenomena requires a profound new theory, **quantum field theory**, to which quantum mechanics is the approximation when particle number is considered to be fixed. It was Dirac himself who arguably took the first step to a quantum field theory with a paper in 1927 concerning the creation and absorption of photons. It would be wrong to think that quantum field theory is only required to explain the creation of particles in the targets of high-energy particle accelerators. The fact that the number of particles is not fixed changes *everything*. For example, the forces between particles can be represented as the result of other particles briefly coming into existence between the two interacting particles. In the case of the Coulomb interaction, these particle are photons. At very short ranges, the Coulomb interaction is modified by the fleeting appearance of electron–positron pairs between the interacting particles … the vacuum is no longer a passive background.

The modification of electromagnetic fields by the fleeting creation of electron–positron pairs does not infringe energy conservation precisely because their existence is so fleeting. Particles of mass m and energy $\Delta E = mc^2$ can exist for time Δt, where $\Delta E \, \Delta t \simeq \frac{1}{2}\hbar$, a form of the uncertainty principle.

The particular part of quantum field theory that is concerned with photons and electrons is known as **quantum electrodynamics** (QED). Two consequences of QED are relevant. You will recall that the Dirac equation predicts that the energies of the states of atomic hydrogen depend only on j but not on which of the two l values go with each value of j. It turns out that this is not quite correct. In 1947, Lamb and Retherford discovered a very small difference between the energies of the $n = 2$, $j = \frac{1}{2}$ levels with $l = 0$ and $l = 1$. This hyperfine energy difference, known as the **Lamb shift**, was too small to be detected by a spectrometer. However, the lower, $l = 0$, state could be excited to the $l = 1$ state by high-frequency radio waves, a process that could be detected. The frequency has been measured to high precision, and is (1057.845 ± 0.003) MHz, which is in agreement with the predictions of QED.

Dirac's relativistic theory predicted the magnetic dipole moment of the electron, Equation 4.28, but that too needed correction. In 1947, Foley and Kusch discovered that it was about one part in a thousand higher than Dirac's prediction.

Both of these corrections reveal the activity of the vacuum and can be calculated with QED to an outstanding level of precision. In 1955, the Nobel prize in physics was shared by Lamb for his discoveries concerning the fine structure of the hydrogen spectrum and Kusch for his precision determination of the magnetic dipole moment of the electron.

More recently quantum field theory has been used to explain the properties of quarks. A key step was made in 1974 when **charmonium** was discovered. Charmonium refers to bound states of heavy **quarks** of a kind known as charmed quarks, and the corresponding anticharmed quarks. The states of charmonium are described using the ideas developed for all the other hydrogen-like systems. Later, similar systems made up of different kinds of quarks were discovered and are known collectively as **quarkonium**. The discovery of systems that could be described as bound states of charmed quarks and their antiparticles was a turning point in the acceptance of what has come to be known as the *standard model* of elementary particles.

Summary of Chapter 4

Section 4.1 Hydrogen-like atoms are systems in which a negatively-charged particle is bound to a positively-charged nucleus by Coulomb attraction. Their properties are similar to those of atomic hydrogen, but with the energy eigenvalues given by

$$E_n = -\frac{E_R^{\text{scaled}}}{n^2},$$

where

$$E_R^{\text{scaled}} = Z^2 \frac{\mu}{\mu_H} E_R,$$

and where Z is the atomic number of the nucleus, μ is the reduced mass of the atom, μ_H is the reduced mass of a hydrogen atom, and E_R is the Rydberg energy. The spatial extent of the eigenfunctions is found by replacing a_0 in the hydrogen atom eigenfunctions by

$$a_0^{\text{scaled}} = \frac{1}{Z} \frac{\mu_H}{\mu} a_0.$$

This model is used to understand properties of ions such as He^+ and Li^{2+}, which have a single electron, and muonic atoms and antiprotonic atoms. Muonic atoms provide information concerning the charge distribution within atomic nuclei, and antiprotonic atoms provide information concerning the distribution of neutrons within nuclei. The same model is the basis for understanding features of X-ray spectra.

Section 4.2 Under closer examination, the states of hydrogen with a particular value of principal quantum number n are not degenerate; their energies are independent of l but depend on the total angular momentum quantum number j. States with lower j have lower energy. Three of the perturbations responsible for the fine structure are (i) a relativistic correction to the kinetic energy, (ii) the spin–orbit interaction, and (iii) the Darwin term. The second of these does not apply to states with $l = 0$, whilst the third applies only to such states. The effect of the three perturbations on the energy of the eigenstates of hydrogen can be calculated using first-order perturbation theory. Yet finer structure, hyperfine structure, of the hydrogen atom spectrum is caused by the interaction of the magnetic dipole moments of the electron and proton. The $n = 1$ state is split into two levels, and transitions between these levels lead to emission of $21\,\text{cm}$ radiation, which is exploited by astronomers.

Section 4.3 Dirac's relativistic wave equation provides a deeper understanding of the hydrogen fine structure. Electron spin and the spin–orbit energy-splitting are correctly predicted, as is the magnetic dipole moment of the electron. Quantum theory and relativity together predict that all subatomic particles with mass have corresponding antiparticles with the same mass. The possibility of creation and annihilation of photons, electrons and all other particles led to the development of quantum field theory. Quantum electrodynamics allows the calculation of further, hyperfine structure, such as the Lamb shift, and corrections to the electron magnetic dipole moment. It is also the fundamental theory to which quantum mechanics is a very good approximation when particle creation and annihilation do not play a major role.

Achievements from Chapter 4

After studying this chapter, you should be able to:

4.1 Explain the meanings of the newly defined (emboldened) terms and symbols, and use them appropriately.

4.2 Calculate the energy eigenvalues of hydrogen-like atoms in situations where the nucleus can be treated as a point charge.

4.3 Calculate the value of $\langle r \rangle$ for a state of a hydrogen-like atom in situations where the nucleus can be treated as a point charge, given the value of $\langle r \rangle$ for the corresponding state of atomic hydrogen.

4.4 Apply the hydrogen-like atom model to muonic atoms, calculate the energies of muonic X-rays, and explain why the energies of the low-lying states are not predicted correctly by a simple scaling of the hydrogen atom expression.

4.5 Explain how muonic atoms can be used to determine the spatial extent of the charge in atomic nuclei, and briefly explain how these studies can be extended to unstable nuclei.

4.6 Briefly explain how antiprotonic atoms contribute to our knowledge of nuclear structure.

4.7 Give an account of how the hydrogen-like atom model can be adapted to explain key features of the X-ray spectra of atoms, calculate X-ray energies for particular elements, and comment on the historical significance of measurements of these energies.

4.8 Give an account of the fine structure of hydrogen and of the various perturbations to the Hamiltonian that give rise to this structure.

4.9 Explain the origin of the hyperfine splitting of the hydrogen atom ground state and the resulting 21 cm radiation.

4.10 Give a general account of the consequences of putting quantum theory and special relativity together in a consistent way, making clear the relevance of the Dirac equation, hyperfine structure, antiparticles and quantum field theory.

Chapter 5 Many-electron atoms

Introduction

In previous chapters, you studied the hydrogen atom in some detail. The study of hydrogen was simplified by neglecting the motion of the atom as a whole and concentrating on the relative motion of the electron and the proton. The problem was therefore reduced to that of a single particle in a spherically-symmetric potential energy well.

We classify all atoms except hydrogen as being many-electron atoms. In these atoms, the electrons interact with one another; it is no longer possible to solve the time-independent Schrödinger equation exactly, and approximations are needed. In addition, a many-electron atom is a system of identical fermions, so its electronic state must be represented by an antisymmetric total wave function.

We will begin by describing the simplest many-electron atom: helium. Using perturbation theory you will see that the indistinguishability of the electrons introduces an effect with no classical counterpart. This effect is called *exchange* and it arises from the symmetry requirements on the wave function. Exchange influences the energy of an atom and the characteristics of its electronic states.

Going beyond helium, we will study how electrons organize themselves in atoms in general. The Pauli exclusion principle is fundamental to this organization. We will use many of the ideas you have learnt about angular momentum, and about the eigenfunctions of the hydrogen atom, and you will see why certain elements have similar properties and can be arranged in a column of the Periodic Table. Although we shall not calculate the energy of all atoms in all possible states, you will see how these states are labelled, and how they are ordered according to their energies.

The chapter is structured as follows. Section 5.1 introduces the Hamiltonian operator for a many-electron atom and uses it in an independent-particle model. In Section 5.2, we go beyond this approximation and calculate the energies of the electronic states of helium in some detail. In Section 5.3, we look at other many-electron atoms, using the central-field approximation, which is based on single-particle atomic orbitals. This leads to a description of atoms in terms of the occupation of atomic orbitals, and to a broad understanding of the Periodic Table. Finally, Section 5.4 goes beyond single-particle orbitals and shows how atomic states can be labelled more precisely by angular momentum quantum numbers that refer to the atom as a whole.

5.1 First steps in describing a many-electron atom

5.1.1 The Hamiltonian operator

You know by now that when a system has a definite energy E, its wave function can be written as a product of a time-dependent phase factor ($e^{-iEt/\hbar}$) and a time-independent energy eigenfunction, a solution of the time-independent

Schrödinger equation. In this chapter we concentrate on finding the energy eigenfunctions.

As with hydrogen, we start by ignoring the spin of the electrons and the spin–orbit interaction. We are not interested in the motion of the atom as a whole, so we work in the centre-of-mass frame. In addition, we shall assume that the nucleus is infinitely massive or, at least, so much more massive than an electron that we can ignore its motion and place it at the origin of coordinates. This approximation is a very good one in most cases. In an oxygen atom, for example, the nuclear mass is 16 times the mass of hydrogen, which in turn is about 1840 times the mass of an electron. Hence, even for this light atom, the nucleus is around 30 000 times more massive than an electron.

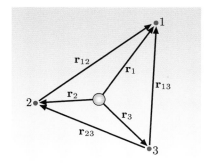

Figure 5.1 An example of the vectors used to specify the Hamiltonian operator of an atom — in this case a lithium atom.

- Write down the Hamiltonian operator for the lithium atom using the coordinates shown in Figure 5.1. A lithium atom has three electrons, each of mass m_e and charge $-e$, and a nucleus of charge $+3e$. Neglect the motion of the centre of mass of the system, assuming an infinite nuclear mass.

○ The Hamiltonian operator depends on the position vector $\mathbf{r}_i$ of each electron relative to the fixed nucleus, and on the relative displacements $\mathbf{r}_{ij} = \mathbf{r}_i - \mathbf{r}_j$ of each pair of electrons. The kinetic energy associated with the motion of electron i is represented by an operator of the form $-(\hbar^2/2m_e)\nabla_i^2$, where the subscript i indicates that the derivatives are taken with respect to the coordinates of electron i.

Each electron is attracted by the positive charge $3e$ of the nucleus. Using the coordinate system of Figure 5.1, the corresponding potential energies are $-3e^2/4\pi\varepsilon_0 r_1$, $-3e^2/4\pi\varepsilon_0 r_2$ and $-3e^2/4\pi\varepsilon_0 r_3$. Finally, the electrons repel one another. This introduces the additional potential energy terms $e^2/4\pi\varepsilon_0 r_{12}$, $e^2/4\pi\varepsilon_0 r_{13}$ and $e^2/4\pi\varepsilon_0 r_{23}$. So the Hamiltonian operator for lithium is

$$\widehat{H} = -\frac{\hbar^2}{2m_e}\nabla_1^2 - \frac{\hbar^2}{2m_e}\nabla_2^2 - \frac{\hbar^2}{2m_e}\nabla_3^2 \qquad \text{electron kinetic energy}$$
$$-\frac{3e^2}{4\pi\varepsilon_0 r_1} - \frac{3e^2}{4\pi\varepsilon_0 r_2} - \frac{3e^2}{4\pi\varepsilon_0 r_3} \qquad \text{electron–nucleus attraction}$$
$$+\frac{e^2}{4\pi\varepsilon_0 r_{12}} + \frac{e^2}{4\pi\varepsilon_0 r_{13}} + \frac{e^2}{4\pi\varepsilon_0 r_{23}} \qquad \text{electron–electron repulsion.}$$

The Coulomb interaction depends only on the distance between the charged particles, so the potential energy terms describing the electron–nucleus attraction and the electron–electron repulsion depend only on r_i and r_{ij}, the magnitudes of the vectors $\mathbf{r}_i$ and $\mathbf{r}_{ij}$.

The above Hamiltonian operator can be generalized to a neutral atom with Z electrons and atomic number Z (and hence nuclear charge Ze):

$$\widehat{H} = \sum_{i=1}^{Z} -\frac{\hbar^2}{2m_e}\nabla_i^2 - \sum_{i=1}^{Z} \frac{Ze^2}{4\pi\varepsilon_0 r_i} + \sum_{i=1}^{Z}\sum_{j>i} \frac{e^2}{4\pi\varepsilon_0 r_{ij}}. \qquad (5.1)$$

This Hamiltonian operator is again the sum of three terms. The first is the operator for the total kinetic energy of all the electrons. The second is the Coulomb potential energy due to the mutual attraction between each electron and the nucleus. Each of these terms is the sum of Z contributions, one for each electron. The third term is the Coulomb potential energy due the mutual repulsion between pairs of electrons. The indices in the double summation are chosen to ensure that there is only one contribution from each pair of electrons, and that there is no contribution due to an electron interacting with itself. That is why the sum over j ranges over values that are strictly greater than i.

Exercise 5.1 Write out the sum $\sum_{i=1}^{4} \sum_{j>i}^{4} A_{ij}$ in full, and satisfy yourself that each pair of distinct indices appears once, and once only. ■

To obtain the energy eigenfunctions and eigenvalues for a many-electron atom, we need to solve the time-independent Schrödinger equation

$$\widehat{H}\,\psi(\mathbf{r}_1, \mathbf{r}_2, \ldots, \mathbf{r}_Z) = E\,\psi(\mathbf{r}_1, \mathbf{r}_2, \ldots, \mathbf{r}_Z) \quad (5.2)$$

with the Hamiltonian operator as in Equation 5.1. The energy eigenfunction $\psi(\mathbf{r}_1, \mathbf{r}_2, \ldots, \mathbf{r}_Z)$ depends on the position vectors of all the electrons relative to the central nucleus.

How difficult is it to solve this time-independent Schrödinger equation? Well, in fact, it is *impossible* to find exact solutions, even for $Z = 2$. The best we can hope for is good approximations. The culprits in this unfortunate situation are the electron–electron repulsion terms. In the next subsection we will introduce a drastic simplification that allows an approximate solution to be found.

5.1.2 The independent-particle model

The simplest approximation is to neglect the electron–electron interaction terms altogether, so that the Hamiltonian operator takes the form

$$\widehat{H} = \sum_{i=1}^{Z} \left[-\frac{\hbar^2}{2m_e}\nabla_i^2 - \frac{Ze^2}{4\pi\varepsilon_0 r_i} \right]. \quad (5.3)$$

This can also be expressed as

$$\widehat{H} = \sum_{i=1}^{Z} \widehat{h}_i, \quad (5.4)$$

where

$$\widehat{h}_i = -\frac{\hbar^2}{2m_e}\nabla_i^2 - \frac{Ze^2}{4\pi\varepsilon_0 r_i} \quad (5.5)$$

depends only on the coordinates of the ith electron. So the many-particle Hamiltonian $\widehat{H}$ has been approximated by a sum of single-particle Hamiltonians, each of which is like that for a single electron in a hydrogen-like atom with nuclear charge Ze.

The time-independent Schrödinger equation corresponding to the Hamiltonian in Equation 5.3 is separable, so we can look for solutions that are products of terms relating to individual electrons:

$$\psi(\mathbf{r}_1, \mathbf{r}_2, \ldots, \mathbf{r}_Z) = \phi_p(\mathbf{r}_1)\,\phi_q(\mathbf{r}_2) \ldots \phi_s(\mathbf{r}_Z). \quad (5.6)$$

Substituting into the many-particle time-independent Schrödinger equation $\widehat{H}\psi = E\psi$, and using the standard method of separation of variables, we find solutions in which the functions $\phi_p(\mathbf{r}_1), \phi_q(\mathbf{r}_2), \ldots, \phi_s(\mathbf{r}_Z)$ obey single-particle time-independent Schrödinger equations of the form

$$\widehat{h}_1 \phi_p(\mathbf{r}_1) = E_p \phi_p(\mathbf{r}_1),$$
$$\widehat{h}_2 \phi_q(\mathbf{r}_2) = E_q \phi_q(\mathbf{r}_2),$$
$$\vdots$$
$$\widehat{h}_Z \phi_s(\mathbf{r}_Z) = E_s \phi_s(\mathbf{r}_Z),$$

and the total energy of the system is the sum of all the single-particle energies: $E = E_p + E_q + \cdots + E_s$.

All the single-particle Schrödinger equations are of the same form, so they have the same set of eigenfunctions. In the context of many-electron atoms, these single-particle normalized eigenfunctions are called **atomic orbitals**. In the independent-particle approximation, the orbitals take the hydrogenic form

$$\phi_{nlm}(r, \theta, \phi) = \left(\frac{r}{a}\right)^l \left(\text{polynomial in } \frac{r}{a}\right) e^{-r/na} Y_{lm}(\theta, \phi), \tag{5.7}$$

where the $Y_{lm}(\theta, \phi)$ are spherical harmonics, and $a = a_0/Z$, where $a_0 = 5.29 \times 10^{-11}$ m is the Bohr radius. The corresponding single-particle energy eigenvalues are

$$E_n = -\frac{Z^2 E_R}{n^2}, \tag{5.8}$$

where $E_R = 13.6\,\text{eV}$ is the Rydberg energy. We shall use the usual **spectroscopic notation** to label these orbitals, as indicated in Table 5.1, and will append the magnetic quantum number m as a subscript. For example, any orbital with $n = 2, l = 1$ and $m = 1$ will be labelled as $2p_1$.

So far, we have not taken the indistinguishability of electrons into account. We know that the total wave function for any system of electrons is antisymmetric. So, taking the total wave function to be a product of a spatial eigenfunction and a spin ket, the spatial eigenfunction must be *either* symmetric (with an antisymmetric spin ket) *or* antisymmetric (with a symmetric spin ket). In general, however, the product eigenfunction in Equation 5.6 is neither symmetric nor antisymmetric.

We can proceed in two ways. The simplest is to ignore the requirements of symmetry, regarding this as just one more approximation in our model. A more satisfactory approach is to construct solutions that are symmetric or antisymmetric by taking linear combinations of functions like that in Equation 5.6, but with the particle labels permuted in all possible ways. In the independent-particle model, all these linear combinations have the same total energy $E = E_p + E_q + \cdots + E_s$, so this procedure does not affect our results for the energy levels of the atom. However, this is a very special result for the independent-particle model. In the next section we shall consider helium, *taking electron–electron repulsion into account*. You will see that the requirement for the spatial energy eigenfunction of helium to be symmetric or antisymmetric, in the presence of electron–electron repulsion, has a significant effect on the energy levels of this atom.

Details of applying the method of separation of variables to many-particle systems were given in Section 4.1 of Book 2. In three-dimensional problems, the labels $p, q, \ldots, s$ can each refer to a set of quantum numbers (e.g. n, l and m).

Throughout this chapter, single-particle eigenfunctions are denoted by the symbol ϕ with one or more subscripts; this is to be distinguished from the azimuthal angle ϕ, which carries no subscripts.

Table 5.1 Spectroscopic notation for atomic orbitals.

n	l	Orbital
1	0	1s
2	0	2s
2	1	2p
3	0	3s
3	1	3p
3	2	3d
4	0	4s
4	1	4p
4	2	4d
4	3	4f
$\vdots$	$\vdots$	$\vdots$

5.2 The helium atom

Helium, with atomic number $Z = 2$, is the simplest atom after hydrogen. Let us apply the independent-particle model to the ground state of this system, and see how well it does. The ground state is the state of lowest energy, which must correspond to both electrons being in the lowest energy orbital (1s), each with energy E_{1s}. In the independent-particle model, the ground-state energy of the whole atom is therefore given by

$$E_1 = E_{1s} + E_{1s} = -\frac{2^2 E_R}{1^2} - \frac{2^2 E_R}{1^2} = -8E_R = -108.8\,\text{eV},$$

where the negative value indicates a bound state because the energy zero is chosen to correspond to stationary particles infinitely far apart.

The experimental value of the ground-state energy of helium is $-78.9\,\text{eV}$. So the independent-particle model, which neglects electron–electron repulsion, does not provide a very good approximation in this case. As you might expect, the energy obtained is too low: electron repulsion will contribute a positive term to the energy, making it higher (i.e. less negative) than the estimate provided by the independent-particle model. We therefore need to take electron–electron repulsion into account.

In this section, we will consider both the ground state and various excited states of helium. You will see that the indistinguishability of the two electrons in a helium atom affects the energies of the excited states. For this reason, we shall begin by reminding you of the rules used to construct the total wave function of a pair of electrons (Book 2, Chapter 4).

5.2.1 The total wave function for two electrons

Electrons are fermions, so a pair of electrons must be described by an *antisymmetric* total wave function. The total electronic wave function is taken to be the product of a spatial part and a spin part, so the following combinations are allowed:

- symmetric spatial × antisymmetric spin = antisymmetric total,
- antisymmetric spatial × symmetric spin = antisymmetric total.

First consider the spatial part of the total wave function. If both electrons are described by the *same* atomic orbital, ϕ_r, the product function

$$\phi_r(\mathbf{r}_1)\,\phi_r(\mathbf{r}_2) \tag{5.9}$$

is automatically symmetric with respect to swapping the particle labels. However, if the electrons are in *different* orbitals, ϕ_r and ϕ_s, we can form either a symmetric combination

$$\frac{1}{\sqrt{2}}\left[\phi_r(\mathbf{r}_1)\,\phi_s(\mathbf{r}_2) + \phi_s(\mathbf{r}_1)\,\phi_r(\mathbf{r}_2)\right], \tag{5.10}$$

or an antisymmetric combination

$$\frac{1}{\sqrt{2}}\left[\phi_r(\mathbf{r}_1)\,\phi_s(\mathbf{r}_2) - \phi_s(\mathbf{r}_1)\,\phi_r(\mathbf{r}_2)\right]. \tag{5.11}$$

We assume that the atomic orbitals are themselves normalized.

In both cases, the factor $1/\sqrt{2}$ ensures that the two-particle spatial wave function is normalized.

Exercise 5.2 Show that the function in Equation 5.10 is an eigenfunction of the independent-particle Hamiltonian $\widehat{h}_1 + \widehat{h}_2$ with eigenvalue $E_r + E_s$, where E_r and E_s are the single-particle energies associated with the orbitals ϕ_r and ϕ_s. ■

Now let's consider the spin part of the total wave function. This is represented by a spin ket $|S, M_S\rangle$, which is an eigenvector of the operators $\widehat{S}^2$ and $\widehat{S}_z$ representing the square of the magnitude, and the z-component, of the total spin of the two-electron system. We have

$$\widehat{S}^2 |S, M_S\rangle = S(S+1)\hbar^2 |S, M_S\rangle \quad \text{and} \quad \widehat{S}_z |S, M_S\rangle = M_S \hbar |S, M_S\rangle.$$

Refer to Book 2, Chapter 4 if you need to refresh your memory on spin kets for a pair of spin-$\frac{1}{2}$ particles.

For a system of two spin-$\frac{1}{2}$ particles, the possible values of S are 0 and 1. For $S = 0$, M_S must be equal to 0. For $S = 1$, M_S can take any of the values $-1, 0$ and 1. So there are four basic spin kets to consider. The spin ket $|0, 0\rangle$ is called a singlet, while the three spin kets $|1, -1\rangle, |1, 0\rangle$ and $|1, 1\rangle$ are said to form a triplet.

It is possible to express these spin kets as follows:

Singlet spin state

$$|0, 0\rangle = \frac{1}{\sqrt{2}} \left(|\uparrow\downarrow\rangle - |\downarrow\uparrow\rangle \right) \qquad S = 0, M_S = 0 \qquad (5.12)$$

Triplet spin states

$$|1, 1\rangle = |\uparrow\uparrow\rangle \qquad S = 1, M_S = +1 \qquad (5.13)$$

$$|1, 0\rangle = \frac{1}{\sqrt{2}} \left(|\uparrow\downarrow\rangle + |\downarrow\uparrow\rangle \right) \qquad S = 1, M_S = 0 \qquad (5.14)$$

$$|1, -1\rangle = |\downarrow\downarrow\rangle \qquad S = 1, M_S = -1 \qquad (5.15)$$

As usual, any arrow without a subscript refers to a spin component along the z-axis. We also use positional notation, so that $|\uparrow\uparrow\rangle = |\uparrow\rangle_1 |\uparrow\rangle_2$, $|\uparrow\downarrow\rangle = |\uparrow\rangle_1 |\downarrow\rangle_2$, etc.

From these equations you can see that the singlet spin ket $|0, 0\rangle$ is *antisymmetric* with respect to swapping particle labels, while the three spin kets $|1, 1\rangle, |1, 0\rangle$ and $|1, -1\rangle$ are all *symmetric*. This is the most important point to remember in the discussion that follows.

Finally, combining the spatial and spin parts of the total wave function, and imposing the requirement that it must be antisymmetric with respect to swapping particle labels, we see that the following possibilities are available:

$$\phi_r(\mathbf{r}_1) \phi_r(\mathbf{r}_2) |0, 0\rangle, \qquad (5.16\text{a})$$

$$\frac{1}{\sqrt{2}} \left[\phi_r(\mathbf{r}_1) \phi_s(\mathbf{r}_2) + \phi_s(\mathbf{r}_1) \phi_r(\mathbf{r}_2) \right] |0, 0\rangle, \qquad (5.16\text{b})$$

$$\frac{1}{\sqrt{2}} \left[\phi_r(\mathbf{r}_1) \phi_s(\mathbf{r}_2) - \phi_s(\mathbf{r}_1) \phi_r(\mathbf{r}_2) \right] |1, 1\rangle, \qquad (5.16\text{c})$$

$$\frac{1}{\sqrt{2}} \left[\phi_r(\mathbf{r}_1) \phi_s(\mathbf{r}_2) - \phi_s(\mathbf{r}_1) \phi_r(\mathbf{r}_2) \right] |1, 0\rangle, \qquad (5.16\text{d})$$

$$\frac{1}{\sqrt{2}} \left[\phi_r(\mathbf{r}_1) \phi_s(\mathbf{r}_2) - \phi_s(\mathbf{r}_1) \phi_r(\mathbf{r}_2) \right] |1, -1\rangle. \qquad (5.16\text{e})$$

The first two have symmetric spatial functions and the antisymmetric singlet spin ket, while the last three have an antisymmetric spatial function and a symmetric triplet spin ket. Note that it is not possible to construct an antisymmetric spatial function when both particles are in the same orbital, so the ground state of helium is described by a *symmetric* spatial function and an *antisymmetric* spin ket; it is a singlet state.

5.2.2 The ground state of helium

In the independent-particle model, the ground state of helium is described by the spatial function $\phi_{1s}(\mathbf{r}_1)\,\phi_{1s}(\mathbf{r}_2)$. We know that this does not give an accurate result for the ground-state energy, so how can we do better? We shall now use perturbation theory to improve our estimate.

Using Equation 5.1, together with Equations 5.4 and 5.5 with $Z = 2$, we see that the Hamiltonian operator for a helium atom is

$$\widehat{H} = \widehat{h}_1 + \widehat{h}_2 + \frac{e^2}{4\pi\varepsilon_0 r_{12}},$$

where $\widehat{h}_1$ and $\widehat{h}_2$ are the single-particle Hamiltonian operators used in the independent-particle approximation. Remember, these have energy eigenfunctions similar to those of a hydrogen atom. The last term, $e^2/4\pi\varepsilon_0 r_{12}$, describes the potential energy due to electron–electron repulsion; this term prevents us from using the method of separation of variables, making the problem hard to solve.

In the perturbation theory approach, we choose $\widehat{h}_1 + \widehat{h}_2$ to be the unperturbed Hamiltonian, and $e^2/4\pi\varepsilon_0 r_{12}$ to be the perturbation. That is, we write

$$\widehat{H} = \widehat{H}^{(0)} + \delta\widehat{H}, \quad \text{where} \quad \widehat{H}^{(0)} = \widehat{h}_1 + \widehat{h}_2 \quad \text{and} \quad \delta\widehat{H} = \frac{e^2}{4\pi\varepsilon_0 r_{12}}.$$

Now, we know the eigenfunctions of the unperturbed Hamiltonian, $\widehat{H}^{(0)}$. In particular, the total wave function describing the ground state of the unperturbed problem is

$$\psi_{gs}^{(0)} = \phi_{1s}(\mathbf{r}_1)\,\phi_{1s}(\mathbf{r}_2)\,|0,0\rangle.$$

The superscript (0) indicates that this is regarded as the zeroth-order approximation in our perturbation calculation. The corresponding zeroth-order approximation to the ground-state energy is

$$E_{gs}^{(0)} = E_{1s} + E_{1s} = 2E_{1s}.$$

Using perturbation theory, the first-order correction to the ground-state energy is

$$E_{gs}^{(1)} = \langle \psi_{gs}^{(0)} | \delta\widehat{H} | \psi_{gs}^{(0)} \rangle$$

$$= \langle 0,0|0,0\rangle \iint \phi_{1s}^*(\mathbf{r}_1)\,\phi_{1s}^*(\mathbf{r}_2)\,\frac{e^2}{4\pi\varepsilon_0 r_{12}}\,\phi_{1s}(\mathbf{r}_1)\,\phi_{1s}(\mathbf{r}_2)\,dV_1\,dV_2,$$

where the integrals are over all space for both particles. The singlet spin ket is normalized, so $\langle 0,0|0,0\rangle = 1$, and we conclude that the first-order correction is

$$E_{gs}^{(1)} = \left\langle \phi_{1s}\phi_{1s} \left| \frac{e^2}{4\pi\varepsilon_0 r_{12}} \right| \phi_{1s}\phi_{1s} \right\rangle, \tag{5.17}$$

where we have used Dirac notation as a shorthand for the integral in the preceding equation. This integral is not at all easy to evaluate, and we shall not attempt to do so here. We just quote the result: 34.0 eV. The ground-state energy of helium calculated within first-order perturbation theory is therefore

$$E_{gs} \simeq -108.8\,\text{eV} + 34.0\,\text{eV} = -74.8\,\text{eV}.$$

This value is much better than the independent-particle model, and is about 5% above the experimental value of -78.9 eV. An even better value could be found by using second-order perturbation theory, but this would involve a great deal of work.

5.2.3 The excited states of helium

The low-lying excited states of helium can be described as states in which one electron is in the lowest-energy orbital (1s), and the other electron is in a higher-energy orbital (2s, 2p, 3s, 3p, 3d, ...). Since these excited states correspond to the two electrons occupying *different* orbitals, the correctly-symmetrized total wave function is given by one of Equations 5.16b–5.16e, rather than by Equation 5.16a, and this leads to interesting effects, as we shall now see.

Let us suppose that the occupied orbitals are ϕ_r and ϕ_s, with energies E_r and E_s, and that the energy of the atom in this excited state is E_{rs}. Then the zeroth-order approximation to the energy is given by the independent-particle result

$$E_{rs}^{(0)} = E_r + E_s. \tag{5.18}$$

See Exercise 5.2.

The corresponding zeroth-order approximation to the total wave function is given by one of Equations 5.16b–5.16e, each of which corresponds to a different allowed state. These four possibilities can be expressed as

$$\psi_{rs}^{(0)} = \frac{1}{\sqrt{2}}\left[\phi_r(\mathbf{r}_1)\phi_s(\mathbf{r}_2) \pm \phi_s(\mathbf{r}_1)\phi_r(\mathbf{r}_2)\right]|S, M_S\rangle, \tag{5.19}$$

where the plus sign goes with the $S = 0$ singlet ket, the minus sign goes with any of the $S = 1$ triplet kets, and the (0) superscript indicates that this is our zeroth-order approximation in perturbation theory.

In perturbation theory, the first-order correction to the excited-state energy E_{rs} is

$$E_{rs}^{(1)} = \langle\psi_{rs}^{(0)}|\delta\widehat{H}|\psi_{rs}^{(0)}\rangle, \quad \text{where } \delta\widehat{H} = \frac{e^2}{4\pi\varepsilon_0 r_{12}}.$$

Using Equation 5.19, we see that

$$E_{rs}^{(1)} = \tfrac{1}{2}\langle S, M_S|S, M_S\rangle \left\langle\phi_r\phi_s \pm \phi_s\phi_r\left|\frac{e^2}{4\pi\varepsilon_0 r_{12}}\right|\phi_r\phi_s \pm \phi_s\phi_r\right\rangle.$$

The spin kets are normalized, so $\langle S, M_S|S, M_S\rangle = 1$. When we expand the remaining bra and ket, we get

$$E_{rs}^{(1)} = \tfrac{1}{2}\Bigg[\left\langle\phi_r\phi_s\left|\frac{e^2}{4\pi\varepsilon_0 r_{12}}\right|\phi_r\phi_s\right\rangle + \left\langle\phi_s\phi_r\left|\frac{e^2}{4\pi\varepsilon_0 r_{12}}\right|\phi_s\phi_r\right\rangle$$
$$\pm \left\langle\phi_r\phi_s\left|\frac{e^2}{4\pi\varepsilon_0 r_{12}}\right|\phi_s\phi_r\right\rangle \pm \left\langle\phi_s\phi_r\left|\frac{e^2}{4\pi\varepsilon_0 r_{12}}\right|\phi_r\phi_s\right\rangle\Bigg].$$

The square brackets on the right-hand side of this equation contain four Dirac brackets, which can be translated back into integrals as follows:

$$\left\langle \phi_r \phi_s \left| \frac{e^2}{4\pi\varepsilon_0 r_{12}} \right| \phi_r \phi_s \right\rangle = \iint \phi_r^*(\mathbf{r}_1)\,\phi_s^*(\mathbf{r}_2)\, \frac{e^2}{4\pi\varepsilon_0 r_{12}} \phi_r(\mathbf{r}_1)\,\phi_s(\mathbf{r}_2)\, \mathrm{d}V_1\, \mathrm{d}V_2,$$

$$\left\langle \phi_s \phi_r \left| \frac{e^2}{4\pi\varepsilon_0 r_{12}} \right| \phi_s \phi_r \right\rangle = \iint \phi_s^*(\mathbf{r}_1)\,\phi_r^*(\mathbf{r}_2)\, \frac{e^2}{4\pi\varepsilon_0 r_{12}} \phi_s(\mathbf{r}_1)\,\phi_r(\mathbf{r}_2)\, \mathrm{d}V_1\, \mathrm{d}V_2,$$

$$\left\langle \phi_r \phi_s \left| \frac{e^2}{4\pi\varepsilon_0 r_{12}} \right| \phi_s \phi_r \right\rangle = \iint \phi_r^*(\mathbf{r}_1)\,\phi_s^*(\mathbf{r}_2)\, \frac{e^2}{4\pi\varepsilon_0 r_{12}} \phi_s(\mathbf{r}_1)\,\phi_r(\mathbf{r}_2)\, \mathrm{d}V_1\, \mathrm{d}V_2,$$

$$\left\langle \phi_s \phi_r \left| \frac{e^2}{4\pi\varepsilon_0 r_{12}} \right| \phi_r \phi_s \right\rangle = \iint \phi_s^*(\mathbf{r}_1)\,\phi_r^*(\mathbf{r}_2)\, \frac{e^2}{4\pi\varepsilon_0 r_{12}} \phi_r(\mathbf{r}_1)\,\phi_s(\mathbf{r}_2)\, \mathrm{d}V_1\, \mathrm{d}V_2.$$

By interchanging the labels 1 and 2, and noting that the distance between the electrons can be written either as r_{12} or as r_{21}, we see that the first two integrals are equal to one another, and the second two integrals are also equal to one another. We therefore have as a first-order correction

$$E_{rs}^{(1)} = \left\langle \phi_r \phi_s \left| \frac{e^2}{4\pi\varepsilon_0 r_{12}} \right| \phi_r \phi_s \right\rangle \pm \left\langle \phi_r \phi_s \left| \frac{e^2}{4\pi\varepsilon_0 r_{12}} \right| \phi_s \phi_r \right\rangle.$$

The $+$ is for the singlet state, and the $-$ is for triplet states.

Adding this to the zeroth-order approximation (Equation 5.18), we conclude that

$$E_{rs} \simeq E_r + E_s + C_{rs} \pm J_{rs}, \tag{5.20}$$

where, with slight rearrangements,

$$C_{rs} = \frac{e^2}{4\pi\varepsilon_0} \iint \frac{|\phi_r(\mathbf{r}_1)|^2\, |\phi_s(\mathbf{r}_2)|^2}{r_{12}}\, \mathrm{d}V_1\, \mathrm{d}V_2 \tag{5.21}$$

and

$$J_{rs} = \frac{e^2}{4\pi\varepsilon_0} \iint \frac{\phi_r^*(\mathbf{r}_1)\,\phi_s^*(\mathbf{r}_2)\,\phi_s(\mathbf{r}_1)\,\phi_r(\mathbf{r}_2)}{r_{12}}\, \mathrm{d}V_1\, \mathrm{d}V_2. \tag{5.22}$$

The first integral, C_{rs}, has a simple classical interpretation. The quantity $|\phi_r(\mathbf{r}_1)|^2$ is the probability density for particle 1 when it is in the orbital ϕ_r; when multiplied by $-e$, this represents the charge distribution of electron 1, and may be interpreted classically in terms of a 'charge cloud'. So C_{rs} represents the potential energy associated with the classical Coulomb repulsion between two charge clouds (those of electrons 1 and 2). It is called the **Coulomb integral**.

The second integral has no classical analogue. It is a quantum-mechanical interference term arising from the fact that the two electrons are indistinguishable, making it impossible to say *which* electron is in *which* orbital. Given that electrons are fermions, this leads to the requirement for the total wave function to be antisymmetric, reversing its sign when the particle labels are exchanged. For this reason, J_{rs} is called the **exchange integral**.

Table 5.2 Coulomb and exchange integrals for the first two excited states of helium.

r, s	C_{rs}	J_{rs}
1s, 2s	11.42 eV	1.19 eV
1s, 2p	13.21 eV	0.93 eV

The Coulomb integral is always positive (because all the terms in the integrand are positive): as expected, the mutual repulsion of the electrons increases the energy of the atom. The exchange integral also turns out to be positive for the excited states of helium, but is roughly an order of magnitude smaller than the Coulomb integral (see Table 5.2). Remembering that the plus sign in Equation 5.20 goes with the singlet state, and that the minus sign goes with the triplet states, we conclude that:

When the electrons in a helium atom occupy two different orbitals, the singlet state is always higher in energy than the triplet states.

One way of interpreting this result is to note that the singlet state is described by a *symmetric* spatial function, in which the electrons tend to huddle together — a consequence of symmetry, unrelated to classical forces. This means that the electrons are likely to be found in close proximity, where they have a large positive mutual potential energy. By contrast, the triplet states are described by *antisymmetric* spatial functions, so the electrons tend to stay further apart and experience a lower mutual potential energy in triplet states. Figure 5.2 summarizes these results for the ground state and low-lying excited states of helium.

The modification of energy levels due to the indistinguishability of electrons is called **exchange**.

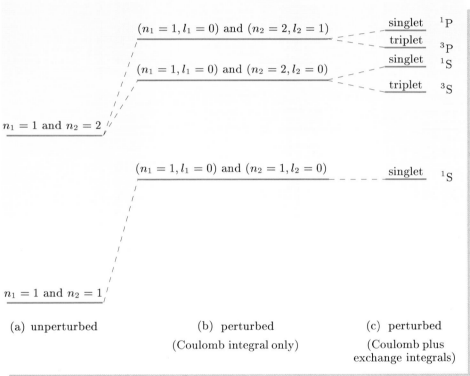

Figure 5.2 Low-lying energy levels in helium: (a) unperturbed energy levels from the independent-particle model; (b) first-order perturbation theory results using the Coulomb integral only; (c) first-order perturbation theory results using both the Coulomb and exchange integrals. The symbols ^{1}P etc. in the column on the far right are part of a labelling convention that will be explained in Section 5.4.

Exercise 5.3 All the bound states of helium have at least one electron in the lowest atomic orbital. All *except the ground state* have separate singlet and triplet levels. Why is the ground state an exception? ∎

5.3 Many-electron atoms

As the number of electrons in the atom increases, so does the number of electron–electron repulsion terms in the Hamiltonian operator. Perturbation theory then becomes very cumbersome. This section describes a more practical way of dealing with atoms containing many electrons.

5.3.1 The central-field approximation

One of the main attractions of the independent-particle model is that it leads to a time-independent Schrödinger equation that is separable, allowing us to use the method of separation of variables to construct solutions from products of orbitals. Unfortunately, this property is lost when we include electron–electron repulsion. We therefore ask whether the effects of electron–electron repulsion can be included an approximate way — one that will still allow us to use separation of variables.

Motivated by these thoughts, we shall replace the exact Hamiltonian of Equation 5.1 by an *effective Hamiltonian*:

$$\widehat{H} = \sum_i \widehat{h}_i \quad \text{with} \quad \widehat{h}_i = -\frac{\hbar^2}{2m_e}\nabla_i^2 + V_i(r_i). \tag{5.23}$$

In this equation,

$$V_i(r_i) = -\frac{Ze^2}{4\pi\varepsilon_0 r_i} + U_i(r_i) \tag{5.24}$$

is an effective potential energy function for electron i. The first term, $-Ze^2/4\pi\varepsilon_0 r_i$, is the potential energy of the electron due to its attraction by the central nucleus. The second term, $U_i(r_i)$, represents in an approximate way the effect of all the other electrons on electron i. This term, whose form still remains to be discussed, *is chosen to depend only on the radial coordinate of electron i*. It is therefore spherically symmetric, and corresponds to a *central field* experienced by electron i. This is what gives the **central-field approximation** its name. Notice that $U_i(r_i)$ may be different for different electrons, so the single-particle Hamiltonian $\widehat{h}_i$ may also be different for different electrons, which is unlike the independent-particle case.

Nevertheless, the time-independent Schrödinger equation is separable, and its solutions can be expressed as products of single-particle atomic orbitals. Because the effective potential energy function $V_i(r_i)$ is spherically symmetric, each orbital can be written as the product of a radial function $R_{nl}(r)$ and a spherical harmonic $Y_{lm}(\theta, \phi)$. Because $V_i(r_i)$ differs from a Coulomb potential energy function, the radial functions will be different to those of hydrogen-like atoms, but the energy eigenfunctions are still labelled by the quantum numbers n, l and m, where l and m are the quantum numbers associated with orbital angular momentum. We can therefore use the same spectroscopic notation as for hydrogen, labelling the orbitals by $1s_0$, $2p_1$, $3d_{-2}$, etc.

This result was used earlier, in Section 4.1.3.

Now let us consider the choice of the effective potential energy function $V_i(r_i)$. A useful result, which emerges from the classical theory of electromagnetism, is that the electric field at a distance r from the centre of a spherically-symmetric charge distribution is the same as that caused by a point charge Q located at the centre of the distribution, where Q is the total amount of charge that is *closer to the centre* than r. This motivates the idea of *screening*.

Consider an electron that is far from the nucleus of an atom with Z electrons. Then we can assume that the other $Z - 1$ electrons will be closer to the nucleus than the distant electron. Modelling these electrons by a spherically-symmetric charge cloud, and using the above result from electromagnetism, the distant electron will be subject to the approximate potential energy function

$$V_i(r_i) \simeq \frac{e^2}{4\pi\varepsilon_0}\left[-\frac{Z}{r_i} + \frac{(Z-1)}{r_i}\right] = -\frac{e^2}{4\pi\varepsilon_0}\frac{1}{r_i}. \quad (5.25)$$

In effect, the repulsion due to the inner electrons partly cancels the attraction due to the nucleus, so it is as if the nucleus had a smaller charge. In atomic physics, this effect is known as **screening**. For an atom with many electrons, most of the nuclear charge will be screened from a distant electron.

Now, let us go to the other extreme and consider an electron that is close to the nucleus. In this case, the other electrons, again modelled by a spherically-symmetric charge cloud, will have practically no effect, and the nuclear charge will be unscreened. The chosen electron will therefore be subject to the approximate potential energy function

$$V_i(r_i) \simeq -\frac{e^2}{4\pi\varepsilon_0}\frac{Z}{r_i}. \quad (5.26)$$

We should therefore choose a function $V_i(r_i)$ that varies smoothly between the limiting tendencies of Equations 5.25 and 5.26. Although, in both limits, $V_i(r_i)$ is approximately proportional to $1/r_i$, the proportionality factor is different for large and small values of r_i, so $V_i(r_i)$ is not proportional to $1/r_i$ over the full range of the radial coordinate. This means that the degeneracy observed in hydrogen for states with the same value of n, but different values of l, is removed. For atoms other than hydrogen, orbitals with the same value of n, but with different values of l, have different energies.

What energy ordering would you expect for orbitals with the same value of n and different values of l? A clue is provided by Equation 5.7, which tells us that the energy eigenfunctions of a hydrogen-like atom are proportional to r^l at small values of r. If you refer back to Chapter 2, you will see that this behaviour arises because the potential energy function in hydrogen is dominated by the centrifugal barrier $l(l+1)/2mr^2$ for small values of r. Exactly the same thing happens in the orbitals of many-electron atoms, so we can safely assume that their orbitals will also vary as r^l close to the nucleus.

It follows that electrons with lower values of l are more likely to be found close to the nucleus than electrons with higher values of l. Consequently, electrons with lower values of l will experience less screening by other electrons, and will be more strongly bound. Since s, p, d and f orbitals correspond to $l = 0, 1, 2$ and 3 respectively, we predict the following ordering of energies for a given value of n:

$$s < p < d < f < \cdots.$$

Figure 5.3 shows the energies of several orbitals for the lithium atom. As you can see, these increase with l. As l increases, the screening becomes more effective and the energy of the lithium orbital becomes closer to that of the corresponding state of hydrogen.

Usually, the screening effect is smaller than the energy differences associated with different values of the principal

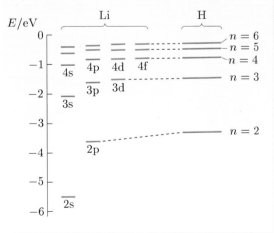

Figure 5.3 Diagram showing the energies of lithium orbitals for the highest-energy electron in the central-field approximation. The other two electrons occupy the 1s orbital. The $n = 1$ orbital is already full, so it is not available to the highest-energy electron. The energy levels of hydrogen are plotted on the right for comparison.

quantum number n, but this is not always so. For example, 4s orbitals are often lower in energy than 3d orbitals, and 5s orbitals are lower in energy than 4f orbitals; in cases like this the reduced screening at low values of l reverses the normal energy-ordering associated with the principal quantum number n. We shall say more about this shortly.

Self-consistency

The choice of the potential energy function $V_i(r_i)$ may have seemed rather arbitrary, but there is a way of making things much more precise.

Starting with a given potential energy function, the central-field model is used to find a set of single-particle orbitals. These orbitals are then filled with electrons, starting with the orbital of lowest energy and working upwards until all Z electrons have been used up. The total wave functions may also be antisymmetrized at this stage. Knowing the shapes of the orbitals, physicists can work out the distribution of electronic charge. They can then use this to find an appropriate potential energy function for each electron that is consistent with the spatial distribution of the other $Z - 1$ electrons. This gives a new central-field model, which can be solved to find a better set of orbitals. The process is repeated again and again until further iterations make no difference. At this point, the solution is said to be **self-consistent**. Provided that the starting point is reasonable, the self-consistent solution should be a good approximation, limited only by the central-field model itself. Moreover, the self-consistent solution is generally insensitive to the precise starting point. Nowadays, this method is implemented using powerful computers, but the calculations were first automated in the 1930s using a mechanical device whose prototype was constructed from Meccano!

5.3.2 Configurations

We shall not discuss the detailed form of the energy eigenfunctions or eigenvalues of a general Z-electron atom. However, we shall describe the way atomic states are classified and labelled. We shall use the word **shell** to describe the set of orbitals that have a specific pair of n and l values. For example, we may refer to the 2p shell or the 3d shell. Then the crudest description of the state of an atom is provided by a list of the occupied shells, together with the numbers of electrons they contain. This list is referred to as the **electronic configuration** of the atom. In a crude approximation, which neglects any corrections to the central-field model, the electronic configuration of an atom determines its energy.

We are particularly interested in the ground-state configuration. For an atom with Z electrons, this can be found by thinking of the shells as being progressively filled by electrons, starting from the shell of lowest energy and working upwards until all the electrons are accounted for.

The possible shells are specified by pairs of n and l values, where $n = 1, 2, \ldots$ and $l = 0, 1, 2, \ldots, n - 1$. To determine which shells are occupied in the ground-state configuration of an atom, we need to know the capacity of each shell (the maximum number of electrons it can contain) and the energy-ordering of the shells. The capacity of a shell is based on the following points:

1. For each value of l, there are $2l + 1$ possible values of m: $0, \pm 1, \ldots, \pm l$.

2. Each electron has two possible values of m_s: $\pm\frac{1}{2}$.
3. By the Pauli exclusion principle, no two electrons in an atom can have the same set of quantum numbers (n, l, m and m_s).

Taken together, these points imply that a shell, characterized by given values of n and l, has a capacity of $2 \times (2l+1)$ electrons. For example, a 2p shell can contain at most $2 \times (2 \times 1 + 1) = 6$ electrons. When a shell is full, it is said to be a **closed shell**; otherwise it is an **open shell**. We can therefore say that a closed nl shell contains $2(2l+1)$ electrons, while an open nl shell contains fewer than $2(2l+1)$ electrons. The electrons in the open shells determine many of the properties of the atoms. They are known as **valence electrons**.

The general pattern of energy-ordering for shells is shown in Figure 5.4, with the 1s shell having the lowest energy, the 2s shell the next-to-lowest energy, and so on. The principal quantum number n increases without limit, but the figure includes all the shells that are occupied in the ground-state configurations of atoms.

The ground-state configuration of an atom is found by filling these shells in order of increasing energy. For example, the ground-state configuration of carbon, with atomic number $Z = 6$, has two electrons in the 1s shell, two electrons in the 2s shell and two electrons in the 2p shell. However f, d and s shells can be very close in energy, and their relative energy-ordering depends sensitively on the screening in individual atoms. Hence, exceptions to the general pattern of shell filling can be found amongst the so-called *transition elements* (in which d shells are being filled) and the *rare-earth elements* (in which f shells are being filled). These exceptional cases are noted in the caption to Figure 5.4.

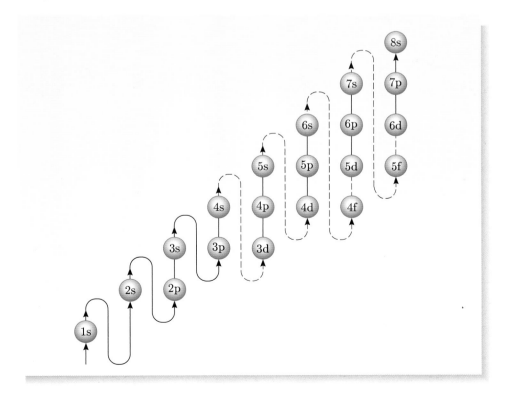

Figure 5.4 The energy-ordering of shells in the central-field approximation, indicated by the snaking line. The line is dashed where the energy differences are so small that the filling of shells may be irregular. (The following elements have ground-state configurations that are exceptions to the normal order of filling: Cr, Cu, Nb, Mo, Tc, Ru, Rh, Ag, Pt and Au have an extra d electron and a missing s electron; Pd has two extra d electrons and two missing s electrons; La, Gd, Ac, Pa, U, Np and Cm have an extra d electron and a missing f electron; Th has two extra d electrons and two missing f electrons.) Symbols of chemical elements are interpreted inside the back cover.

To indicate the number of electrons in a given shell, we use a superscript (omitted in the case of '1'). So, for example, the ground-state configurations of hydrogen and helium are 1s and $1s^2$. The ground-state configurations of the first six elements are listed in Table 5.3.

Table 5.3 The ground-state electronic configurations of the first six elements.

Symbol	Name	Z	Configuration
H	hydrogen	1	1s
He	helium	2	$1s^2$
Li	lithium	3	$1s^2\,2s$
Be	beryllium	4	$1s^2\,2s^2$
B	boron	5	$1s^2\,2s^2\,2p$
C	carbon	6	$1s^2\,2s^2\,2p^2$

Configurations of excited states

The configurations of low-lying excited states can be written down in a similar way. In general, the configuration describing the first excited state of a many-electron atom is one in which an electron in the *highest* energy orbital in the ground-state configuration is excited to the next orbital in energy. For example, the ground-state configuration of lithium is $1s^2\,2s$, so the first excited configuration is $1s^2\,2p$. Continuing this pattern, the next two excited configurations of lithium are $1s^2\,3s$ and $1s^2\,3p$. Configurations such as $1s\,2s^2$ or $1s\,2s\,2p$, in which an electron is excited from the low-lying energy of the 1s shell, turn out to have much higher energies.

Essential skill
Writing down a ground-state configuration

Exercise 5.4 Write down the ground-state configurations of oxygen ($Z = 8$), neon ($Z = 10$), sodium ($Z = 11$), argon ($Z = 18$) and iron ($Z = 26$).

Exercise 5.5 Show that the maximum number of electrons that can have principal quantum number n in an atom is $2n^2$. You may use the fact that $\sum_{i=0}^{k} i = \frac{1}{2}k(k+1)$.

5.3.3 The Periodic Table

In 1869, Dmitri Mendeleev constructed a table in which chemical elements with similar properties appear in the same column, and atomic mass increases from left to right and from top to bottom of the table. The table has been refined since Mendeleev's day, and is now called the **Periodic Table** of the elements. A modern version is shown on the inside back cover of this book. Elements in the same column are said to belong to the same **group**, and elements in the same row are said to belong to the same **period**.

It is remarkable that Mendeleev was able to sketch the Periodic Table almost thirty years before the electron was discovered, without any inkling of the electronic structure of atoms. Nowadays we can use our knowledge of the ground-state configurations of atoms to gain a deep understanding of the Periodic Table. It turns out that the valence electrons — electrons in open shells — are responsible for the chemical properties of elements. Elements in the same column of the Periodic Table (i.e. the same group) have the same number of valence electrons, and the principal quantum number n of these electrons increases as we go down the column. As we go from left to right along a row, open shells fill up. For example, the 2s shell fills as we go from Li to Be, and the 2p shell fills as we go from B to Ne.

The group of elements in the column on the extreme right of the Periodic Table is called the **noble gases**. The ground-state configurations of the noble gas atoms are shown in Table 5.4. In these atoms, all the nl shells up to a given value of n are closed. This makes them very non-reactive, with little tendency to form molecules with other atoms.

Table 5.4 The ground-state electronic configurations of the noble gases.

Symbol	Z	Configuration
He	2	$1s^2$
Ne	10	$1s^2\, 2s^2\, 2p^6$
Ar	18	$1s^2\, 2s^2\, 2p^6\, 3s^2\, 3p^6$
Kr	36	$1s^2\, 2s^2\, 2p^6\, 3s^2\, 3p^6\, 4s^2\, 3d^{10}\, 4p^6$
Xe	54	$1s^2\, 2s^2\, 2p^6\, 3s^2\, 3p^6\, 4s^2\, 3d^{10}\, 4p^6\, 5s^2\, 4d^{10}\, 5p^6$
Rn	86	$1s^2\, 2s^2\, 2p^6\, 3s^2\, 3p^6\, 4s^2\, 3d^{10}\, 4p^6\, 5s^2\, 4d^{10}\, 5p^6\, 6s^2\, 4f^{14}\, 5d^{10}\, 6p^6$

Elements in the first group of the Periodic Table — from Li to Fr — are called **alkali atoms**, and those in the penultimate group — from F to At — are called **halogens**. The ground-state configurations of the alkali atoms and halogens are shown in Tables 5.5 and 5.6, using a shorthand notation in which [Ne(10)], for example, means that the first 10 electrons are arranged in the ground-state electronic configuration of neon ($1s^2\, 2s^2\, 2p^6$).

Table 5.5 The ground-state electronic configurations of the alkali atoms.

Symbol	Z	Configuration
Li	3	[He(2)] 2s
Na	11	[Ne(10)] 3s
K	19	[Ar(18)] 4s
Rb	37	[Kr(36)] 5s
Cs	55	[Xe(54)] 6s
Fr	87	[Rn(86)] 7s

Table 5.6 The ground-state electronic configurations of the halogens.

Symbol	Z	Configuration
F	9	[He(2)] $2s^2\, 2p^5$
Cl	17	[Ne(10)] $3s^2\, 3p^5$
Br	35	[Ar(18)] $4s^2\, 3d^{10}\, 4p^5$
I	53	[Kr(36)] $5s^2\, 4d^{10}\, 5p^5$
At	85	[Xe(54)] $6s^2\, 4f^{14}\, 5d^{10}\, 6p^5$

Alkali atoms have a single valence electron in an s shell. The other $Z-1$ electrons form a tightly bound core in closed shells around the nucleus, so the valence electron sees a well-screened nuclear charge and is well described by the central-field model. All alkali atoms display similar spectra and form similar types of molecule when combining with other atoms. The halogen atoms also bear a family resemblance to one another; they all lack one electron in an open p shell. Closed shells tend to be low in energy, so alkali atoms and halogens react strongly with one another to form stable molecules in which an alkali atom donates its valence electron to a halogen atom; examples are NaCl (sodium chloride, table salt) and KBr (potassium bromide).

The electronic configuration of an element can help us understand many of its properties. One of these is the **first ionization energy**. This is the minimum energy required to remove one electron from the atom to produce a charged ion and an unbound electron. More energy is required to remove a second electron, having removed the first. This is because the nuclear charge remains unchanged, and with fewer electrons, the screening of the nuclear charge is less effective. For example, the first and second ionization energies for helium are 24.5 eV and 54.4 eV, respectively.

> These trends in the Periodic Table are only a general guide; exceptional cases which go against the general trends can be found.

In the Periodic Table, the first ionization energy generally increases from left to right along a row and decreases as we go down a column (Figure 5.5a). We can understand this in the following way. As we go along a row, the nuclear charge increases, and so does the number of valence electrons. However, these electrons are entering an open shell and are described by orbitals with similar spatial distributions. Such orbitals provide only weak screening for one another. So, as we go along a row, screening does not fully compensate for the increasing nuclear charge; the valence electrons become more tightly bound and the first ionization energy tends to increase. As we go down a column, the nuclear charge increases, and there are more electrons in closed shells to provide screening. Far from the nucleus, these two effects largely cancel out. However, as n increases, the average distance of the valence electrons from the nucleus increases; this means that the valence electrons are less tightly-bound, leading to a decrease in the first ionization energy.

A similar analysis can be given for the sizes of atoms (Figure 5.5b). Atoms are not hard spheres with defined edges, but an atomic radius can be defined as the radial distance within which a given percentage (say 90%) of the total electronic charge of the atom is located, on average. Alkali atoms, with a loosely-bound valence electron, have large radii. As we go from the left along a row, the valence electrons become more strongly bound and the atomic radius tends to decrease. Also, as noted above, the atomic radius tends to increase as we go down a column. However, this increase in radius is not dramatic; cesium (Cs) contains nearly 20 times as many electrons as lithium, but its radius is less than twice as large. This is because the greater nuclear charge in cesium keeps the electrons more densely packed.

Exercise 5.6

(a) Which is expected to have the larger first ionization energy: sulphur (S) or magnesium (Mg)?

(b) Which is expected to have the larger atomic radius: magnesium (Mg) or barium (Ba)?

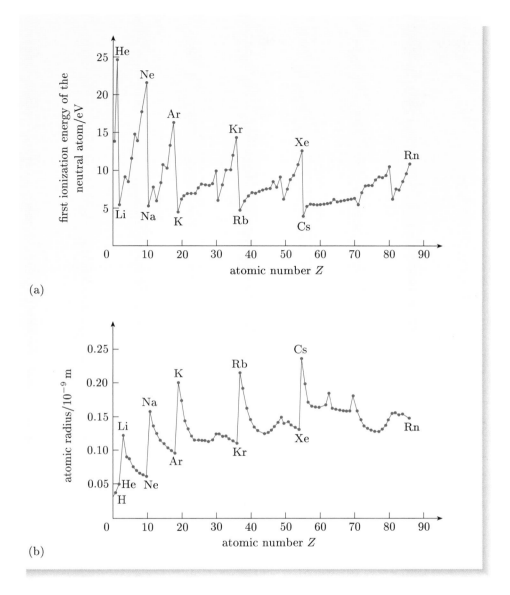

Figure 5.5 (a) First ionization energy as a function of atomic number. (b) Atomic radius as a function of atomic number (technically, 'covalent atomic radii' have been used in this figure).

5.4 Terms and levels

Figure 5.6 shows the visible parts of the spectra of hydrogen and helium. Each of the lines in the spectra corresponds to a transition between two energy levels in the atom. There are many more lines in the helium spectrum, so one might guess that, in a given energy interval, there are more energy levels in the helium atom than in hydrogen.

Figure 5.6 The visible spectra of atomic hydrogen and helium.

A clue as to how this happens is given in Figure 5.7. Each horizontal line in the figure indicates an energy level for the carbon atom. Looking at the three lowest lines, you can see that *a single configuration corresponds to more than one energy level.* This is not surprising. The configuration of an atom only tells us how the electrons occupy shells of given n and l; to completely specify the state of the atom, we must supply further information.

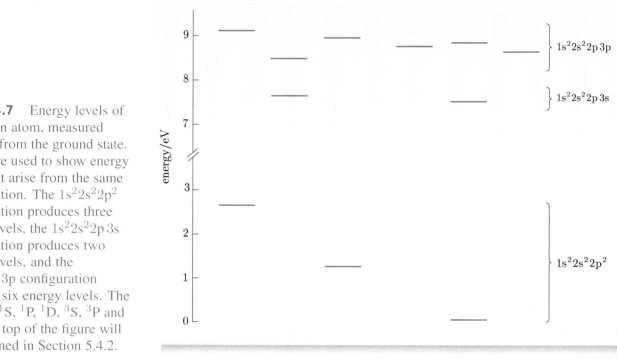

Figure 5.7 Energy levels of the carbon atom, measured upwards from the ground state. Braces are used to show energy levels that arise from the same configuration. The $1s^2 2s^2 2p^2$ configuration produces three energy levels, the $1s^2 2s^2 2p\,3s$ configuration produces two energy levels, and the $1s^2 2s^2 2p\,3p$ configuration produces six energy levels. The symbols ^1S, ^1P, ^1D, ^3S, ^3P and ^3D at the top of the figure will be explained in Section 5.4.2.

Within the central-field model, atomic states with the same configuration (differing only in the m and m_s quantum numbers of individual electrons) remain degenerate. However, the central-field model is only an approximation. In reality, electron–electron repulsion cannot be *exactly* modelled by a central field, and additional effects such as the spin–orbit interaction also come into play. Consequently, the electrons in an atom interact in a variety of ways. You have already seen an example of this in the helium atom, where the 1s 2s configuration corresponds to one singlet and three triplet states. Electron–electron repulsion causes the singlet and triplet states to split apart by a couple of electronvolts. Although this does not destroy the broad pattern of energy levels determined by the various configurations, it does lead to a refinement of it.

5.4.1 Good quantum numbers

We shall not calculate the energy levels in a complicated atom. Instead, we shall explain how these energy levels are classified and labelled. We rely on the following general principle.

5.4 Terms and levels

Good quantum numbers

If observables $A, B, \ldots, C$ are represented by operators $\widehat{A}, \widehat{B}, \ldots, \widehat{C}$ that commute with the Hamiltonian operator $\widehat{H}$ of the system and with each other, it is possible to find a set of simultaneous eigenfunctions for $\widehat{H}, \widehat{A}, \widehat{B}, \ldots, \widehat{C}$.

These eigenfunctions describe states in which the energy, and observables $A, B, \ldots, C$, all have definite constant values. They can be labelled by the quantum numbers that specify the values of these observables. These quantum numbers are said to be **good quantum numbers** for the system.

Within the central-field model, the operators $\widehat{L}_i^2$, $\widehat{L}_{zi}$ and $\widehat{S}_{zi}$ for the ith electron all commute with the Hamiltonian operator, so the single-particle quantum numbers l_i, m_i and m_{s_i} are all good quantum numbers in this approximation. But beyond the central-field model, we must think again. The exact Hamiltonian contains electron–electron interaction terms that depend on the separations of the electrons, rather than just their radial coordinates. It also contains spin–orbit interaction terms that couple together the orbital and spin angular momenta of the particles. When these terms are included, l_i, m_i and m_{s_i} do not remain good quantum numbers, and they cannot be used to label the exact energy eigenfunctions of the atom.

It is clear what needs to be done: we must construct operators that commute with the full Hamiltonian of the many-electron atom. The energy eigenfunctions of the atom can then be labelled by the good quantum numbers associated with these operators. There are two cases to consider: atoms for which the spin–orbit interaction can be ignored, and atoms for which the spin–orbit force plays a significant role. We start with the first case.

5.4.2 Spin–orbit interaction neglected

For atoms of low atomic number, the energy associated with the spin–orbit interaction is much smaller than the energy associated with electron–electron repulsion, so it is a reasonable approximation to ignore the spin–orbit interaction altogether. The central-field approximation already takes some account of electron–electron repulsion, but only in an approximate way. The difference between the exact electron–electron repulsion of Equation 5.1 and the central-field approximation to it is called the *residual electron–electron interaction*. In this subsection, we investigate the effects of this residual interaction in the absence any spin–orbit forces.

The residual electron–electron interaction prevents the quantum numbers l_i and m_i (associated with the orbital angular momentum of the ith electron) from being good quantum numbers. At first sight, you might suppose that m_{s_i} (associated with the z-component of the spin of the ith electron) would remain a good quantum number, provided that there are no spin-dependent terms in the Hamiltonian. But this is not true: we must form an antisymmetric total wave function, and this automatically links the electron spin to the electrostatic energy. You have seen an example of this in the helium atom, where the ground state involves the singlet spin ket $|0,0\rangle$ in which one electron is spin-up and the other is spin-down, but it is impossible to say which electron has which spin. It is the *total* spin quantum numbers, S and M_S, that have definite values. Taking this as a clue, we construct the operators

$$\widehat{\mathbf{L}} = \sum_{i=1}^{Z} \widehat{\mathbf{L}}_i \quad \text{and} \quad \widehat{\mathbf{S}} = \sum_{i=1}^{Z} \widehat{\mathbf{S}}_i, \tag{5.27}$$

where $\widehat{\mathbf{L}}_i$ and $\widehat{\mathbf{S}}_i$ are the orbital angular momentum and spin angular momentum operators for electron i, and $\widehat{\mathbf{L}}$ and $\widehat{\mathbf{S}}$ are the *total* orbital and spin angular momentum operators for all Z electrons in the atom. Taking the z-components of these operators gives

$$\widehat{\mathbf{L}}_z = \sum_{i=1}^{Z} \widehat{\mathbf{L}}_{zi} \quad \text{and} \quad \widehat{\mathbf{S}}_z = \sum_{i=1}^{Z} \widehat{\mathbf{S}}_{zi}, \tag{5.28}$$

while taking the squares of the magnitudes gives

Just for the record: $\widehat{\mathbf{L}}^2 \neq \sum_{i=1}^{Z} \widehat{\mathbf{L}}_i^2$ because there are cross-product terms.

$$\widehat{\mathbf{L}}^2 = \widehat{\mathbf{L}} \cdot \widehat{\mathbf{L}} \quad \text{and} \quad \widehat{\mathbf{S}}^2 = \widehat{\mathbf{S}} \cdot \widehat{\mathbf{S}}. \tag{5.29}$$

It is possible to show that the operators in Equations 5.28 and 5.29 all commute with the Hamiltonian operator $\widehat{\mathbf{H}}$ of Equation 5.1, which includes the exact electron–electron repulsion terms, but ignores the spin–orbit interaction. That is,

$$[\widehat{\mathbf{L}}^2, \widehat{\mathbf{H}}] = 0, \quad [\widehat{\mathbf{L}}_z, \widehat{\mathbf{H}}] = 0, \quad [\widehat{\mathbf{S}}^2, \widehat{\mathbf{H}}] = 0, \quad [\widehat{\mathbf{S}}_z, \widehat{\mathbf{H}}] = 0.$$

Following the general pattern for angular momentum operators, the eigenvalues of $\widehat{\mathbf{L}}^2, \widehat{\mathbf{L}}_z, \widehat{\mathbf{S}}^2$ and $\widehat{\mathbf{S}}_z$ are $L(L+1)\hbar^2$, $M_L\hbar$, $S(S+1)\hbar^2$ and $M_S\hbar$. It follows that the quantum numbers L, M_L, S and M_S are *good quantum numbers* for the atom in the absence of spin–orbit interactions. This means that the energy eigenfunctions of the atom can be labelled by these quantum numbers (and by additional quantum numbers if necessary).

Note the use of capital letters to denote quantum numbers for total angular momenta. Here, L is the the **total orbital angular momentum quantum number**; it is *not* the magnitude of the orbital angular momentum. Similarly, in this context, S is the **total spin quantum number**.

In general, the energy of an isolated atom depends on its electronic configuration, and on the quantum numbers L and S, but is independent of the quantum numbers M_L and M_S. Each set of states, arising from a given configuration and specified by a given pair of L and S values is called an **atomic term**, or a **term** for short. So, when specifying an energy level in an atom, we must give the electronic configuration *and* the atomic term.

To find the possible atomic terms, we need to know the values of L and S that are consistent with a given electronic configuration, and to find the degeneracy of an atomic term we need to know which values of M_L and M_S are possible for given values of L and S. The following rules (no proofs given) apply in simple cases:

> **Rules for the possible values of L, S, M_L and M_S**
>
> For L and S derived from a given configuration:
>
> 1. Closed shells in the configuration do not contribute to the values of S and L; only the valence electrons in open shells are relevant.
>
> 2. An open shell containing a single valence electron with quantum numbers l_i and s_i produces a single term with $L = l_i$ and $S = s_i$.

3. A pair of valence electrons that occupy different open shells and have quantum numbers (l_1, s_1) and (l_2, s_2) produces terms in which the possible values of L and S are

$$L = |l_1 - l_2|, |l_1 - l_2| + 1, \ldots, l_1 + l_2 - 1, l_1 + l_2, \quad (5.30)$$
$$S = |s_1 - s_2|, |s_1 - s_2| + 1, \ldots, s_1 + s_2 - 1, s_1 + s_2. \quad (5.31)$$

The case in which two valence electrons occupy the same open shell is discussed on page 136.

4. For fixed values of L and S, the quantum numbers M_L and M_S are restricted to the values

$$M_L = -L, -L + 1, \ldots, L - 1, L, \quad (5.32)$$
$$M_S = -S, -S + 1, \ldots, S - 1, S, \quad (5.33)$$

giving $(2L + 1)$ values of M_L and $(2S + 1)$ values of M_S.

The following worked example shows how these rules are applied.

Worked Example 5.1

What are the possible values of L, S, M_S and M_L for the carbon atom configuration $1s^2 2s^2 2p\,3s$?

Solution

$1s^2$ and $2s^2$ are closed shells, so we only need to consider the electrons in the open 2p and 3s shells. These electrons have $l_1 = 1$ and $l_2 = 0$, so the minimum and maximum values of L are

$$L_{\min} = |l_1 - l_2| = |1 - 0| = 1,$$
$$L_{\max} = l_1 + l_2 = 1 + 0 = 1.$$

Hence the only possible value of L is 1, and the corresponding values of M_L are -1, 0 and 1.

For spin, $s_1 = s_2 = \tfrac{1}{2}$, and the minimum and maximum values of S are

$$S_{\min} = |s_1 - s_2| = |\tfrac{1}{2} - \tfrac{1}{2}| = 0,$$
$$S_{\max} = s_1 + s_2 = \tfrac{1}{2} + \tfrac{1}{2} = 1.$$

The values of S increase in steps of 1, so the possible values are $S = 0$ and $S = 1$. This result is familiar: the total spin state of a pair of electrons is either singlet ($S = 0$) or triplet ($S = 1$). When $S = 0$, M_S is 0. When $S = 1$, M_S is equal to -1, 0 or 1.

Essential skill

For a given configuration with two electrons in different open shells, determine the possible values of the quantum numbers L, S, M_S and M_L

Exercise 5.7 What are the possible values of L and S for the lithium atom configuration $1s^2\,2s$?

Exercise 5.8 What are the possible values of L, S, M_S and M_L for the carbon atom configuration $1s^2\,2s^2\,2p\,3p$? ∎

A special **spectroscopic notation** is used to label atomic terms. For a term with quantum numbers L and S, we write

$$^{2S+1}L, \quad (5.34)$$

Table 5.7 Spectroscopic notation for atomic terms.

L	L
0	S
1	P
2	D
3	F
4	G
⋮	⋮

where, for L, we adopt the same notation as for orbitals, but using capital letters, as indicated in Table 5.7.

The quantity $2S+1$ is called the **multiplicity** of the term, and represents the $2S+1$ different values of M_S that are associated with a given value of S. For $S = 0, \frac{1}{2}, 1, \frac{3}{2}, \ldots$, we have $2S + 1 = 1, 2, 3, 4, \ldots$, and the terms are called *singlet, doublet, triplet, quartet*, and so on. There are also $2L + 1$ values of M_L associated with a given value of L. In the absence of spin–orbit effects, the energy of a term is independent of M_S and M_L, so each term has a degeneracy of $(2S + 1)(2L + 1)$.

We can now see what the labels ^{1}P, ^{3}P, etc. signify in Figures 5.2 and 5.7; they label different atomic terms arising from given configurations.

● Use spectroscopic notation to specify the terms arising from the carbon atom configurations $1s^2 2s^2 2p\,3s$ and $1s^2 2s^2 2p\,3p$.

○ According to Worked Example 5.1, the configuration $1s^2\,2s^2\,2p\,3s$ gives rise to $L = 1$ and $S = 0$ or $S = 1$, so its possible atomic terms are ^{1}P and ^{3}P. According to Exercise 5.8, the configuration $1s^2\,2s^2\,2p\,3p$ gives rise to $L = 0$, $L = 1$ or $L = 2$ and $S = 0$ or $S = 1$, so its possible atomic terms are ^{1}S, ^{3}S, ^{1}P, ^{3}P, ^{1}D and ^{3}D. All of these terms are marked in Figure 5.7.

Exercise 5.9 Use spectroscopic notation to specify the terms arising from the 1s 2s and 1s 2p helium configurations, and state the degeneracies of these terms. ■

A limitation

Note that Rule 3 in the box on pages 134–5 refers to electrons that are in *different* open shells. For example, the configuration could be $1s^2\,2s\,2p$ or $1s^2\,2s\,3s$ as, in both cases, the two valence electrons have different sets of nl quantum numbers. An added complication arises in configurations such as $1s^2\,2p^2$, where the two valence electrons occupy the *same* shell. In such cases, it turns out that the indistinguishability of the electrons and the Pauli exclusion principle imply that some of the predicted atomic terms are not realized in practice. We mention this only for completeness. Electrons in the same open shell are said to be **equivalent electrons**; you will only be asked to find atomic terms for configurations involving *non-equivalent* electrons (electrons in *different* open shells), for which Rule 3 applies.

5.4.3 Spin–orbit interaction included

The spin–orbit interaction increases with the atomic number Z. We now consider the case where the spin–orbit interaction is not neglected, but is still small compared to the residual electron–electron interaction. This is true for atoms with $Z \leq 35$.

In the presence of the spin–orbit interaction, L and S remain good quantum numbers, but $\widehat{L}_z$ and $\widehat{S}_z$ do not commute with the Hamiltonian operator, so M_L and M_S are not good quantum numbers. However, we can form a **total atomic angular momentum**, **J**, where 'total' now refers to a sum over all the electrons *and* to the sum of orbital and spin contributions. The operators $\hat{J}^2$ and $\hat{J}_z$ have eigenvalues $J(J+1)\hbar^2$ and $M_J\hbar$, and the quantum numbers J and M_J are good quantum numbers for the whole atom.

When the spin–orbit interaction is included, the energy of the atom depends on J, but is independent of M_J. Each atomic term, associated with a given electronic configuration and with given values of L and S, splits into separate energy levels called **atomic levels**, or **levels** for short. This splitting produces fine structure in atomic spectra. In general, the splitting between different levels arising from the same term is much smaller than the spacing between different terms arising from the same configuration. Because the energy does not depend on M_J, each level corresponds to a set of $2J+1$ degenerate states.

If the spin–orbit interaction is small, the appropriate way to construct **J** is to add the total orbital angular momentum to the total spin:

$$\mathbf{J} = \mathbf{L} + \mathbf{S}.$$

For obvious reasons, this is called the **LS-coupling scheme**. Within this scheme, it turns out that fixed values of L and S, labelling an atomic term, give rise to the following allowed values of J:

$$J = |L-S|, |L-S|+1, \ldots, L+S-1, L+S, \qquad (5.35)$$

and, as for all angular momenta, the possible values of M_J are

$$M_J = -J, -J+1, \ldots, J-1, J. \qquad (5.36)$$

For $L \geq S$, the number of different J values, and hence the number of levels, is equal to the multiplicity, $2S+1$.

Exercise 5.10 What are the possible values of the total angular momentum quantum numbers J and M_J for atomic terms in which: (a) $L = S = 0$; (b) $L = 2$, $S = 0$; (c) $L = 1$, $S = 1$? ■

Because the splitting depends on the total angular momentum, the value of J is now added to the labelling. To specify a level in **spectroscopic notation**, we write

$$^{2S+1}L_J.$$

When the LS-coupling scheme holds, the possible values of J are determined by Equation 5.35. Figure 5.8 illustrates how a configuration leads to several terms which then split into different levels when a weak spin–orbit interaction is taken into account.

Exercise 5.11 Write down the levels that the following terms split into: (a) ^{3}P; (b) ^{1}F; (c) ^{2}D.
What are the degeneracies of these levels? ■

Strong spin–orbit coupling

For atoms with $Z \geq 80$, the spin–orbit interaction becomes stronger than the residual electron–electron repulsion. Under these circumstances, it is more appropriate to combine the orbital and spin angular momenta of each electron to find its total

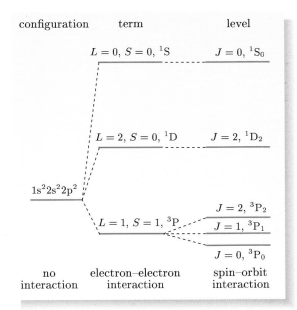

Figure 5.8 The splitting of the configuration $1s^2\,2s^2\,2p^2$ of a carbon atom into terms with different values of L and S (see Figure 5.7). When the spin–orbit interaction is taken into account, the terms split into levels with different values of J (see Exercise 5.10).

angular momentum, $\mathbf{j}_i$, and then express the total angular momentum of the atom as a sum of contributions from each electron: $\mathbf{J} = \sum_i \mathbf{j}_i$. This is known as the **jj-coupling scheme**, but we will not go into its details here. For atoms with Z between 35 and 80, neither the LS-coupling scheme nor the jj-coupling scheme is completely satisfactory, although they are often used for pragmatic reasons.

5.4.4 Hund's rules

In general, different terms and levels arising from the same configuration have different energies, but what is their energy ordering? In many cases, the answer is provided by a set of rules known as **Hund's rules**. Strictly speaking, Hund's rules apply to ground-state configurations, but they can also be applied to some excited states. Hund's rules can be summarized as follows.

> **Hund's rules for atomic terms and levels**
>
> For terms and levels arising from a given configuration:
>
> 1. The term with the largest value of S has the lowest energy. The energy of other terms increases with decreasing S.
>
> 2. For a given value of S, the term with the largest value of L has the lowest energy. The energy of other terms increases with decreasing L.
>
> 3. For a given term, the levels are ordered as follows:
>
>   - If the shell is less than half-full, the level with the smallest value of J has the lowest energy.
>
>   - If the shell is more than half-full, the level with the largest value of J has the lowest energy.
>
>   - If the shell is exactly half-full, there is no splitting due to the spin–orbit interaction.

Hund's rules do not apply to all excited states. For example, they fail for the $1s^2\, 2s^2\, 2p\, 3p$ configuration in carbon (Figure 5.7).

The following example shows how these rules are applied.

Essential skill:
Applying Hund's rules to order terms and levels arising from a given configuration

> **Worked Example 5.2**
>
> Use Hund's rules to confirm the order of the levels shown in Figure 5.8.
>
> **Solution**
>
> The levels shown in Figure 5.8 are $^1S_0$, $^1D_2$, $^3P_2$, $^3P_1$ and $^3P_0$. Hund's first rule shows that the triplets $^3P_2$, $^3P_1$ and $^3P_0$ are lower in energy than the singlets $^1S_0$ and $^1D_2$. Hund's second rule shows that $^1D_2$ is lower in energy than $^1S_0$. Finally, the 2p shell is less than half-full, so Hund's third rule shows that $^3P_0$ is lower in energy than $^3P_1$, which is lower in energy than $^3P_2$. Hence the energy ordering is
>
> $$^3P_0 < {}^3P_1 < {}^3P_2 < {}^1D_2 < {}^1S_0,$$
>
> as shown in Figure 5.8.

Exercise 5.12 The following terms arise from the same configuration: ^{2}P, ^{4}P, ^{2}D, ^{2}F, ^{4}F, ^{2}G, ^{2}H. Assuming that Hund's rules apply, write these terms in order of increasing energy. ∎

Summary of Chapter 5

Section 5.1 Taking the mass of the nucleus to be infinite, the Hamiltonian operator of an Z-electron atom can be written as

$$\widehat{H} = \sum_{i=1}^{Z} \left[-\frac{\hbar^2}{2m_e} \nabla_i^2 - \frac{Ze^2}{4\pi\varepsilon_0 r_i} \right] + \frac{1}{4\pi\varepsilon_0} \sum_{i=1}^{Z} \sum_{j>i}^{Z} \frac{e^2}{r_{ij}},$$

where motion of the whole atom has been neglected. The corresponding time-independent Schrödinger equation is non-separable, and cannot be solved exactly. In the independent-particle approximation, we neglect electron–electron repulsions. Energy eigenfunctions can then be found that are products of single-particle orbitals, and the energy of the atom is the sum of the energy eigenvalues associated with individual electrons.

Section 5.2 Electrons are identical fermions, so the electrons in an atom must be described by an antisymmetric total wave function. In first-order perturbation theory, an excited state of helium, in which the electrons occupy different orbitals ϕ_r and ϕ_s with energies E_r and E_s, has total energy

$$E_{rs} \simeq E_r + E_s + \frac{e^2}{4\pi\varepsilon_0} \int \frac{|\phi_r(\mathbf{r}_1)|^2 |\phi_s(\mathbf{r}_2)|^2}{r_{12}} \, dV_1 \, dV_2$$

$$\pm \frac{e^2}{4\pi\varepsilon_0} \int \frac{\phi_r^*(\mathbf{r}_1) \phi_s^*(\mathbf{r}_2) \phi_r(\mathbf{r}_2) \phi_s(\mathbf{r}_1)}{r_{12}} \, dV_1 \, dV_2,$$

where the plus sign is for a singlet state and the minus sign is for a triplet state. The last integral on the right-hand side is the exchange integral; this makes triplet states lower in energy than singlet states.

Section 5.3 In the central-field approximation, electron–electron repulsion is modelled by a spherically-symmetric effective potential energy function. This gives a separable time-independent Schrödinger equation, and solutions can be obtained as products of single-particle orbitals. Each orbital is the product of a radial function and a spherical harmonic. For a given n, orbitals with low l-values have lower energies than orbitals with high l-values. Self-consistency can be achieved by an iterative procedure which ensures that the potential energy function is consistent with the distribution of electrons.

A shell is a set of orbitals with a given pair of n and l values. The configuration of an atom specifies how electrons occupy shells. Ground-state configurations explain the gross structure of the Periodic Table, in which elements with similar properties are organized into columns.

Section 5.4 When electron–electron and spin–orbit interactions are taken into account, the quantum numbers associated with a single electron are not good quantum numbers for the whole atom.

When the spin–orbit interaction is neglected, L and S, the total orbital angular momentum and total spin quantum numbers of all the electrons in the atom, are good quantum numbers, along with the corresponding M and M_S. Because of the residual electron–electron interaction, a single configuration splits into a number of energy levels, called atomic terms, labelled by ^{2S+1}L. Closed shells do not contribute to L and S. Two non-equivalent valence electrons with quantum numbers (l_1, s_1) and (l_2, s_2) gives rise to terms labelled by

$$L = |l_1 - l_2|, |l_1 - l_2| + 1, \ldots, l_1 + l_2 - 1, l_1 + l_2,$$
$$S = |s_1 - s_2|, |s_1 - s_2| + 1, \ldots, s_1 + s_2 - 1, s_1 + s_2.$$

When the spin–orbit interaction is included, the good quantum numbers are L, S, J and M_J, where J and M_J, refer to the total angular momentum (orbital plus spin) of all the electrons in the atom. If the spin–orbit interaction is weak, the LS-coupling scheme can be used and the allowed values of J are

$$J = |L - S|, |L - S| + 1, \ldots, L + S - 1, L + S.$$

Because of the spin–orbit interaction, states of different J have different energies, but have a $(2J + 1)$-fold degeneracy because the energy does not depend on M_J. Thus each *term* is split into a number of *levels* defined by different values of J. Levels are labelled by $^{2S+1}L_J$.

Hund's rules often allow us to determine the energy ordering of the terms and levels arising from a given configuration.

Achievements from Chapter 5

After studying this chapter, you should be able to:

5.1 Explain the meanings of the newly defined (emboldened) terms and symbols, and use them appropriately.

5.2 Describe the independent-particle model of atoms.

5.3 Write down singlet and triplet total wave functions for helium.

5.4 Give an account of the calculation of the energy of bound states of helium using perturbation theory; explain how the exchange interaction arises.

5.5 Explain the basic ideas behind the central-field approximation.

5.6 Write down the ground-state electronic configuration for given atoms.

5.7 Explain general characteristics of the Periodic Table, in particular why atoms in the same column have similar behaviour, why noble gases are non-reactive, and general trends in the first ionization energy and atomic radius of elements.

5.8 Use the spectroscopic notation for a many-electron atom, and describe how L, S and J arise within the LS-coupling scheme.

5.9 Obtain the terms and levels arising from a given configuration in the case of a single valence electron or a pair of non-equivalent electrons, and use Hund's rules to predict the energy-ordering of terms and levels.

Chapter 6 Diatomic molecules

Introduction

You have seen how the time-independent Schrödinger equation is used to describe atoms. However, most of the matter on Earth is not in the form of isolated atoms; instead it consists of atoms joined together in molecules or in solids, and these have very different properties from isolated atoms. Even the air we breathe consists mainly of molecules: nitrogen molecules N_2 (two nitrogen atoms bound together), oxygen molecules O_2 (two oxygen atoms bound together) and carbon dioxide molecules CO_2 (two oxygen atoms bound to a carbon atom). These molecules remain intact when they collide with one another in a gas, and a considerable amount of energy is required to split them into their constituent atoms. For example, 9.8 eV is required to split a nitrogen molecule into two nitrogen atoms.

The aim of this chapter is to apply the time-independent Schrödinger equation to simple molecules, and hence to show how quantum mechanics explains chemical bonding, that is, how atoms bind together to form molecules.

We shall start by studying the hydrogen molecule ion, H_2^+, which consists of two protons and one electron. This is not a molecule you would come across in everyday life, nor can you store it in the laboratory because it would rapidly gain an electron to form a hydrogen molecule, H_2. However, the hydrogen molecule ion has been observed experimentally and its properties have been measured. It was discovered by J. J. Thompson in cathode rays and it is formed fleetingly in a tube of hydrogen gas through which an electrical discharge is passed. Measurements show that H_2^+ is a molecule in which the two protons are bound together by the electron; it is not a hydrogen atom plus a free proton. Early studies of the hydrogen molecule ion demonstrated that the time-independent Schrödinger equation can explain how chemical bonds arise, and this opened up the whole new area of quantum chemistry.

An early triumph for molecular quantum mechanics was the explanation of the unusual magnetic properties of the oxygen molecule. When a stream of liquid oxygen (which consists of oxygen molecules) is poured between the poles of a magnet, it is held by the field, as shown in Figure 6.1. This behaviour could not be explained by pre-quantum mechanical descriptions of chemical bonding, but we will show that it can be explained using results obtained from the time-independent Schrödinger equation.

Figure 6.1 A stream of liquid oxygen is held in the inhomogeneous magnetic field between the poles of a magnet.

Section 6.1 starts by setting up the time-independent Schrödinger equation for a diatomic molecule, and explaining how it can be simplified using an approximation due to Born and Oppenheimer, which allows the motions of electrons and nuclei to be studied separately. Using the Born–Oppenheimer approximation, it is possible to solve the electronic part of the problem for H_2^+ exactly. Rather than doing this, Section 6.2 uses the variational method (introduced in Chapter 3) to obtain an estimate of the ground-state energy and the corresponding eigenfunction for the hydrogen molecule ion. The advantage is that the variational method can be applied to *any* molecule, whereas exact solutions can only be found for H_2^+. We shall also see how the first excited state of H_2^+ emerges from the variational calculation. Section 6.3 explains how molecular orbitals can be constructed from linear combinations of atomic orbitals. It also introduces a spectroscopic notation for labelling molecular orbitals in diatomic molecules. Section 6.4 then shows how to find the ground-state electronic configurations of a number of diatomic molecules. You will see why the nitrogen atoms in a nitrogen molecule are so strongly bound together, and how quantum mechanics explains the unusual magnetic properties of the oxygen molecule. Finally, Section 6.5 describes ways of improving the approximations used in this chapter.

6.1 The time-independent Schrödinger equation

6.1.1 Setting up the equation

As for any quantum system, the energy eigenfunctions needed to describe the stationary states of a diatomic molecule are found by solving the time-independent Schrödinger equation. We shall simplify this task by ignoring the spin of the particles, just as we did for atoms initially. The first step is to find an expression for the Hamiltonian operator, $\widehat{H}$.

As an example, consider a two-electron diatomic molecule, with the variables defined as in Figure 6.2. We can write the Hamiltonian operator for this molecule as

$$\widehat{H} = -\frac{\hbar^2}{2M_A}\nabla_A^2 - \frac{\hbar^2}{2M_B}\nabla_B^2$$
$$+ \sum_{i=1}^{2}\left[-\frac{\hbar^2}{2m_e}\nabla_i^2 - \frac{Z_A e^2}{4\pi\varepsilon_0 r_{iA}} - \frac{Z_B e^2}{4\pi\varepsilon_0 r_{iB}}\right]$$
$$+ \frac{e^2}{4\pi\varepsilon_0 r_{12}} + \frac{Z_A Z_B e^2}{4\pi\varepsilon_0 R_{AB}}. \tag{6.1}$$

In this expression, ∇_A^2 is a shorthand for
$$\frac{\partial^2}{\partial X_A^2} + \frac{\partial^2}{\partial Y_A^2} + \frac{\partial^2}{\partial Z_A^2},$$
where (X_A, Y_A, Z_A) are the coordinates of nucleus A, with a similar interpretation for ∇_B^2.

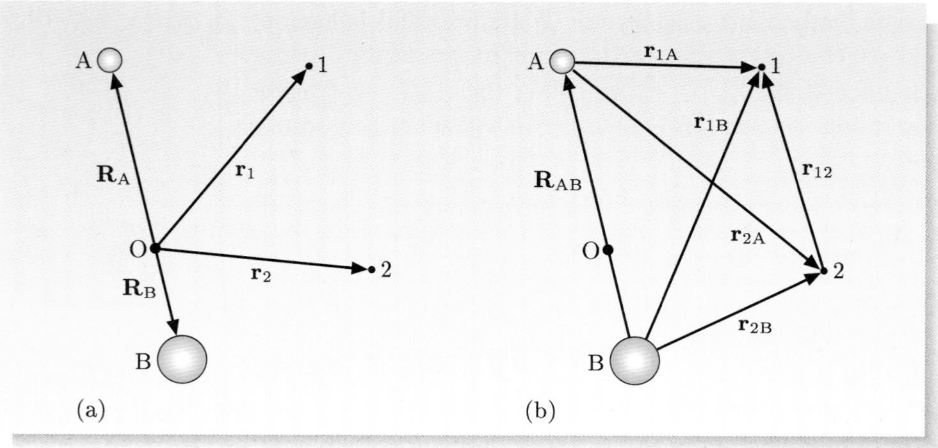

Figure 6.2 Position vectors used to define the Hamiltonian operator of a diatomic molecule with two nuclei labelled A and B and two electrons labelled 1 and 2: (a) positions relative to an origin O at the centre of mass of the molecule; (b) relative positions.

The first line on the right-hand side represents the total kinetic energy of the two nuclei, A and B, of masses M_A and M_B. The second line represents the kinetic energies of the two electrons, each of mass m_e, plus the negative potential energies of the electrons due to their attractions by the nuclei, which have atomic numbers Z_A and Z_B and are separated from the ith electron by $r_{iA} = |\mathbf{r}_i - \mathbf{R}_A|$ and $r_{iB} = |\mathbf{r}_i - \mathbf{R}_B|$. The third line adds the potential energies due to Coulomb repulsion between the two electrons, separated by $r_{12} = |\mathbf{r}_1 - \mathbf{r}_2|$, and Coulomb repulsion between the two nuclei, separated by $R_{AB} = |\mathbf{R}_A - \mathbf{R}_B|$.

As with atoms, we are not interested in translational motion of the molecule as a whole. We therefore work in the centre-of-mass frame of the molecule, where the overall translational motion is absent. When defining the centre of mass, we can safely neglect the masses of the electrons compared to the masses of the nuclei. Omitting the details, this approximation allows us to write the Hamiltonian operator as

$$\widehat{H} = -\frac{\hbar^2}{2\mu}\nabla^2_{AB} + \sum_{i=1}^{2}\left[-\frac{\hbar^2}{2m_e}\nabla^2_i - \frac{Z_A e^2}{4\pi\varepsilon_0 r_{iA}} - \frac{Z_B e^2}{4\pi\varepsilon_0 r_{iB}}\right]$$
$$+ \frac{e^2}{4\pi\varepsilon_0 r_{12}} + \frac{Z_A Z_B e^2}{4\pi\varepsilon_0 R_{AB}}. \tag{6.2}$$

Compared with Equation 6.1, the kinetic energy terms of the two nuclei have been replaced by a single term $(-\hbar^2/2\mu)\nabla^2_{AB}$. This includes the reduced mass μ of the nuclei and involves derivatives with respect to the components of the *relative* position vector $\mathbf{R}_{AB} = \mathbf{R}_A - \mathbf{R}_B$, rather than the coordinates of the individual nuclei.

$$\mu = \frac{M_A M_B}{M_A + M_B}$$

With the centre-of-mass motion discounted, the energy eigenfunctions describing a two-electron diatomic molecule are the solutions of the time-independent Schrödinger equation

$$\widehat{H}\psi(\mathbf{r}_1, \mathbf{r}_2, \mathbf{R}_{AB}) = E\psi(\mathbf{r}_1, \mathbf{r}_2, \mathbf{R}_{AB}), \tag{6.3}$$

where the Hamiltonian operator $\widehat{H}$ is given by Equation 6.2. The energy eigenfunction ψ depends on the position vectors $\mathbf{r}_1$ and $\mathbf{r}_2$ of electrons 1 and 2, and also on the relative position $\mathbf{R}_{AB}$ of the nuclei. It does not depend separately on the individual nuclear positions $\mathbf{R}_A$ and $\mathbf{R}_B$ because, in the centre-of-mass frame, these can be expressed directly in terms of M_A, M_B and $\mathbf{R}_{AB}$, for example:

$$\mathbf{R}_A = \frac{M_B}{M_A + M_B}\mathbf{R}_{AB}. \tag{6.4}$$

Unfortunately, the Hamiltonian in Equation 6.2 contains terms $-Z_A e^2/(4\pi\varepsilon_0 r_{iA})$ and $-Z_B e^2/(4\pi\varepsilon_0 r_{iB})$ that depend on the coordinates of an electron and a nucleus. These terms prevent us from using the method of separation of variables and obtaining completely independent equations for the electronic and nuclear coordinates. This means that the time-independent Schrödinger equation cannot be solved exactly; we need to find suitable approximations.

6.1.2 The Born–Oppenheimer approximation

The key approximation used in the quantum-mechanical description of nearly all molecules was proposed by Max Born and Robert Oppenheimer in 1927. It is called the **Born–Oppenheimer approximation**, and is based on the fact that the nuclei in a molecule are much more massive than the electrons, and therefore move much more slowly than them. This leads to:

> **The Born–Oppenheimer approximation**
>
> 1. In order to study the behaviour of the electrons in a molecule, we can initially treat the nuclei as being static.
>
> 2. In order to study the nuclear motions in a molecule, we can assume that the electrons adapt instantaneously to each new set of positions of the nuclei.

The Born–Oppenheimer approximation does not achieve a complete separation of the electronic and nuclear motions, but it does provide a framework within which we can understand how the nuclear positions affect the electronic states, and how these electronic states then influence the more sluggish motion of the nuclei.

In this chapter, we are mainly concerned with the states of electrons in molecules. From this perspective, the most important point about the Born–Oppenheimer approximation is that $\mathbf{R}_{AB}$ can be treated as a constant vector. This means that we can solve the electronic part of the problem by neglecting the first term in Equation 6.2 (which describes the motion of the nuclei) and also neglecting the last term (which just adds a constant to the energy, without affecting the electronic energy eigenfunctions). We therefore arrive at the following equation for the electrons:

Electronic time-independent Schrödinger equation

$$\sum_{i=1}^{2} \left[-\frac{\hbar^2}{2m_e} \nabla_i^2 - \frac{Z_A e^2}{4\pi\varepsilon_0 r_{iA}} - \frac{Z_B e^2}{4\pi\varepsilon_0 r_{iB}} \right] \psi_{el}(\mathbf{r}_1, \mathbf{r}_2)$$

$$+ \frac{e^2}{4\pi\varepsilon_0 r_{12}} \psi_{el}(\mathbf{r}_1, \mathbf{r}_2) = E_{el} \psi_{el}(\mathbf{r}_1, \mathbf{r}_2). \tag{6.5}$$

The solutions ψ_{el} are the **electronic energy eigenfunctions** of the molecule; they describe the probability distribution of the electrons for a fixed internuclear separation $\mathbf{R}_{AB}$. The corresponding eigenvalues are the **electronic energies**, which also depend on $\mathbf{R}_{AB}$.

A different choice of $\mathbf{R}_{AB}$ would lead to different electronic energy eigenfunctions. It is therefore fair to say that these eigenfunctions depend on $\mathbf{R}_{AB}$. However, $\mathbf{R}_{AB}$ is fixed in the Born–Oppenheimer approximation. For this reason, we shall regard $\mathbf{R}_{AB}$ as a parameter (like M_A or Z_A) that implicitly determines the form of the electronic energy eigenfunctions, but is not displayed as an argument of these functions. We say that $\psi_{el}(\mathbf{r}_1, \mathbf{r}_2)$ *depends parametrically* on $\mathbf{R}_{AB}$. The energy eigenvalues also depend on $\mathbf{R}_{AB}$. In the discussion that follows, it will be helpful to make this dependence explicit by writing $E_{el} = E_{el}(\mathbf{R}_{AB})$.

The Born–Oppenheimer approximation assumes that the energy eigenfunction of the whole molecule (electrons plus nuclei) can be expressed as a product:

$$\psi(\mathbf{r}_1, \mathbf{r}_2, \mathbf{R}_{AB}) = \psi_{el}(\mathbf{r}_1, \mathbf{r}_2) \chi_{nuc}(\mathbf{R}_{AB}). \tag{6.6}$$

Here, $\psi_{el}(\mathbf{r}_1, \mathbf{r}_2)$ is the electronic energy eigenfunction, whose modulus squared describes the probability density of the electrons for a particular internuclear separation $\mathbf{R}_{AB}$, and $\chi_{nuc}(\mathbf{R}_{AB})$ is the nuclear part of the energy eigenfunction, whose modulus squared describes the probability density of the nuclei. To find the form of $\chi_{nuc}(\mathbf{R}_{AB})$, we substitute Equation 6.6 into the time-independent Schrödinger equation for the whole molecule (Equations 6.2 and 6.3). Using the fact that ψ_{el} satisfies the electronic time-independent Schrödinger equation (Equation 6.5), we then see that $\chi_{nuc}(\mathbf{R}_{AB})$ satisfies the following equation:

Nuclear time-independent Schrödinger equation

$$\left[-\frac{\hbar^2}{2\mu} \nabla^2_{AB} + \frac{Z_A Z_B e^2}{4\pi\varepsilon_0 R_{AB}} + E_{el}(\mathbf{R}_{AB}) \right] \chi_{nuc}(\mathbf{R}_{AB}) = E \chi_{nuc}(\mathbf{R}_{AB}), \tag{6.7}$$

where E is the total energy of the whole molecule. This equation contains $E_{el}(\mathbf{R}_{AB})$, an energy eigenvalue of the electronic time-independent Schrödinger equation at the nuclear separation $\mathbf{R}_{AB}$. Because this is a function of $\mathbf{R}_{AB}$, the electronic energy contributes to the effective potential energy function in which the nuclei move. This effective potential energy function represents the total energy of the molecule minus the kinetic energies of the nuclei; we shall call it the **total static energy** of the molecule and denote it by E_{stat}. Thus

$$E_{stat} = \frac{Z_A Z_B e^2}{4\pi\varepsilon_0 R_{AB}} + E_{el}(\mathbf{R}_{AB}). \tag{6.8}$$

Note that E_{stat} depends on the precise electronic state that is being considered.

In practise, the Born–Oppenheimer approximation splits the task of finding the solutions of Equation 6.3 into that of finding the solutions of Equations 6.5, and then finding the solutions of Equation 6.7. You might think that it would be more work to solve two equations rather than one, but this is not the case because the new equations are much simpler to solve than the original one.

6.1.3 Three different energy contributions

If we record the absorption spectrum of a diatomic molecule, we observe sets of lines in three distinct regions of the electromagnetic spectrum. One group lies in the ultraviolet or visible region, a second group is in the infrared region and there is a third group in the far infrared or microwave region. Figure 6.3 schematically illustrates this with the spectrum for the hydrogen chloride molecule, HCl. The fact that these lines show up in different regions of the spectrum indicates that the separation between the energy levels producing the lines varies greatly (largest for the lines in the ultraviolet region and smallest in the microwave region).

Chapter 6 Diatomic molecules

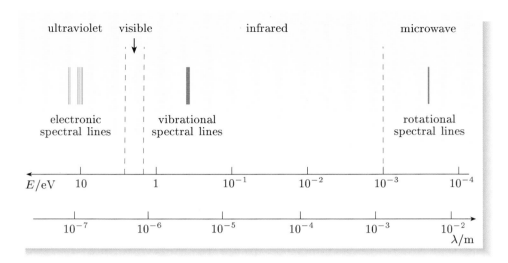

Figure 6.3 Schematic representation of the lines found in the absorption spectrum of HCl in terms of the photon energy $E = hf = hc/\lambda$, where f is the frequency and λ is the wavelength.

The lines in the ultraviolet region are associated with changes in *electronic energy*; how to calculate the energy levels associated with these lines is the major topic in the rest of this chapter. The other lines, in the infrared and microwave regions, are associated with motion of the nuclei, and the fact that they occur in two distinct regions suggests that there are two distinct types of nuclear motion. This is indeed the case.

To a good approximation, the nuclear time-independent Schrödinger equation (Equation 6.7) can be separated into two time-independent Schrödinger equations: one that describes the *vibrational* motion of the molecule (how the distance between the nuclei changes), and another that describes the *rotational* motion of the molecule about various axes. This allows us to define vibrational and rotational energies for the molecule. Spectral lines in the infrared region are associated with changes in vibrational energy and lines in the microwave region are associated with changes in rotational energy. The total energy of the molecule can be written as a sum of electronic, vibrational and rotational contributions.

Exercise 6.1 Would you expect the vibrational energy levels of a molecule to depend on its electronic state? ∎

6.2 The hydrogen molecule ion

We now concentrate on the electronic part of the problem. For simplicity, we shall look at the hydrogen molecule ion, H_2^+, which consists of two protons and one electron. The reason for starting here, rather than with more familiar molecules such as oxygen or nitrogen, is that the hydrogen molecule ion provides the simplest possible example of a chemical bond. The two protons are somehow bound together by the electron, and we shall try to understand how this works.

We start by writing down the electronic time-independent Schrödinger equation for the hydrogen molecule ion:

$$\left[-\frac{\hbar^2}{2m_e}\nabla^2 - \frac{e^2}{4\pi\varepsilon_0 r_A} - \frac{e^2}{4\pi\varepsilon_0 r_B}\right]\psi_{el}(\mathbf{r}) = E_{el}\psi_{el}(\mathbf{r}). \quad (6.9)$$

There is no electron–electron repulsion term in this case because the hydrogen molecule ion has only one electron. Since H_2^+ has only one electron, we have dropped the index i in ∇_i, r_{iA}, r_{iB} and the vectors that define the electron's position. Thus $\mathbf{r}$ is the position vector for the electron relative to the centre of mass O, and $\mathbf{r}_A$ and $\mathbf{r}_B$ are its position vectors relative to protons A and B. We shall also denote the magnitude of the proton–proton separation R_{AB} by R. For hydrogen atoms, we set $Z_A = Z_B = 1$.

We choose a coordinate system with its origin at the centre of mass of the molecule, midway between the two protons, and with both protons on the z-axis (Figure 6.4). The Cartesian coordinates of the two protons are therefore $(0, 0, +R/2)$ and $(0, 0, -R/2)$, where R is distance between the protons, which is taken to be constant according to the Born–Oppenheimer approximation.

Unlike the hydrogen atom, the hydrogen molecule ion is not spherically symmetric. Nevertheless, there are advantages to using spherical coordinates, as you will see. The spherical coordinates for a hydrogen molecule ion are shown in Figure 6.5.

The absence of spherical symmetry is marked by the fact that the potential energy of the electron depends on θ as well as r. For example, the potential energy of the electron at $r = 0.6R$, $\theta = 0$, $\phi = 0$, is very different from that at $r = 0.6R$, $\theta = \pi/2$, $\phi = 0$. However, the potential energy function *is* symmetric with respect to rotations about the z-axis, and so does not depend on the angle ϕ in spherical coordinates. As a consequence, the operator for the z-component of the orbital angular momentum of the electron, $\widehat{L}_z = -i\hbar\,\partial/\partial\phi$, commutes with the Hamiltonian operator in Equation 6.9. (For diatomic molecules with many electrons, the total $\widehat{L}_z$ also commutes with the Hamiltonian operator.)

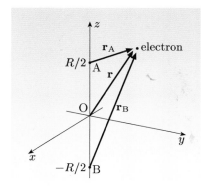

Figure 6.4 The Cartesian coordinate system for the hydrogen molecule ion, with proton A fixed at $(0, 0, +R/2)$ and proton B at $(0, 0, -R/2)$.

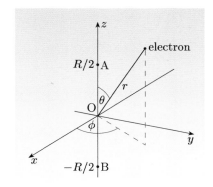

Figure 6.5 Spherical coordinates r, θ, ϕ for the electron in the hydrogen molecule ion.

- What do you conclude about the solutions $\psi_{el}(\mathbf{r})$ of Equation 6.9 from the fact that $\widehat{L}_z$ commutes with the Hamiltonian operator $\widehat{H}$?

○ The solutions $\psi_{el}(\mathbf{r})$ can be chosen to be eigenfunctions of $\widehat{L}_z$ as well as of the Hamiltonian.

Although the operator $\widehat{L}_z$ commutes with the Hamiltonian, the same cannot be said for $\widehat{L}_x, \widehat{L}_y$ or $\widehat{L}^2$. You may find this rather puzzling because it means that the total orbital angular momentum of the electron is not a conserved quantity and yet the total angular momentum of an isolated hydrogen molecule ion must always be conserved, as in classical physics. The answer to this puzzle lies in the Born–Oppenheimer approximation. In fixing the positions of the protons, we have ignored any contribution they make to the angular momentum, and it is the total angular momentum of the whole molecule (both the electron and the protons) that must be conserved.

Because the operator $\widehat{L}_z = -i\hbar\,\partial/\partial\phi$ commutes with the Hamiltonian operator within the Born–Oppenheimer approximation, the quantum number m, where $L_z = m\hbar$, is a good quantum number for this system, and can be used to label the electronic energy eigenfunctions. In more detail, these eigenfunctions can be written as

$$\psi_{\text{el}}(\mathbf{r}) = u(r, \theta)\, e^{im\phi}, \tag{6.10}$$

where m is an integer (zero, positive or negative) known as the *magnetic quantum number*, or the *azimuthal quantum number*. The part of the energy eigenfunction that depends on ϕ, namely $e^{im\phi}$, is similar to that for the hydrogen atom. For a hydrogen molecule ion, however, the z-direction is defined by the molecular axis, whereas for the hydrogen atom it is arbitrary, or is defined with reference to something external, such as an applied magnetic field.

The allowed energy levels do not depend on whether the z-axis points in the direction from proton B to proton A (as in Figure 6.5) or in the opposite direction, from proton A to proton B. Reversing the direction of the z-axis has the effect of changing the sense in which ϕ is measured around the molecular axis, and therefore changes $e^{im\phi}$ into $e^{-im\phi}$. So, if two energy eigenvalues differ only in being labelled by m and $-m$, they will have the same energy, and be degenerate with one another. The electronic energies in any diatomic molecule depend on $|m|$, but they do not depend on the sign of m. This fact will be important when we discuss bonding in molecules such as nitrogen and oxygen.

6.2.1 Linear combination of atomic orbitals

We shall now use the variational method, which you met in Chapter 3, to find an approximate solution to Equation 6.9 for a hydrogen molecule ion in its electronic ground state, for a particular value of the proton–proton distance. To do this, we must first choose a suitable trial function.

There are two common methods for choosing trial functions for molecules. The first method chooses a trial function that incorporates any molecular symmetries that are expected to be present. However, this method requires a different starting point for each molecular shape.

The second method regards molecules as being made up of combinations of atoms, and therefore uses atomic orbitals to build a trial function. Such an approach has the advantage that we have a set of known functions that we can use as building blocks to form trial functions for any molecule, and this is the approach usually adopted. The trial function is taken to be a linear combination of atomic orbitals centred on each of the nuclei that form the molecule. This method is known as the **linear combination of atomic orbitals**, frequently abbreviated to **LCAO**. The resulting one-electron eigenfunctions for the molecule are called **molecular orbitals**. We shall now apply the LCAO method to the electronic ground state of the hydrogen molecule ion.

The hydrogen molecule ion in its electronic ground state dissociates into a proton and a hydrogen atom. This means that when R is very large, we expect the electronic ground-state eigenfunction to resemble that for a hydrogen atom with the electron in a 1s atomic orbital. The orbital can be centred on either proton, but we cannot have a solution that 'prefers' one over the other. This suggests that the appropriate choice will be to take the electron in a H_2^+ molecule to be in a linear combination of two 1s hydrogen atomic orbitals — one centred on proton A and the other centred on proton B.

Let us denote the 1s atomic orbital centred on proton A by $\phi_{1s}^A(\mathbf{r})$ and the 1s atomic orbital centred on proton B by $\phi_{1s}^B(\mathbf{r})$, where $\mathbf{r}$ is the position vector of the electron. Then we have

$$\phi_{1s}^A(\mathbf{r}) = \frac{1}{\sqrt{\pi}} \left(\frac{1}{a_0}\right)^{3/2} e^{-|\mathbf{r}-\mathbf{R}_A|/a_0} \qquad (6.11a)$$

See Tables 1.1 and 2.1.

$$\phi_{1s}^B(\mathbf{r}) = \frac{1}{\sqrt{\pi}} \left(\frac{1}{a_0}\right)^{3/2} e^{-|\mathbf{r}-\mathbf{R}_B|/a_0}, \qquad (6.11b)$$

and our trial function can be written as:

$$\psi_{el}(\mathbf{r}) = c_1 \phi_{1s}^A(\mathbf{r}) + c_2 \phi_{1s}^B(\mathbf{r}), \qquad (6.12)$$

where the coefficients c_1 and c_2 remain to be determined.

At the heart of the variational method is the idea that the expectation value of the energy, computed with the exact Hamiltonian and our trial function, obeys the inequality

$$\langle E \rangle = \frac{\langle \psi_{el} | \widehat{H} | \psi_{el} \rangle}{\langle \psi_{el} | \psi_{el} \rangle} \geq E_{gs}, \qquad (6.13)$$

where E_{gs} is the exact ground-state energy. We shall now use our trial function to calculate $\langle E \rangle$, and then vary the parameters c_1 and c_2 to find the smallest possible answer; this will be our variational estimate of the ground-state energy.

Substituting ψ_{el} from Equation 6.12 into Equation 6.13, we get

$$\langle E \rangle = \frac{\langle c_1 \phi_{1s}^A + c_2 \phi_{1s}^B | \widehat{H} | c_1 \phi_{1s}^A + c_2 \phi_{1s}^B \rangle}{\langle c_1 \phi_{1s}^A + c_2 \phi_{1s}^B | c_1 \phi_{1s}^A + c_2 \phi_{1s}^B \rangle}.$$

If we expand the bra and ket vectors, and remember that the linear operator $\widehat{H}$ has no effect on the constant coefficients c_1 and c_2, we obtain

$$\langle E \rangle = \frac{c_1^* c_1 \langle \phi_{1s}^A | \widehat{H} | \phi_{1s}^A \rangle + c_2^* c_2 \langle \phi_{1s}^B | \widehat{H} | \phi_{1s}^B \rangle + c_1^* c_2 \langle \phi_{1s}^A | \widehat{H} | \phi_{1s}^B \rangle + c_2^* c_1 \langle \phi_{1s}^B | \widehat{H} | \phi_{1s}^A \rangle}{c_1^* c_1 \langle \phi_{1s}^A | \phi_{1s}^A \rangle + c_2^* c_2 \langle \phi_{1s}^B | \phi_{1s}^B \rangle + c_1^* c_2 \langle \phi_{1s}^A | \phi_{1s}^B \rangle + c_2^* c_1 \langle \phi_{1s}^B | \phi_{1s}^A \rangle}.$$

Now we can make some simplifications. The atomic orbitals $\phi_{1s}^A(\mathbf{r})$ and $\phi_{1s}^B(\mathbf{r})$ are normalized, so we have

$$\langle \phi_{1s}^A | \phi_{1s}^A \rangle = \langle \phi_{1s}^B | \phi_{1s}^B \rangle = 1.$$

We cannot give the values of $\langle \phi_{1s}^A | \phi_{1s}^B \rangle$ or $\langle \phi_{1s}^B | \phi_{1s}^A \rangle$ because these refer to hydrogen atom orbitals centred on *different* atoms, and the overlap between these orbitals depends on the separation of the protons. However, the orbitals in Equations 6.11 have been chosen to be real, and this allows us to say that $\langle \phi_{1s}^A | \phi_{1s}^B \rangle = \langle \phi_{1s}^B | \phi_{1s}^A \rangle$. It is convenient to introduce the shorthand notation

$$S = \langle \phi_{1s}^A | \phi_{1s}^B \rangle = \langle \phi_{1s}^B | \phi_{1s}^A \rangle, \qquad (6.14)$$

where the real quantity S is called the **interatomic overlap integral**.

- Use Equations 6.11 to write down an explicit integral for S. Do not attempt to simplify your answer.

○ Using Equations 6.11, we have
$$S = \langle \phi_{1s}^A | \phi_{1s}^B \rangle$$
$$= \frac{1}{\pi a_0^3} \int_0^{2\pi} \int_0^{\pi} \int_0^{\infty} e^{-|\mathbf{r}-\mathbf{R}_A|/a_0} e^{-|\mathbf{r}-\mathbf{R}_B|/a_0} r^2 \sin\theta \, dr \, d\theta \, d\phi,$$

where r, θ and ϕ are the spherical coordinates in Figure 6.5. Further work could relate $|\mathbf{r} - \mathbf{R}_A|$ and $|\mathbf{r} - \mathbf{R}_B|$ to r, R and $\cos\theta$, but such simplification is not required by the question.

Exercise 6.2 Use physical arguments to deduce the limiting values of S as the distance R between the two protons tends to (a) zero and (b) infinity. ■

It is also helpful to introduce two more symbols for other terms that appear in our expression for $\langle E \rangle$. We define

> These terms (often called *matrix elements*) cannot depend on our choice of labels for the two protons.

$$H_{AA} = \langle \phi_{1s}^A | \widehat{H} | \phi_{1s}^A \rangle = \langle \phi_{1s}^B | \widehat{H} | \phi_{1s}^B \rangle \tag{6.15}$$

$$H_{AB} = \langle \phi_{1s}^A | \widehat{H} | \phi_{1s}^B \rangle = \langle \phi_{1s}^B | \widehat{H} | \phi_{1s}^A \rangle. \tag{6.16}$$

Finally, we assume that the coefficients c_1 and c_2 are real. This assumption is not essential, and it will not affect our final results, but we make it in order to simplify the working. Putting all these simplifications and symbols together, we obtain the following expression for $\langle E \rangle$:

$$\langle E \rangle = \frac{(c_1^2 + c_2^2)H_{AA} + 2c_1 c_2 H_{AB}}{c_1^2 + c_2^2 + 2c_1 c_2 S}. \tag{6.17}$$

● Show that, when the proton–proton separation R becomes very large, $\langle E \rangle$ becomes equal to the ground-state energy of an isolated hydrogen atom.

○ When R becomes very large, S and H_{AB} become vanishingly small because atomic orbitals of different atoms scarcely overlap. In this limit Equation 6.17 reduces to

$$\langle E \rangle = \frac{(c_1^2 + c_2^2)H_{AA}}{c_1^2 + c_2^2} = H_{AA}.$$

When R is large, the contribution of the potential energy of one or other of the protons to the Hamiltonian in Equation 6.9 is negligible (in other words, if $e^2/4\pi\varepsilon_0 r_A$ is significant, then $e^2/4\pi\varepsilon_0 r_B$ is negligible and vice versa). The Hamiltonian operator therefore reduces to that for an isolated hydrogen atom. Thus at large R,

$$\langle E \rangle = H_{AA} = \langle \phi_{1s}^A | \widehat{H} | \phi_{1s}^A \rangle = \langle \phi_{1s}^A | E_{1s} | \phi_{1s}^A \rangle = E_{1s} \langle \phi_{1s}^A | \phi_{1s}^A \rangle = E_{1s},$$

where E_{1s} is the ground-state energy of an isolated hydrogen atom.

6.2.2 The ground state of the hydrogen molecule ion

We now determine the electronic ground state of the hydrogen molecule ion by minimizing the expression for $\langle E \rangle$ in Equation 6.17 with respect to the coefficients c_1 and c_2. To do this, we need to find the partial derivatives of $\langle E \rangle$ with respect to c_1 and c_2 and equate each of these partial derivatives to zero.

It is helpful to note that the derivative of any quotient function, $f(x) = T(x)/B(x)$ is given by

$$\frac{df}{dx} = \frac{B(x)\frac{dT}{dx} - T(x)\frac{dB}{dx}}{B(x)^2} = \frac{1}{B(x)}\left[\frac{dT}{dx} - f(x)\frac{dB}{dx}\right],$$

so the extrema of any quotient function, $f(x) = T(x)/B(x)$ satisfy the condition

$$\frac{dT}{dx} - f(x)\frac{dB}{dx} = 0. \tag{6.18}$$

We shall now apply this condition to $\langle E \rangle$. Calculating the partial derivatives with respect to c_1 of the top T and bottom B of Equation 6.17, and substituting them into Equation 6.18, we obtain

$$2c_1 H_{AA} + 2c_2 H_{AB} - \langle E \rangle (2c_1 + 2c_2 S) = 0,$$

and collecting terms in c_1 and c_2 gives

$$c_1(H_{AA} - \langle E \rangle) + c_2(H_{AB} - S\langle E \rangle) = 0. \tag{6.19}$$

Differentiating $\langle E \rangle$ with respect to c_2 leads to a similar expression in which c_1 and c_2 are interchanged:

$$c_1(H_{AB} - S\langle E \rangle) + c_2(H_{AA} - \langle E \rangle) = 0. \tag{6.20}$$

In order for Equations 6.19 and 6.20 to be satisfied non-trivially (that is, without setting $c_1 = c_2 = 0$) we require that

$$\begin{vmatrix} H_{AA} - \langle E \rangle & H_{AB} - S\langle E \rangle \\ H_{AB} - S\langle E \rangle & H_{AA} - \langle E \rangle \end{vmatrix} = 0. \tag{6.21}$$

It would be legitimate to obtain Equation 6.20 by interchanging c_1 and c_2 in Equation 6.19, on the grounds that c_1 and c_2 appear symmetrically in Equation 6.17.

This is called the **secular equation** and the determinant on the left-had side is called the **secular determinant**. Expanding the determinant, we obtain

$$(H_{AA} - \langle E \rangle)^2 - (H_{AB} - S\langle E \rangle)^2 = 0,$$

so

$$H_{AA} - \langle E \rangle = \pm(H_{AB} - S\langle E \rangle).$$

There are two solutions:

$$\langle E \rangle = \frac{H_{AA} + H_{AB}}{1 + S} \quad \text{and} \quad \langle E \rangle = \frac{H_{AA} - H_{AB}}{1 - S}. \tag{6.22}$$

By considering the integrals implicit in S, H_{AA} and H_{AB}, it is possible to show that the first of these solutions has the lower energy. This solution therefore gives us our estimate of the electronic ground-state energy:

$$E_{gs} \simeq \langle E \rangle_{\min} = \frac{H_{AA} + H_{AB}}{1 + S}. \tag{6.23}$$

Remember that the integrals S, H_{AA} and H_{AB} all depend on the proton–proton separation R. So, within the Born–Oppenheimer approximation, we have found an approximation to the electronic ground-state energy of a molecule with a fixed value of R. We will relate this to the total energy of the molecule in the next section. First, we investigate the wave function that corresponds to the minimum value we have just found.

Exercise 6.3 Show that we must have $c_1 = c_2$ in order for the minimum expectation value $\langle E \rangle = (H_{AA} + H_{AB})/(1 + S)$ to satisfy Equations 6.19 and 6.20. ∎

The value of c_1 is determined by the normalization condition, $\langle \psi_{el} | \psi_{el} \rangle = 1$. We have already calculated $\langle \psi_{el} | \psi_{el} \rangle$ as the denominator in Equation 6.17. So we require that

$$\langle \psi_{el} | \psi_{el} \rangle = c_1^2 + c_2^2 + 2c_1 c_2 S = 1.$$

Putting $c_1 = c_2$, this gives $2c_1^2(1 + S) = 1$, so normalization is achieved by setting

$$c_1 = \frac{1}{\sqrt{2(1+S)}}. \tag{6.24}$$

Our approximate eigenfunction for the electronic ground state of the hydrogen molecule ion is therefore

$$\psi_{gs}(\mathbf{r}) = \frac{1}{\sqrt{2(1+S)}} \left(\phi_{1s}^A(\mathbf{r}) + \phi_{1s}^B(\mathbf{r}) \right), \tag{6.25}$$

where we use the subscript gs to indicate that this function refers to the electronic ground state of the molecule. The electron is in a linear superposition of two states, one described by a 1s atomic orbital centred on proton A, and the other by a 1s atomic orbital centred on proton B. Not surprisingly, given the symmetry of the situation, the coefficients of these two states have equal magnitudes.

- Can we specify to which proton the electron is bound in the limit of large R?

○ No — not until we make a measurement that decides between these two possibilities. Such a measurement only makes sense in the limit of large R. When it is made, the eigenfunction in Equation 6.25 collapses into either that of a hydrogen atom at $z = R/2$ or that of a hydrogen atom at $z = -R/2$. Since $S \to 0$ in the limit of large R, the probabilities for these two outcomes are both equal to $\frac{1}{2}$.

It is instructive to write down an expression for the probability density, $|\psi_{gs}(\mathbf{r})|^2$. Using Equation 6.25, and remembering that the 1s atomic orbitals are described by real functions, we obtain

$$|\psi_{gs}(\mathbf{r})|^2 = \frac{[\phi_{1s}^A(\mathbf{r})]^2}{2(1+S)} + \frac{[\phi_{1s}^B(\mathbf{r})]^2}{2(1+S)} + \frac{\phi_{1s}^A(\mathbf{r})\phi_{1s}^B(\mathbf{r})}{(1+S)}. \tag{6.26}$$

The first two terms represent the probability densities due to separate atomic eigenfunctions centred on the two protons. Since $S > 0$, the factor $1/2(1+S)$ makes each of their contributions less than half of that for an isolated hydrogen atom. The final term is only significant in regions where ϕ_{1s}^A and ϕ_{1s}^B are both non-negligible. It is an interference term between the two atomic wave functions, centred on different protons. Since $\phi_{1s}^A(\mathbf{r})$ and $\phi_{1s}^B(\mathbf{r})$ in Equations 6.11 are positive everywhere, this interference term leads to an enhanced probability density between the protons where the overlap is largest. The interference between the two atomic wave functions is *constructive*.

This is the key to chemical bonding: the increased electron probability density between the protons partially screens the protons from one another, reducing their effective electrostatic repulsion. When all the energies are added together, the total energy of the hydrogen molecule ion is smaller than that of an isolated atom plus a proton, so the hydrogen molecule ion remains bound together: it is stable.

Although the build-up of electron probability density between the protons is a key factor, it is not the only one. In classical physics, it is impossible to achieve stable equilibrium for a static distribution of charge interacting via electrostatic forces alone. A classical molecule with a concentration of electron charge between the nuclei would not be stable. Stability is only achieved because the energy-accounting must be done using quantum mechanics. One factor promoting the stability of molecules in quantum mechanics is that the electronic eigenfunctions in a molecule are more spread out than in an atom, and this tends to reduce the average kinetic energy of the electrons.

Finally, we ask how accurate our variational approximation to the ground-state eigenfunction is. In fact, it is possible to obtain an exact solution for the hydrogen molecule ion within the Born–Oppenheimer approximation, using a coordinate system based on the variables $\xi = (r_A + r_B)/R$ and $\eta = (r_A - r_B)/R$. The lengthy calculation is not included here because it cannot be extended to other molecules, but it does provide a useful 'gold standard' against which our results can be compared.

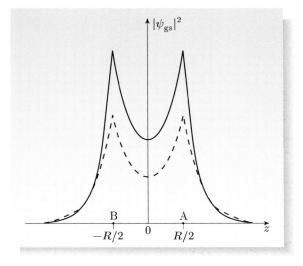

Figure 6.6 The ground-state electron probability density $|\psi_{gs}|^2$ along the z-axis of a hydrogen molecule ion: variational solution for the exact proton–proton separation of $R = 1.06 \times 10^{-10}$ m (dashed line); exact solution (solid line).

Figure 6.6 shows the ground-state electron probability density $|\psi_{gs}|^2$ both for the variational solution and for the exact solution. The probability density for the variational solution is close to the exact solution for $|z| > 0.6R$. However, for small $|z|$, the variational solution underestimates the probability density, and between the two nuclei the difference becomes quite significant.

6.2.3 The ground-state energy curve

The Born–Oppenheimer approximation, gives us the energy eigenfunctions and eigenvalues of the electrons in the centre-of-mass frame, *assuming that the nuclei are fixed in position*.

The same approximation assumes that, when the nuclei in a molecule move, the wave functions of the electrons adjust instantaneously. So, for the hydrogen molecule ion, the electron is assumed at each instant to be in a stationary state described by a solution of the electronic time-independent Schrödinger equation (Equation 6.9) with the appropriate instantaneous value of R. As the protons move, the electron eigenfunction changes to match the new proton positions.

The eigenvalues $E_{el}(R)$ of Equation 6.9 give us the electronic contribution to the total energy of the molecule as a function of R, but they are not the only contribution to the total energy. So far, we have omitted the potential energy $+e^2/4\pi\varepsilon_0 R$ due to the Coulomb repulsion of the two protons. This was sensible when calculating the electronic states because, with R fixed, this term just adds a constant to all energies. However, when discussing the total energy of the whole molecule, we must include this term, along with the kinetic energies of the protons. Equation 6.7, which gives the energy eigenvalues of the whole molecule, includes both the electronic energies $E_{el}(R)$ and the proton–proton repulsion term, $e^2/4\pi\varepsilon_0 R$. The effective potential energy in which nuclei move is given by the *total static energy*

$$E_{\text{stat}} = E_{el}(R) + \frac{e^2}{4\pi\varepsilon_0 R}, \qquad \text{(Eqn 6.8)}$$

and we shall concentrate on this energy now.

We shall take our energy zero to correspond to a hydrogen atom in its ground state separated from a free proton by an infinite distance, with both the hydrogen atom and the proton being at rest. Then, a hydrogen molecule ion whose total static energy at a particular value of R is *negative* will be more stable than a separated hydrogen atom and proton. With our choice of energy zero, the electronic contribution to the ground-state energy is negative, and the proton–proton repulsion energy is positive. For any value of R, the total static energy is a balance between these two contributions.

Figure 6.7 shows how the total static energy of a hydrogen molecule ion in its electronic ground state depends on the proton–proton separation, R. We shall refer to a graph like this, which plots the total static energy as a function of the internuclear distance, as an **energy curve**. Several points should be noted:

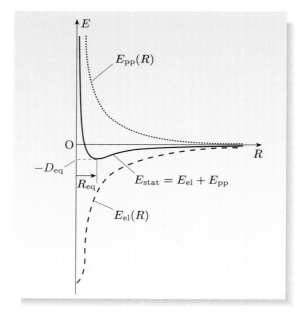

Figure 6.7 The total static energy E_{stat} of the hydrogen molecule ion (solid line) is the sum of the negative electronic energy $E_{el}(R)$ and the positive proton–proton repulsion energy, $E_{pp}(R) = e^2/4\pi\varepsilon_0 R$, where R is the proton–proton separation.

- At large values of R, the total static energy approaches zero, so there is practically no difference between the energy of the molecule and that of a free proton and an isolated hydrogen atom in its ground state.
- At very small values of R, the proton–proton repulsion energy becomes very large and the total static energy becomes large and positive.
- There is a minimum in the energy curve. The value of R at which this occurs is the **equilibrium separation**, R_{eq}, for the ground state.
- The hydrogen molecule ion is most stable when $R = R_{eq}$. The depth of the potential energy well at this equilibrium separation is called the **spectroscopic**

dissociation energy and is labelled D_{eq}. Even the lowest vibrational state has some zero-point vibrational energy. Allowing for this, the directly-observed ground-state dissociation energy of the molecule is slightly less than the spectroscopic dissociation energy, and is given by D_{eq} minus the zero-point energy.

How do the predictions made using our trial function compare with the exact solution, and with experiment? Figure 6.8 compares the approximate ground-state energy curve obtained using our trial function with that obtained using the exact ground-state eigenfunction. Both curves are based on the Born–Oppenheimer approximation.

As expected, the variational approximation to the energy is always above the exact ground-state energy. Although our variational solution correctly predicts that there is a minimum in the energy curve, it places it at a proton–proton separation of $R_{eq} = 1.32 \times 10^{-10}$ m, whereas both the exact solution and experimental measurements place it at 1.06×10^{-10} m. The approximate solution also underestimates the spectroscopic dissociation energy, predicting a value of $D_{eq} = 1.76$ eV compared with the experimental value of 2.79 eV.

The conclusion is that our trial function gives the right qualitative picture, and gives a reasonable explanation of the origins of chemical bonding, but is inaccurate in the quantitative details. At the end of the chapter, we shall see how trial functions can be improved in order to make better predictions for both R_{eq} and D_{eq}.

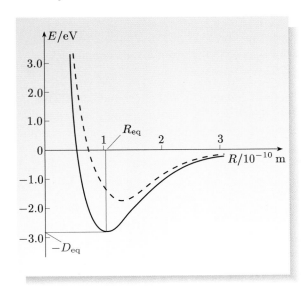

Figure 6.8 The total static energy for the ground-state variational energy eigenfunction of Equation 6.25 (dashed line) compared with the exact total static energy (solid line). The minimum value of the total static energy is $-D_{eq}$, where $D_{eq} > 0$.

6.2.4 First excited state of the hydrogen molecule ion

So far we have concentrated on the ground state of H_2^+. However, experimental measurements show that other states exist. Can we describe these excited states using the LCAO method? It turns out that we can, as we shall now show.

When using the variational principle to obtain an estimate for the electronic ground-state energy, we obtained a secular equation with two solutions, and we selected the solution of lower energy to describe the ground state. But what about the other solution?

You saw in Chapter 3 that, if we could search among all possible functions, any function ψ that produces a stationary value in $\langle\psi|\widehat{H}|\psi\rangle/\langle\psi|\psi\rangle$ is an exact eigenfunction of the Hamiltonian $\widehat{H}$. Of course, we cannot search among all possible functions, but assuming that it is reasonable to approximate molecular electronic eigenfunctions by a linear combination of atomic orbitals, the rejected solution to the secular equation should provide an approximate description of an excited state of the hydrogen molecule ion. In fact (although we cannot prove it) it approximates the *first* excited state. Referring back to Equation 6.22, we therefore have the following estimate for the electronic energy of the first excited state:

$$E_{\text{exc}} = \frac{H_{\text{AA}} - H_{\text{AB}}}{1 - S}. \tag{6.27}$$

As you can confirm in the following exercises, the corresponding normalized energy eigenfunction is

$$\psi_{\text{exc}}(\mathbf{r}) = \frac{1}{\sqrt{2(1-S)}} \left(\phi_{1s}^{A}(\mathbf{r}) - \phi_{1s}^{B}(\mathbf{r}) \right). \tag{6.28}$$

Exercise 6.4 By substituting $\langle E \rangle = E_{\text{exc}}$ from Equation 6.27 into Equation 6.19, show that this value of the energy corresponds to a linear combination of atomic orbitals of the form $c_1 \left(\phi_{1s}^{A}(\mathbf{r}) - \phi_{1s}^{B}(\mathbf{r}) \right)$, where c_1 is a constant.

Exercise 6.5 Confirm that the wave function ψ_{exc} in Equation 6.28 is normalized. Show also that ψ_{exc} is orthogonal to the ground state eigenfunction ψ_{gs} of Equation 6.25. ∎

The first excited state has a very different character from the ground state. This can be seen calculating the probability density $|\psi_{\text{exc}}|^2$. Using Equation 6.28 and remembering that the 1s atomic orbitals are described by real functions, we obtain

$$|\psi_{\text{exc}}(\mathbf{r})|^2 = \frac{\left[\phi_{1s}^{A}(\mathbf{r})\right]^2}{2(1-S)} + \frac{\left[\phi_{1s}^{B}(\mathbf{r})\right]^2}{2(1-S)} - \frac{\phi_{1s}^{A}(\mathbf{r})\phi_{1s}^{B}(\mathbf{r})}{(1-S)}. \tag{6.29}$$

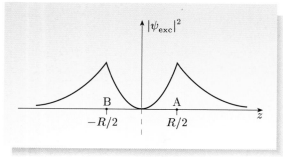

Figure 6.9 Probability density along the z-axis for an electron in the first excited state ψ_{exc} of a hydrogen molecule ion in the LCAO approximation.

Figure 6.9 shows a plot of this probability density. The last term in Equation 6.29 is an interference term between the two atomic wave functions, centred on different protons. It makes a negative contribution to the electron probability density, especially in the region between the protons so there is only a small electron probability density there. This interference is *destructive*.

Without a significant electron probability density between the protons, screening is slight and the effective proton–proton repulsion contributes a large positive term to the total energy. In practice, this means that the energy of H_2^+ in its first excited state is always higher than for an isolated hydrogen atom in its ground state and a free proton.

The energy curve for the first excited state is also quite different from that for the ground state, as you can see in Figure 6.10. The lower curve E_{gs} in this figure is the ground-state energy that you met in Figure 6.8. The upper curve E_{exc} is that for the excited-state eigenfunction in Equation 6.28. At large values of R, both curves tend to the energy of an isolated hydrogen atom in its ground state. However there is no minimum in the energy curve for the excited state. This means that the hydrogen molecule ion in its first excited state is not stable and will dissociate into a hydrogen atom and a free proton.

6.3 Molecular orbitals

The hydrogen molecule ion is the simplest molecule. In this section, we shall continue to use H_2^+ as an example, but will introduce concepts and terminology that are appropriate for all diatomic molecules.

In Chapter 5 you saw that atoms can be described using single-electron states called atomic orbitals. We shall now see that a similar description can be given for molecules. The single-electron states describing molecules are called **molecular orbitals**.

In the LCAO approximation, molecular orbitals are linear combinations of atomic orbitals centred on different atoms. You have seen how this works for the 1s orbitals centred on the two protons in a hydrogen molecule ion. Many more molecular orbitals can be produced by combining other pairs of atomic orbitals of the same type, centred on different atoms: (2s, 2s), (2p, 2p), (3s, 3s) and so on.

A molecular orbital with a minimum in its energy curve at a value that is lower than the energies of either of the atomic orbitals from which it is formed is called a **bonding orbital**. A molecular orbital for which the energy is always higher than the energies of the contributing atomic orbitals is called an **antibonding orbital**. Thus the orbital describing the ground state of H_2^+ is a bonding orbital, while that describing the first excited state is an antibonding orbital. The electron probability density between the nuclei is enhanced for bonding orbitals but reduced for antibonding orbitals. Also, energy curves for bonding orbitals always have a minimum, while those for antibonding orbitals do not.

Before discussing molecular orbitals in more detail, it will be helpful to introduce the standard notation used to label them.

6.3.1 Spectroscopic notation for molecular orbitals

The notation used to label molecular orbitals in diatomic molecules is loosely based on that used to label atomic orbitals, but there are some important differences. **Spectroscopic notation** for molecular orbitals involves a number, a lower-case Greek letter and sometimes a subscript g or u. For example, a molecular orbital might be labelled $2\sigma_u$, or $3\delta_g$. The meaning of the different parts of this notation is as follows.

- The *Greek letter* refers to the value of $|m|$ for the molecular orbital, where m is the magnetic (or azimuthal) quantum number. We use the Greek equivalents of the letters s, p, d, f, ... Orbitals with $|m| = 0$ are labelled with the Greek equivalent of s, which is σ. Orbitals with $|m| = 1$ are labelled with the Greek equivalent of p, which is π, and so on as indicated in Table 6.1.

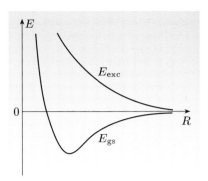

Figure 6.10 Energy curves showing the total static energies of the electronic ground state and first excited state of H_2^+ as a function of the internuclear distance R.

Table 6.1 Spectroscopic notation for molecular orbitals.

$|m|$	symbol	word
0	σ	sigma
1	π	pi
2	δ	delta
3	ϕ	phi
4	γ	gamma
5	η	eta
$\vdots$	$\vdots$	$\vdots$

Chapter 6 Diatomic molecules

Symmetric orbitals are said to have *even parity* and antisymmetric orbitals are said to have *odd parity*.

- The *subscript* g or u stands for **gerade** and **ungerade**, the German words for even and odd. Diatomic molecules with two identical nuclei, such as H_2 or N_2, are called **homonuclear diatomic molecules**. These molecules have a **centre of symmetry** — an origin about which the operation of *inversion* ($\mathbf{r} \to -\mathbf{r}$) produces no change in the Hamiltonian. It turns out that all the molecular orbitals in homonuclear diatomic molecules are either symmetric or antisymmetric under inversion. The symmetric orbitals are labelled by the subscript g (e.g. σ_g) and the antisymmetric orbitals are labelled by the subscript u (e.g. σ_u).

- The *number* labelling a molecular orbital is allocated by starting with the lowest-energy molecular orbital of any particular type, such as σ_u or π_g, and numbering all the orbitals of that type in sequence, from 1 upwards. For example, successive σ_u orbitals are labelled $1\sigma_u$, $2\sigma_u$, $3\sigma_u$ and so on.

Note carefully that neither the Greek letter nor the number labelling a molecular orbital tell us directly about the l and n quantum numbers of the parent atomic orbitals in the LCAO approximation. A σ molecular orbital is produced by combining two atomic orbitals with $m = 0$, but this can be done by combining two atomic s orbitals ($l = 0, m = 0$), or two atomic p_0 orbitals ($l = 1, m = 0$) for example. Moreover, an orbital like $3\sigma_g$ may be derived from two atomic orbitals with quantum number $n = 3$, but this is not necessarily the case. All we can say for sure is that it is the third molecular orbital of type σ_g in the energy ordering. The number labelling a molecular orbital is *not* a quantum number.

6.3.2 Forming molecular orbitals from atomic orbitals

Figure 6.11 shows a series of energy curves corresponding to the first few molecular orbitals of the hydrogen molecule ion. The lowest two curves are those for the ground state and the first excited state, which were shown in Figure 6.10. Some of the energy curves, like that for the ground state, have a minimum and therefore correspond to bonding orbitals. Other curves do not have a minimum, and these correspond to antibonding orbitals.

The curve labelled $2\sigma_g$ has a distinct minimum, but its energy never falls below that of the ground state. This is a bonding orbital, but at large R the energy of this state is that of a free proton and a hydrogen atom in a state with $n = 2$ rather than $n = 1$. The energy of this bonding orbital is always higher than that of a free proton and a hydrogen atom in its ground state, but lower for a range of R than that of the systems it dissociates into: a proton and a hydrogen atom with $n = 2$.

In the LCAO approximation we obtain such molecular orbitals by applying the variational method to a trial function built from an appropriate linear combination of atomic orbitals.

- Which molecular orbitals can be approximated by a linear combination of two 2s orbitals centred on different nuclei?

○ Atomic s orbitals have $m = 0$, so the resulting molecular orbitals must have $m = 0$, and therefore be σ orbitals. We have already seen that the $1\sigma_g$ and $1\sigma_u$ molecular orbitals are produced by combining 1s atomic orbitals. The 2s orbitals are expected to produce molecular orbitals of higher energy than this, so it is reasonable to suppose that they will produce $2\sigma_g$ and $2\sigma_u$ molecular orbitals, where the 2 prefix indicates that these are the next σ orbitals in the energy ordering after $1\sigma_g$ and $1\sigma_u$.

The $2\sigma_g$ molecular orbital is described by the wave function $C\left(\phi_{2s}^A(\mathbf{r}) + \phi_{2s}^B(\mathbf{r})\right)$, where $\phi_{2s}^A(\mathbf{r})$ and $\phi_{2s}^B(\mathbf{r})$ are 2s orbitals centred on atoms A and B, and C is a normalization constant. There is also an antibonding orbital $2\sigma_u$, described by the wave function $D\left(\phi_{2s}^A(\mathbf{r}) - \phi_{2s}^B(\mathbf{r})\right)$, where D is a different normalization constant.

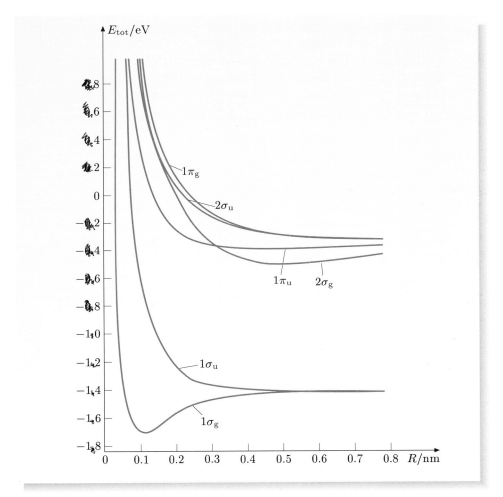

Figure 6.11 Some energy curves for the molecular orbitals of H_2^+. The energy zero is defined in this case as the energy of two protons and an electron, all infinitely far apart.

Combining s orbitals is quite straightforward, but when we combine 2p orbitals we need to take account of the fact that there are three atomic 2p orbitals, each with a different azimuthal quantum number m. We noted in Section 6.2 that the energy of the hydrogen molecular ion depends only on $|m|$, and as a consequence states with $m = +|m|$ and $m = -|m|$ are degenerate. The same considerations apply to molecular orbitals in all diatomic molecules and affects their degeneracies.

In the LCAO approximation, we always take linear combinations of atomic orbitals with the same values of $|m|$. We can combine a $2p_0$ orbital on one atom with a $2p_0$ orbital on the other atom to produce bonding and antibonding σ orbitals. Ignoring spin, each of these molecular orbitals is non-degenerate. However, when we combine $2p_{+1}$ and $2p_{-1}$ orbitals on one atom with $2p_{+1}$ and

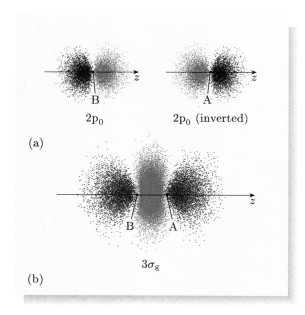

Figure 6.12 (a) Probability densities for two $2p_0$ atomic orbitals on different atoms. Blue indicates a positive real part and red a negative real part, so the atomic orbital on the right is inverted relative to that on the left. (b) Probability density for the bonding σ molecular orbital obtained by combining the two $2p_0$ atomic orbitals in (a).

$2p_{-1}$ orbitals on another atom, we get a total of four π-molecular orbitals; two degenerate bonding orbitals and two degenerate antibonding orbitals.

Figure 6.12b shows the probability density for the bonding $3\sigma_g$ molecular orbital that results from combining $2p_0$ orbitals on each proton. As for any bonding orbital, the electron probability density is high between the nuclei. In this case, however, the combination that produces this result is that with $c_1 = -c_2$ in the wave function. This is because the increased probability density between the nuclei can only be achieved by overlapping parts of atomic orbitals with the same sign. Figure 6.12a shows that this is achieved by combining orbitals with their positive lobes pointing in different directions, corresponding to a wave function of the form $C\left(\phi_{2p_0}^A(\mathbf{r}) - \phi_{2p_0}^B(\mathbf{r})\right)$.

The π molecular orbitals produced by combining $2p_{+1}$ and $2p_{-1}$ atomic orbitals have completely different probability densities from the σ molecular orbitals produced by combinations of $2p_0$. These π orbitals have a probability density that vanishes along the line joining the two protons. In spite of this, a bonding π orbital like that shown in Figure 6.13a has sufficient electron probability density around the centre of the molecule to stabilize the structure. By contrast, the antibonding π orbital in Figure 6.13b has a nodal plane perpendicular to the axis of the molecule and this structure is not stable.

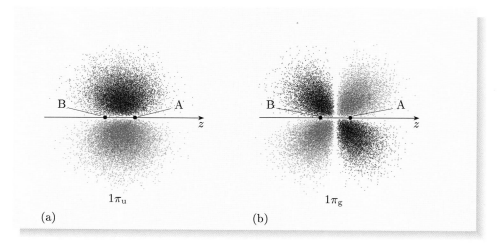

Figure 6.13 Probability densities for combinations of $\phi_{2p_{+1}}$ and $\phi_{2p_{-1}}$. (a) A bonding molecular orbital, $1\pi_u$. (b) An antibonding orbital, $1\pi_g$. The colours have the same meanings as in Figure 6.12.

All of the energy curves for the molecular orbitals formed from 2p atomic orbitals tend to the energy of a atom in a state with $n = 2$ at large values of R. For higher excited orbitals, the energy curves tend to the energy of an isolated atom in states with $n = 3, 4, 5, \ldots$. For each pair of atomic orbitals, one centred around each nucleus, with particular values of n, l, and m, we can use the LCAO approximation to form a bonding and an antibonding molecular orbital. However, molecular orbitals formed from atomic orbitals with $m = +|m|$ and $m = -|m|$, where $m \neq 0$, are doubly degenerate.

Exercise 6.6 Enumerate the bonding and antibonding orbitals that can be formed by appropriate pairs of atomic orbitals with $n = 3$. Classify the molecular orbitals as being σ, π or δ, and state their degree of degeneracy. ∎

6.4 Homonuclear diatomic molecules

Table 6.2 gives ground-state experimental data for the spectroscopic dissociation energy D_{eq} and the equilibrium separation R_{eq} of diatomic molecules X_2 formed from atoms in the first two rows of the Periodic Table (H_2^+ to Ne_2). We consider only homonuclear molecules — those formed from two identical atoms. Some of these molecules such as He_2 have very small dissociation energies, whereas others, such as N_2, have very large dissociation energies. In this section we shall discuss the ground-state electronic configurations of the molecules and show they help to explain some of the data in this table. As with atoms, the **electronic configuration** of a molecule is given by listing the different types of occupied molecular orbital and stating how many electrons are contained in each type.

Table 6.2 Ground-state properties of homonuclear diatomic molecules: equilibrium internuclear separation R_{eq}, spectroscopic dissociation energy D_{eq}, and the total spin quantum number S.

Molecule	$R_{eq}/10^{-10}$ m	D_{eq}/eV	S
H_2	0.74	4.75	0
H_2^+	1.06	2.79	$\frac{1}{2}$
He_2	3.0	0.0009	0
He_2^+	1.08	2.5	$\frac{1}{2}$
Li_2	2.67	1.07	0
Be_2	2.45	0.10	0
B_2	1.59	3.1	1
C_2	1.24	6.3	0
N_2	1.10	9.91	0
N_2^+	1.12	8.85	$\frac{1}{2}$
O_2	1.21	5.21	1
F_2	1.41	1.66	0
Ne_2	3.1	0.0036	0

6.4.1 Molecules of hydrogen and helium

For atoms, you have seen that the Pauli exclusion principle restricts the occupation of atomic orbitals, so that each atomic orbital holds a maximum of two electrons, one spin-up and the other spin-down. The ground-state electronic configuration of an atom is obtained by filling the atomic orbitals in order of increasing energy. Some atomic orbitals are degenerate, so a given atomic shell may contain more than two electrons. The electronic configuration of the atom specifies the number of electrons in each shell.

Very similar ideas apply to molecules. Each molecular orbital contains a maximum of two electrons, and the ground-state electronic configuration of the molecule is obtained by filling the molecular orbitals in order of increasing energy. We must take account of the fact that some molecular orbitals are degenerate, so some molecular energy levels can hold more than two electrons. The electronic configuration of a molecule specifies the number of electrons in each molecular energy level.

The hydrogen molecule

The hydrogen molecule, H_2, is very much like the hydrogen molecular ion, except that it has two electrons, one from each hydrogen atom. The lowest-energy molecular orbital is $1\sigma_g$, and this can hold both electrons in opposite spin states. So the ground-state electronic configuration for H_2 is written as $1\sigma_g^2$.

Because there are two electrons in this bonding orbital, the protons in H_2 are held together more strongly than in H_2^+, giving a shorter equilibrium separation and a higher dissociation energy, as reflected in Table 6.2. However, the dissociation energy is less than twice that of H_2^+. This is because the two electrons in a hydrogen molecule repel one another, increasing the ground-state energy and reducing the dissociation energy.

The next molecular orbital above $1\sigma_g$ is $1\sigma_u$. This is empty in the ground-state configuration of H_2, but the first excited configuration of H_2 would have one electron in $1\sigma_g$ and the other in $1\sigma_u$.

The diatomic helium molecule

The He_2 molecule has four electrons, so you might think that the helium nuclei would be held together even more strongly than the protons in H_2. However, the He_2 molecule is unknown under normal conditions of temperature and pressure: at room temperature, helium gas contains only helium atoms.

We need to consider how the electrons occupy the available molecular orbitals. As with H_2, the two orbitals of lowest energy are $1\sigma_g$ and $1\sigma_u$. In He_2, two electrons fill the bonding $1\sigma_g$ orbital, and the other two fill the antibonding $1\sigma_u$ orbital. The two electrons in the antibonding orbital do not help to bind the molecule together; on the contrary, they practically cancel out the effect of the electrons in the bonding orbital. *Stable molecules generally have more electrons in bonding orbitals than in antibonding orbitals.*

Accurate calculations indicate that there is a very shallow minimum in the energy curve of He_2 at $R_{eq} = 3 \times 10^{-10}$ m, with a dissociation energy of $D_{eq} = 0.0009$ eV. This very small dissociation energy is close to the energy of the lowest vibrational state, so detection of He_2 molecules requires very low temperatures, and has only been achieved for a beam of helium atoms cooled to 10^{-3} K. Because the energy curve has a very shallow minimum, the molecule samples a range of interatomic distances that are far from the equilibrium value. Because the energy curve is asymmetric, the average separation of the two nuclei is much greater than the equilibrium separation, and has been estimated to be about 50×10^{-10} m.

6.4.2 Molecules from lithium to neon

The 2s and 2p energy levels are degenerate in a hydrogen atom, but for heavier atoms the 2s level has a lower energy than the 2p levels. In the LCAO model for diatomic molecules formed from elements other than hydrogen, we might therefore expect molecular orbitals formed from 2s orbitals to be lower in energy than those formed from 2p orbitals.

If the separation between the atomic 2s and 2p orbitals is large, the general energy-ordering of molecular orbitals is as shown in Figure 6.14a. If the separation between the atomic 2s and 2p orbitals is small, a slightly different pattern applies, as shown in Figure 6.14b.

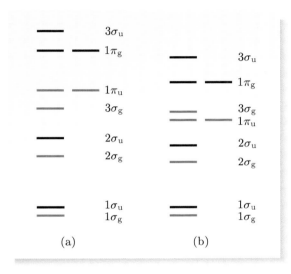

Figure 6.14 The general pattern of energy-ordering of molecular orbitals for a homonuclear diatomic molecule: (a) large gap between the 2s and 2p atomic orbitals and (b) small gap. Bonding orbitals are red and antibonding black. Each molecular orbital can hold a maximum of two electrons, so doubly-degenerate molecular energy levels (indicated by a pair of horizontal lines) can hold a maximum of four electrons. In the second period of the Periodic Table, Li to N follow scheme (b), while O to Ne follow scheme (a).

The lithium molecule

At normal temperatures and pressures, lithium is a metal, but Li_2 molecules have been observed in the vapour above the heated metal. The lithium molecule has six electrons, so its ground-state electronic configuration is

$$1\sigma_g^2 \, 1\sigma_u^2 \, 2\sigma_g^2.$$

Since $1\sigma_g$ and $2\sigma_g$ are bonding orbitals, and $1\sigma_u$ is an antibonding orbital, Li_2 has four electrons in bonding orbitals and two in antibonding orbitals and is therefore stable with respect to dissociation to two isolated Li atoms.

For any series of molecules that use similar orbitals for bonding, such as the series from Li_2 to Ne_2, a quantity called the **formal bond order** (or just *bond order*) can be defined. It is given by

$$\text{formal bond order} = \frac{n_b - n_a}{2}, \tag{6.30}$$

where n_b is the number of electrons in bonding orbitals and n_a the number of electrons in antibonding orbitals. Thus Li_2 has a formal bond order of $(4-2)/2 = 1$. A formal bond order of 1 is referred to as a **single bond**, a formal bond order of 2 is a **double bond**, and so on.

Chapter 6 Diatomic molecules

- What are the formal bond orders of H_2^+ and H_2?
- H_2^+ has one electron in a bonding orbital and none in antibonding orbitals. Its formal bond order is therefore $\frac{1}{2}$. H_2 has two electrons in bonding orbitals and none in antibonding orbitals so its formal bond order is 1.

If you look at Table 6.2, you will see that the equilibrium separation for H_2 is shorter than that for H_2^+ and its dissociation energy is higher than for H_2^+. This is an example of a general principle:

> When comparing diatomic molecules composed of atoms in the same row of the Periodic Table, a higher bond order generally implies stronger bonding, with a higher dissociation energy and a shorter equilibrium distance between the nuclei.

- Does the fact that H_2^+ has a higher dissociation energy and a shorter internuclear distance than Li_2 contravene this principle?
- No, the principle does not apply because hydrogen and lithium are in different rows of the Periodic Table.

The nitrogen molecule

A nitrogen atom has $Z = 7$ and contains 7 electrons.

In nitrogen, the ordering of molecular orbitals is that shown in Figure 6.14b. When filling these orbitals, it is important to remember that π_u orbitals are degenerate so π_u energy levels can hold 4 electrons, and the same is true for π_g energy levels. Hence, the ground-state electronic configuration of nitrogen is

$$1\sigma_g^2 \, 1\sigma_u^2 \, 2\sigma_g^2 \, 2\sigma_u^2 \, 1\pi_u^4 \, 3\sigma_g^2.$$

Referring to Figure 6.14b, we see that there are 10 electrons in bonding orbitals ($1\sigma_g, 2\sigma_g, 1\pi_u$ and $3\sigma_g$), and 4 in antibonding orbitals, giving a formal bond order of $(10-4)/2 = 3$. The nitrogen molecule therefore has a **triple bond**. The two nitrogen atoms are very strongly bound together, with a large dissociation energy and a small internuclear spacing. About 80% of air is composed of nitrogen molecules, but our bodies cannot split N_2 molecules apart to use them to build nitrogen-containing molecules such as proteins (although some bacteria can).

Table 6.2 also shows that forming a N_2^+ ion by removing an electron from a bonding orbital of the N_2 molecule leads to a small decrease in the dissociation energy and a small increase in the equilibrium separation, as you would expect.

The oxygen molecule

An oxygen atom has $Z = 8$ and contains 8 electrons.

The oxygen molecule, O_2, follows the pattern of Figure 6.14a. It has two more electrons than N_2 and we allocate these to the $1\pi_g$ orbital, giving the ground-state electron configuration

$$1\sigma_g^2 \, 1\sigma_u^2 \, 2\sigma_g^2 \, 2\sigma_u^2 \, 3\sigma_g^2 \, 1\pi_u^4 \, 1\pi_g^2.$$

According to Figure 6.14a, the $1\pi_g$ orbital is an antibonding orbital so the oxygen molecule is bound more weakly, and has a lower dissociation energy, than a nitrogen molecule,

Exercise 6.7 What is the formal bond order of O_2? ∎

Now, at last, we can give an explanation for the behaviour of a stream of liquid oxygen in a magnetic field (Figure 6.1). Liquid oxygen is composed of oxygen molecules in their ground-state electronic configurations. Most of the electrons are in full molecular orbitals, with each spin-up electron partnered by a spin-down electron. However, the π_g molecular energy level consists of two degenerate π_g molecular orbitals, and so can hold a maximum four electrons. In an oxygen molecule, this energy level contains only 2 electrons and is not full.

In the first excited configuration of a helium atom (Chapter 5), you saw that various different energy levels can arise, associated with triplet or singlet spin states. The triplet spin state is symmetric and is accompanied by an antisymmetric spatial function in which the electrons tend to stay apart; this reduces the electron–electron repulsion energy and so leads to a state of low energy. Something very similar occurs in the oxygen molecule, but in the ground state. This is possible because the $1\pi_g$ molecular orbitals are degenerate, so one electron can go in each $1\pi_g$ orbital; we can then form an antisymmetric spatial function which must be accompanied by a symmetric (i.e. triplet, $S = 1$) spin state. This is the lowest energy state of the molecule. In an external magnetic field, the $S = 1$, $M_S = -1$ spin state develops a lower energy than the other $S = 1$ spin states. In this state, each oxygen molecule has a magnetic moment and so experiences a force in an inhomogeneous magnetic field, as witnessed by Figure 6.1.

Exercise 6.8 Use the LCAO model to explain why the dissociation energy of O_2^- is less than that of O_2.

Exercise 6.9 (a) Write down the ground-state electron configuration for the carbon molecule C_2 and explain the reasoning behind your answer. The ordering of molecular orbitals in this case is as in Figure 6.14b. (b) What is the formal bond order of C_2?

A carbon atom has $Z = 6$ and contains 6 electrons. C_2 molecules can be produced in flames.

6.5 Beyond the LCAO approximation

In this section we briefly look at ways of improving the LCAO solutions.

First, consider what happens to the hydrogen molecule ion at very small proton–proton separations. As $R \to 0$, we effectively have a helium ion, He^+. So, we would expect the wave function for the electron in H_2^+ to resemble that for the electron in He^+ in this limiting case. However, the ground-state wave function for He^+ is proportional to $\exp(-2r/a_0)$, and this falls off more rapidly with distance than the hydrogen atom orbitals used in the LCAO approximation for H_2^+, which are proportional to $\exp(-r/a_0)$.

One way ahead is to use 1s orbitals centred around each nucleus as before, but with an effective atomic number Z_{eff} so that each orbital takes the form $\exp(-Z_{\text{eff}} r/a_0)$. We then treat Z_{eff} as an adjustable parameter in a variational calculation. This method gives the optimum value of Z_{eff} as 1.24, which lies between 1 and 2 as expected. The corresponding values for the equilibrium proton–proton separation and the dissociation energy are $R_{\text{eq}} = 1.07 \times 10^{-10}$ m and $D_{\text{eq}} = 2.35$ eV. Comparing these results with experimental data in Table 6.2, you can see that they are considerable improvements on our previous estimates ($R_{\text{eq}} = 1.32 \times 10^{-10}$ m and $D_{\text{eq}} = 1.76$ eV), given in Section 6.2.3.

Figure 6.15 compares the electron probability density for the variational solution including Z_{eff} with that for the LCAO solution obtained earlier (Figure 6.6), and also with that for the exact solution. The probability density with $Z_{\text{eff}} = 1.24$ is much closer to the exact solution, although there are still some discrepancies.

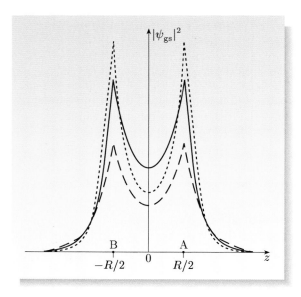

Figure 6.15 Probability density along the z-axis for the hydrogen molecule ion: variational solution with $Z_{\text{eff}} = 1.24$ (short-dashed line), LCAO solution (long-dashed line) and the exact solution (full line).

Another way to improve our approximation is to include additional terms in the trial function. In calculations on diatomic molecules, the additional terms usually involve pairs of atomic orbitals, one centred on each nucleus. For the ground state of H_2^+ we could add a 2s orbital centred on proton A, and a 2s orbital centred on proton B, to the original function giving a new trial function

$$\psi_{\text{el}}(\mathbf{r}) = C_1 \left[\phi_{1s}^A(\mathbf{r}) + \phi_{1s}^B(\mathbf{r})\right] + C_2 \left[\phi_{2s}^A(\mathbf{r}) + \phi_{2s}^B(\mathbf{r})\right].$$

Applying the variational method will then produce two improved molecular orbitals, one for $1\sigma_g$ and one for $2\sigma_g$. In general, it is possible to combine any atomic orbitals with the same values of $|m|$, and it is most important to include orbitals that have similar energies. For example, the ordering of molecular orbitals in Figure 6.14b is only obtained by including both 2s and $2p_0$ in the $2\sigma_g$ and $3\sigma_g$ molecular orbitals.

Lastly, it should be realized that the electron configurations we have described here are only a first approximation to the electronic structure of molecules. Just as for atoms, there are additional effects arising from the need to symmetrize the total wave function, electron–electron repulsion and spin–orbit interactions. (The magnetic properties of an oxygen molecule are one example of this.) These effects generally split electronic configurations into terms and levels, but do not completely overturn the gross structure we have described. For many purposes, the electronic configuration provides an adequate description. The methods described here, including the improvements outlined in this section, have been successfully applied to many molecules, not just diatomic ones, but also molecules like carbon dioxide, benzene, insulin and many, many more.

Summary of Chapter 6

Section 6.1 The Born–Oppenheimer approximation is based on the fact that nuclei are much heavier than electrons and therefore move much more slowly. It allows us to study the behaviour of the electrons in a molecule by treating the nuclei as being in fixed positions. The nuclear motions can then be predicted by assuming that the electrons adapt instantaneously to each new position of the nuclei. The time-independent Schrödinger equation effectively splits into two parts, one for the electrons and the other for the nuclei, but these equations are coupled by the fact that the energy eigenvalues of the electronic equation depend on the nuclear positions, and these eigenvalues contribute to the effective potential energy function in which the nuclei move.

Section 6.2 The hydrogen molecule ion H_2^+ has a single electron. With the molecular axis in the z-direction, $\hat{L}_z$ commutes with the electronic Hamiltonian and m is a good quantum number. The energy eigenvalues depend on the magnitude of m, but not on its sign.

The LCAO approximation to the ground state of H_2^+ is obtained by a variational calculation that uses a linear combination of two hydrogen-atom 1s orbitals, centred on the two protons, as the trial function. The optimized minimum-energy solution is

$$\psi_{\text{el}}(\mathbf{r}) = \frac{1}{\sqrt{2(1+S)}} \left(\phi_{1s}^A(\mathbf{r}) + \phi_{1s}^B(\mathbf{r}) \right),$$

where the interatomic overlap integral $S = \langle \phi_{1s}^A | \phi_{1s}^B \rangle$. This solution corresponds to constructive interference, resulting in a high electron probability density in the region between the protons, and this helps to stabilize the molecule.

The first excited state is orthogonal to the ground state and is proportional to $\phi_{1s}^A(\mathbf{r}) - \phi_{1s}^B(\mathbf{r})$. This solution corresponds to destructive interference, resulting in a low electron probability density in the region between the protons, so the molecule is unstable in this state.

A plot of the electronic energy E as a function of the internuclear separation R is called an energy curve. The energy curve for a bonding molecular orbital has a minimum at the equilibrium nuclear separation; the energy curve for an antibonding orbital has no minimum.

Section 6.3 The LCAO approximation can be applied to any diatomic molecule by taking linear combinations of atomic orbitals, centred on the two different nuclei, with the same value of $|m|$. This results in a series of molecular orbitals.

In spectroscopic notation, molecular orbitals are labelled according to the value of $|m|$. Orbitals with $|m| = 0$ are labelled σ; those with $|m| = 1$ are labelled π, etc. The σ orbitals are non-degenerate, but all other types are doubly-degenerate. Orbitals of homonuclear diatomic molecules have a subscript g (for even parity) or u (for odd parity). The Greek letter is prefaced by a number, which orders orbitals of the same type by increasing energy.

Section 6.4 The electronic configurations of diatomic molecules are obtained by allocating electrons to molecular orbitals in order of increasing energy until all the electrons in the molecule are accounted for. Two electrons with opposite spin can occupy each molecular orbital.

The formal bond order is defined as $(n_b - n_a)/2$, where n_b is the number of electrons in bonding orbitals and n_a is the number of electrons in antibonding orbitals. For elements in the same row of the Periodic Table, a higher formal bond order generally implies a greater dissociation energy and a shorter internuclear distance.

Section 6.5 LCAO solutions can be improved by taking the nuclear charge to be an adjustable parameter or by using more atomic orbitals in the trial function. Additional effects such as symmetrization, electron–electron repulsion and the spin–orbit interaction cause electron configurations to split into terms and levels, but this does not undermine the gross description provided by the electronic configuration.

Achievements from Chapter 6

After studying this chapter, you should be able to:

6.1 Explain the meanings of the newly defined (emboldened) terms and symbols, and use them appropriately.

6.2 Write down the time-independent Schrödinger equation for a two-electron diatomic molecule, and explain how the Born–Oppenheimer approximation is used to separate this equation into one equation dependent on electron coordinates and another equation dependent on nuclear coordinates.

6.3 Apply the variational method to determine the ground-state and first-excited state eigenfunctions and eigenvalues for the hydrogen molecule ion H_2^+, using a linear combination of hydrogen atom 1s orbitals as the trial function.

6.4 Explain the meaning of the energy curves for diatomic molecules, and distinguish between energy curves for bonding and antibonding orbitals.

6.5 Interpret and use spectroscopic notation for orbitals of diatomic molecules, and recall the degeneracy and symmetry of the orbitals.

6.6 Explain how molecular orbitals are constructed from atomic orbitals.

6.7 Use an energy level diagram for molecular orbitals to write down the configurations of homonuclear diatomic molecules formed from atoms in the second row of the Periodic Table.

6.8 Determine the formal bond orders of homonuclear diatomic molecules.

6.9 Describe ways of improving LCAO results for molecular orbitals and electronic energies.

Chapter 7 Solid state physics

Introduction

Solids have an astonishing range of properties. For example, carbon, silicon and lead all belong to the same group in the Periodic Table. Carbon can exist as diamond, transparent and hard, which scarcely conducts electricity at all, but conducts heat so well that diamonds are cold to touch. Silicon is a brittle semiconducting material of great importance in electronic devices. Its electrical conductivity depends on the concentration and type of impurities it contains. Lead is a soft metal which melts at a low temperature and conducts electricity well.

The characteristic properties of solids cannot be understood by scaling up the properties of individual atoms. In the solid state, new properties emerge, such as hardness or conductivity, which only make sense when we consider the solid as a whole. The subject of solid state physics seeks to explain these properties — and in many cases the explanations involve quantum mechanics. At the start of the course, we mentioned that there are strong links between quantum mechanics and industry. Many of these links are found in electronics and communication companies, where the electrical properties of solids in general, and of semiconductors in particular, are of paramount importance. We cannot give a detailed account of solid state physics here — that would take a whole course. Instead, we pick out two areas where quantum mechanics gives deep insight into the behaviour of solids: bonding and the conduction of electricity.

The chapter is organized as follows. Section 7.1 discusses the structure and bonding of solids and extends the LCAO method for molecules to a line of atoms. This method does not take proper account of the symmetry inherent in a crystal, and this is remedied in Section 7.2 which states and proves a key result known as Bloch's theorem. The section also introduces the tight-binding approximation and uses it to describe energy bands in solids. Finally, Section 7.3 gives an overview of electrical conduction in solids, explaining why metals, insulators and semiconductors behave so differently, and exploring the origins of electrical resistance and the special properties of semiconductors.

7.1 Structure and bonding

7.1.1 The arrangement of atoms in solids

We start with a basic question, essential for an understanding of the solid state. How are atoms arranged in a solid? Are they arranged regularly, like soldiers on parade, or are they in more haphazard arrangements, like people on a beach?

In fact, both kinds of arrangement exist. When the atoms are arranged regularly, the solid is said to be **crystalline**; when they are arranged more haphazardly, the solid is said to be **amorphous**. Diamond has a crystalline structure, while glass is amorphous.

In this chapter we shall consider only crystalline solids. One reason is that they are simpler to discuss than amorphous solids, and more is known about them.

Another reason is that all solid elements, and a large number of compounds, are crystalline in their lowest energy state. In some cases the crystals are large enough to see, but they can also be very small and pass unnoticed. A typical metal, for example, consists of *crystalline grains* that are about 0.1 mm across; different grains, oriented in different directions, butt onto one another. Ceramics are also crystalline, with microscopic grains fused together.

A crystal is a solid structure in which a small group of atoms is arranged in a basic pattern, and this pattern is distributed repeatedly over a regular three-dimensional grid called a **lattice**.

To avoid complications due to surfaces, we shall frequently consider crystals that are infinite in extent. The properties we wish to describe do not depend sensitively on the size of the crystal, so this is a fair simplification. Suppose we sit at a particular point in an infinite crystal and observe the atoms around us. The crucial feature of a crystalline structure is that there are many other points, extending throughout the crystal, with exactly similar environments. The set of all such points defines the lattice of the crystal, and each point in the lattice is called a **lattice point**.

A displacement vector joining two lattice points is called a **lattice vector**. It turns out that any lattice vector can be written in the form

$$\mathbf{R} = n_1 \mathbf{a}_1 + n_2 \mathbf{a}_2 + n_3 \mathbf{a}_3,$$

where $\mathbf{a}_1$, $\mathbf{a}_2$ and $\mathbf{a}_3$ are constant non-coplanar vectors, and n_1, n_2 and n_3 are integers (positive, negative or zero). For example, Figure 7.1 shows the arrangement of atoms in crystals of copper and tungsten, with appropriate choices of $\mathbf{a}_1$, $\mathbf{a}_2$ and $\mathbf{a}_3$.

In each of these structures, all the atoms have similar environments in an infinite crystal.

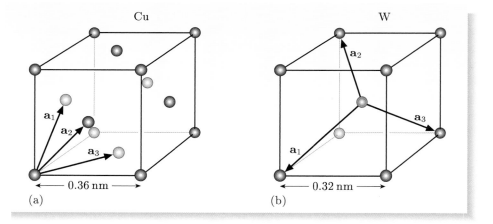

Figure 7.1 The arrangements of atoms in crystals of (a) copper (Cu) and (b) tungsten (W). You should picture these structures as extending indefinitely in three dimensions.

None of these details is vital for our purposes, but one very important concept must be understood: a crystal has a restricted sort of **translational symmetry**. This is not the full translational symmetry of empty space, where any displacement makes no difference, but in a crystal we can say that *any displacement through a lattice vector* makes no difference. This has profound consequences, as you will soon see.

7.1.2 Types of bonding

Why do the individual atoms in a solid remain bound together in a rigid structure? In one way or another, the answer lies with the *valence electrons* — electrons that occupy open shells. Inner electrons in closed shells screen the valence electrons from the atomic nuclei, but do not participate directly in bonding. We shall regard an atom as a positively-charged *core* (the nucleus plus the filled inner shells) plus the valence electrons, which provide a sort of glue, binding the atomic cores together. There are the three main types of bonding found in solids: ionic, covalent and metallic.

Ionic bonding occurs when one or more electrons transfer from one type of atom to another, and the solid is held together by electrostatic forces. A good example is provided by ordinary salt (NaCl) where each sodium atom transfers one of its electrons to a chlorine atom, and the resulting sodium ions (Na^+) and chlorine ions (Cl^-) pack together in a way that minimizes their total potential energy. The crystal structure depends on the relative sizes of the ions. For sodium chloride, the minimum energy is achieved by the structure shown in Figure 7.2, where each positive ion is surrounded by six negative ion nearest neighbours and vice versa.

Ionic bonding necessarily involves two different types of atom, one of which has a tendency to lose electrons, while the other has a tendency to gain electrons. Clearly, this type of bonding cannot occur in *elements*, where all atoms are the same.

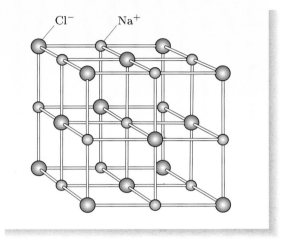

Figure 7.2 The sodium chloride structure.

Covalent bonding occurs when electrons from neighbouring atoms form bonding orbitals and antibonding orbitals, just like those you met for diatomic molecules. Bonding orbitals have a high probability density of electrons between the atoms, and this helps to stabilize the structure. The solid holds together because more electrons occupy bonding orbitals than antibonding orbitals. The great difference between ionic bonding and covalent bonding is illustrated in the electron probability density contour maps of Figure 7.3. In an ionic solid, the electrons are concentrated on the two ions, but in a covalently-bonded solid there is a significant concentration of electrons between neighbouring atoms.

The covalent bonds joining an atom to its nearest neighbours generally have preferred angles relative to one another. This leads to open structures in which each atom has only a few nearest neighbours (four in diamond, silicon and germanium). Also, since covalent bonds can be very strong, the structures can be very rigid.

Metallic bonding is the type of bonding found in metals. The valence electrons in metals have wave functions that extend far beyond the closest atoms. As a result, many atoms contribute to bonding, not just nearest neighbours, and the valence electrons have a probability density that spreads out over the whole crystal, rather than being concentrated along lines joining neighbouring atoms. In a metal, we can picture the atom cores as being bathed in a sea of electrons, which are free to flow throughout the metal. These electrons are responsible for many characteristic properties of metals, including their shiny appearance and high conductivities.

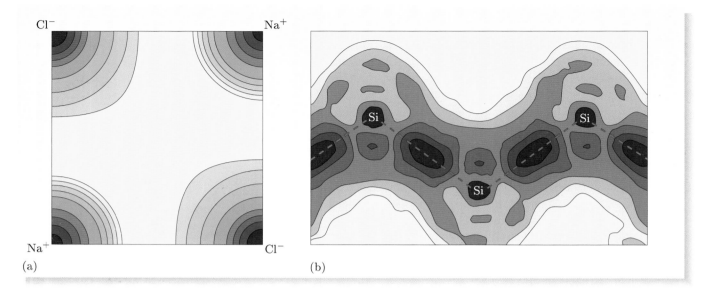

Figure 7.3 Contour maps showing the probability density of electrons in: (a) an ionic solid, sodium chloride, and (b) a covalent solid, silicon. Darker shades of blue indicate higher probability densities of electrons.

Metallic bonding is non-directional. To minimize energy, atoms in metals stack together in ways that fill space efficiently. This leads to structures similar to those you may have seen when spheres such as oranges or cannon balls are stacked together. In a given layer, each sphere has six neighbours and the spheres in the next layer fit neatly into the hollows between spheres in the first layer. This produces a dense structure in which each atom has twelve nearest neighbours. Because metallic bonding is non-directional, metals are malleable and can be beaten into shape in ways that would be inconceivable for diamond.

Exercise 7.1 Explain why metals tend to be denser than covalently-bonded solids such as diamond. ∎

7.1.3 An LCAO approach to bonding in solids

To gain some insight into bonding in solids, we shall begin by treating a solid as a sort of giant molecule. In the preceding chapter we had some success applying the LCAO method to diatomic molecules; now we will try to extend this method to solids.

First, we adopt the Born–Oppenheimer approximation: in order to find the possible states of the valence electrons, we treat the atomic cores as being frozen in position. This approximation relies on the valence electrons moving much faster than the atomic cores, and works just as well for solids as for molecules. It is used throughout solid state physics.

To illustrate the LCAO method, we consider a line of six lithium atoms, labelled A, B, C, D, F and G. Each lithium atom has a single valence electron in a 2s atomic orbital, and we can construct a *trial function* that is a linear combination of six 2s atomic orbitals, one centred on each atom. This gives

$$\psi(\mathbf{r}) = c_1\phi_{2s}^{A}(\mathbf{r}) + c_2\phi_{2s}^{B}(\mathbf{r}) + c_3\phi_{2s}^{C}(\mathbf{r}) + \cdots + c_6\phi_{2s}^{G}(\mathbf{r}), \tag{7.1}$$

where the coefficients c_1 to c_6 are initially unknown parameters and ϕ_{2s}^{A}, for example, is a 2s atomic orbital centred on atom A.

We then apply the variational method, which asks us to use our trial function to calculate

$$\langle E \rangle = \frac{\langle \psi | \widehat{H} | \psi \rangle}{\langle \psi | \psi \rangle}, \tag{7.2}$$

where $\widehat{H}$ is the Hamiltonian operator for the 2s electrons, including their interaction with the atomic cores.

When Equation 7.1 is substituted into Equation 7.2 and the result expanded out, we get many different terms. If we anticipate covalent bonding, it would be reasonable to neglect contributions such as $\langle \phi_{2s}^A | \widehat{H} | \phi_{2s}^C \rangle$ or $\langle \phi_{2s}^A | \phi_{2s}^C \rangle$, since these terms involve atomic orbitals based on non-adjacent atoms, but we would retain these terms if metallic bonding were suspected.

Just as for molecules, we impose the requirement that $\langle E \rangle$ should be an extremum with respect to variations of the coefficients c_1 to c_6. The only difference is that our trial function now leads to a 6×6 secular determinant, leading to six different solutions for $\langle E \rangle$. These are our estimates for the lowest energy levels of the six-atom 'molecule'. For each energy level, we get a particular set of coefficients c_1 to c_6, which we can substitute back into Equation 7.1 to get the corresponding energy eigenfunction.

Figure 7.4 shows the results obtained for the energy levels and the eigenfunctions. Each eigenfunction extends over all six atoms and would be called a molecular

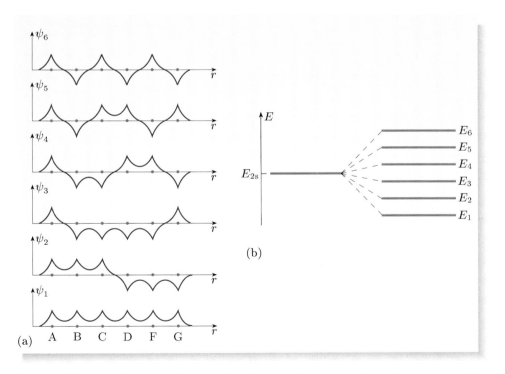

Figure 7.4 A schematic diagram showing (a) the energy eigenfunctions and (b) the eigenvalues, found by applying the LCAO approximation to a line of six lithium atoms, A, B, C, D, F, G, each with one valence electron. The six atomic states produce six molecular orbitals. The ground state has no nodes, and the number of nodes increases by one for each successive increase in energy.

orbital in the context of molecules. The three lowest energy levels are below the energy of the atomic orbital that gave rise to them, and the three highest energy levels are above it. You might think that these two effects would cancel out, giving no net binding. But, remember, each molecular orbital can hold two electrons, one with spin-up and the other with spin-down. This means that all the valence electrons from the six lithium atoms can go into the three lowest molecular orbitals, giving a net lowering of the energy of the system. The six-atom cluster is bound together.

Our simple model has taken one atomic 2s orbital per atom, but we can also consider shells with degenerate atomic orbitals. The 3d shell, for example, has five degenerate atomic orbitals. In this case a similar calculation produces five molecular orbitals per atom, each capable of holding two electrons. Half of the molecular orbitals are bonding, and half are antibonding.

As we increase the number of atoms, the energy between the highest level and the lowest level does not change dramatically, but the number of states is proportional to the number of atoms. In a macroscopic crystal, there are of order 10^{23} atoms, and the energy levels practically form a continuum, *but only within a restricted range*. As shown in Figure 7.5, each energy level in an atom broadens out into an **energy band**. Because the bands derive from atomic shells, we can label them by the quantum numbers of the parent atomic states. For example, we talk of a 2p band (capacity 6 electrons per atom) or a 3d band (capacity 10 electrons per atom). Neighbouring energy bands may overlap, but they are often separated by **band gaps**, in which there are no allowed states. The pattern of energy bands and band gaps will be of vital importance in understanding electrical conductivity of solids.

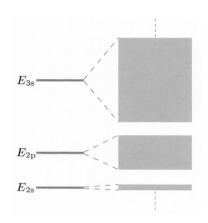

Figure 7.5 Energy bands in a solid separated by band gaps.

The width of an energy band depends on the extent to which atomic orbitals centred on different atoms overlap. Electrons in low-lying states are closely bound to the nucleus of their parent atoms, so two of their atomic orbitals, centred on neighbouring atoms, scarcely overlap at all. That is why the lowest energy band in Figure 7.5 is narrow. In these low-lying states, there is little chance of an electron moving between atoms, so the energy levels in the solid are essentially the same as in an isolated atom. However, electrons in the outer shells of an atom are less tightly bound to their atomic cores; there may then be considerable overlap between atomic orbitals centred on different atoms, leading to relatively broad energy bands.

Although the LCAO method gives a reasonable picture of the emergence of energy bands in a solid, it does not describe the energy eigenfunctions very well, since it takes no account of the translational symmetry. The functions in Figure 7.4 do not give a realistic description of electrons in crystals because they imply that there is a high electron probability density between some atoms, and a small electron probability density between others. This does not agree with the fundamental symmetry requirement that all the lattice points in a crystal should be equivalent. This problem will be addressed in the next section.

The cohesive energy of a solid is the minimum energy needed to split the solid into separate atoms.

Exercise 7.2 Figure 7.6 shows how the cohesive energy per atom varies with the number of d electrons per atom for transition elements in two periods of the Periodic Table. Given that transition elements have open d shells, identify and explain the main features of these data. ∎

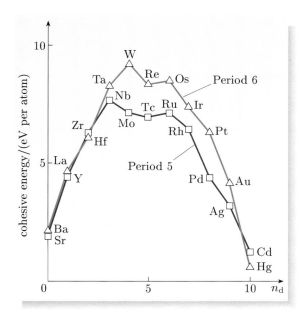

Figure 7.6 Cohesive energies per atom for transition elements in Period 5 (blue line) and Period 6 (red line), plotted against the number n_d of d electrons per atom.

7.2 Bloch's theorem and energy bands

7.2.1 Bloch's theorem

We have seen that the LCAO method ignores the fundamental symmetry requirement that all the lattice points in a crystal should be equivalent. The true nature of electron states in crystals was discovered in 1928 by Felix Bloch, a postgraduate student working under Heisenberg's supervision. Bloch's discovery is regarded as a cornerstone of solid state physics and is known as **Bloch's theorem**.

It is important to remember that *any displacement by a lattice vector* produces no discernable change in an infinite crystal. By virtue of this symmetry property, the potential energy function of an electron in a crystal obeys the condition

$$V(\mathbf{r} + \mathbf{R}) = V(\mathbf{r}) \quad \text{for any lattice vector } \mathbf{R}. \tag{7.3}$$

Any function that repeats itself in this way is said to have the **periodicity of the lattice**. The fact that the potential energy function (and hence the Hamiltonian) of an infinite crystal has this periodicity is the key fact behind Bloch's theorem, which we now state.

> **Bloch's theorem**
>
> In an infinite crystal, the energy eigenfunctions of electrons can be written in the form
>
> $$\psi(\mathbf{r}) = e^{i\mathbf{k}\cdot\mathbf{r}} u_{\mathbf{k}}(\mathbf{r}), \tag{7.4}$$
>
> where $\mathbf{k}$ is a real constant vector and $u_{\mathbf{k}}(\mathbf{r})$ is a periodic function, which may depend on $\mathbf{k}$, and has the periodicity of the lattice:
>
> $$u_{\mathbf{k}}(\mathbf{r} + \mathbf{R}) = u_{\mathbf{k}}(\mathbf{r}) \quad \text{for any lattice vector } \mathbf{R}. \tag{7.5}$$

The function $e^{i\mathbf{k}\cdot\mathbf{r}}$ describes the spatial variation of a plane wave of wavelength $\lambda = 2\pi/k$, propagating in the direction of $\mathbf{k}$. The constant vector $\mathbf{k}$ is called the **wave vector**; it points in the direction of propagation of the electron wave, and its magnitude k is equal to $2\pi/\lambda$. So Bloch's theorem tells us that each energy eigenfunction of an electron in an infinite crystal is the product of a plane wave $e^{i\mathbf{k}\cdot\mathbf{r}}$ and a periodic function $u_\mathbf{k}(\mathbf{r})$. In the language used to describe radio waves, we can say that each energy eigenfunction is a plane wave *modulated by* a periodic function $u_\mathbf{k}(\mathbf{r})$. The resulting function is called a **Bloch wave** (see Figure 7.7).

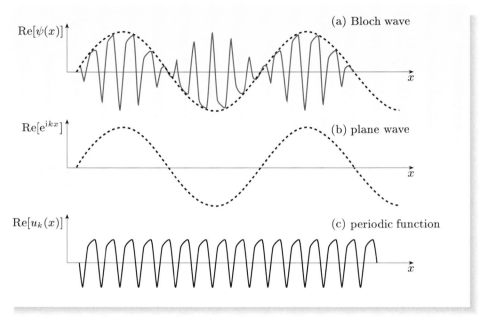

Figure 7.7 An illustration of Bloch's theorem in one dimension. The energy eigenfunction of an electron in an infinite crystal (real part shown in (a)) is the product of a plane wave (real part shown in (b)) and a function with the periodicity of the lattice (real part shown in (c)). The resulting eigenfunction in (a) is a plane wave modulated by a periodic function, and is called a Bloch wave.

Bloch relates: 'I found to my delight that the wave [of an electron in a crystal] differed from the plane wave of free electrons by only a periodic modulation. This was so simple that I didn't think it could be much of a discovery, but when I showed it to Heisenberg he said right away "That's it!".'

Heisenberg saw immediately that Bloch's theorem solved a puzzle that had been worrying physicists for some time: how can electrons move so freely along conducting wires when there are so many atoms blocking their way? Bloch's theorem effectively removes this problem. In perfect crystals, it shows that the eigenfunctions of electrons extend indefinitely, so electron waves have no difficulty in travelling past the atoms in a crystal, just as light has no difficulty in passing through a diffraction grating. It turns out that the origins of electrical resistance lie in the *imperfections* that are present in real crystals, a point we shall return to later.

Exercise 7.3 Use Bloch's theorem to show that the probability density in any

Bloch wave has the periodicity of the lattice. That is,
$$|\psi(\mathbf{r}+\mathbf{R})|^2 = |\psi(\mathbf{r})|^2 \quad \text{for any lattice vector } \mathbf{R}. \tag{7.6}$$

■

Equation 7.6 shows that Bloch's theorem avoids the problem encountered with LCAO eigenfunctions. In any Bloch wave, if there is a given electron probability density between two atoms, this will be reproduced at all equivalent sites in the crystal. As we go to higher energy levels, the energy cost paid for failing to have the maximum possible electron probability density between atoms is *shared equally* between all equivalent sites in the crystal. This contrasts with the LCAO method which incorrectly predicts a very low electron probability density between *some* of the atoms, where a disproportionately high energy cost is paid.

Is Bloch's theorem trivial?

At first sight, you might think that Bloch's theorem is a trivial result, expressed in a needlessly complicated way. Given that the Hamiltonian operator has the periodicity of the lattice, you might expect that its eigenfunctions would have the same periodicity. In other words, you might expect that $\psi(\mathbf{r}) = \psi(\mathbf{r} + \mathbf{R})$ for any lattice vector $\mathbf{R}$, which is consistent with Bloch's theorem in the special case $\mathbf{k} = 0$. If so, you would be falling into a trap — but one that is so inviting that it merits a special warning:

> It is not safe to assume that the solutions of an equation have the same symmetry as the equation itself.

You have seen examples of this previously in the course. For example, the time-independent Schrödinger equation for a hydrogen atom is spherically symmetric, but only some of its solutions (the s states) are spherically symmetric; the others are not. Here is a simpler example:

- ● The equation $x^2 = a^2$ is unchanged by reversing the sign of a. Are its solutions also unchanged by reversing the sign of a?
- ○ Of course not! The solutions $x = a$ and $x = -a$ both change sign when the sign of a is reversed, so the solutions do not have the symmetry of the equation they satisfy.

Justifying Bloch's theorem

We now give a sketch of the proof of Bloch's theorem. Although this proof will not be assessed, Bloch's theorem is too important to take on trust. The first step will be to show that an energy eigenfunction for an electron in a crystal obeys
$$\psi(\mathbf{r}+\mathbf{R}) = e^{i\mathbf{k}\cdot\mathbf{R}}\psi(\mathbf{r}) \tag{7.7}$$
for any lattice vector $\mathbf{R}$, where $\mathbf{k}$ is a real-valued constant vector, independent of $\mathbf{R}$. We will then show that Equation 7.7 leads directly to Bloch's theorem.

We start with a definition. For any lattice vector $\mathbf{R}$, we define the **lattice translation operator** $\widehat{T}(\mathbf{R})$ to be an operator that has the effect
$$\widehat{T}(\mathbf{R})\psi(\mathbf{r}) = \psi(\mathbf{r}+\mathbf{R}) \tag{7.8}$$

when applied to any function $\psi(\mathbf{r})$. This just corresponds to a shift of the origin of coordinates by $-\mathbf{R}$. There are many different lattice translation operators, corresponding to different lattice vectors. It is not difficult to show that all these operators commute with one another, and that they also commute with the Hamiltonian operator $\widehat{H}$, which has the periodicity of the lattice.

As you have seen many times before, the existence of a set of mutually commuting operators allows us to choose the eigenfunctions of one operator to be simultaneous eigenfunctions of all the other operators in the set. In the present context, this means that we can choose the energy eigenfunctions to be eigenfunctions of the lattice translation operators. But what is an eigenfunction of $\widehat{T}(\mathbf{R})$? It is a function that obeys the eigenvalue equation

$$\widehat{T}(\mathbf{R})\psi(\mathbf{r}) = \lambda(\mathbf{R})\,\psi(\mathbf{r}), \tag{7.9}$$

where the eigenvalue $\lambda(\mathbf{R})$ is a scalar that may depend on $\mathbf{R}$. Combining Equations 7.8 and 7.9, we see that the energy eigenfunctions obey the condition

$$\psi(\mathbf{r}+\mathbf{R}) = \lambda(\mathbf{R})\,\psi(\mathbf{r}) \tag{7.10}$$

for any lattice vector $\mathbf{R}$.

We now focus our attention on the form of $\lambda(\mathbf{R})$. The key idea here is that the combined effect of two displacements $\mathbf{R}_1$ and $\mathbf{R}_2$ is the same as that of a single displacement $\mathbf{R}_1+\mathbf{R}_2$. So Equation 7.10 gives

$$\psi(\mathbf{r}+\mathbf{R}_1+\mathbf{R}_2) = \lambda(\mathbf{R}_1)\,\lambda(\mathbf{R}_2)\,\psi(\mathbf{r}) = \lambda(\mathbf{R}_1+\mathbf{R}_2)\,\psi(\mathbf{r}),$$

from which it follows that

$$\lambda(\mathbf{R}_1)\,\lambda(\mathbf{R}_2) = \lambda(\mathbf{R}_1+\mathbf{R}_2) \tag{7.11}$$

for all pairs of lattice vectors $\mathbf{R}_1$ and $\mathbf{R}_2$ in the crystal. It turns out that the only way of satisfying this condition is for $\lambda(\mathbf{R})$ to be an *exponential function* of some sort. The two possibilities are

$$\lambda(\mathbf{R}) = \mathrm{e}^{\mathbf{k}\cdot\mathbf{R}} \quad \text{or} \quad \lambda(\mathbf{R}) = \mathrm{e}^{\mathrm{i}\mathbf{k}\cdot\mathbf{R}},$$

where $\mathbf{k}$ is a real constant vector, independent of $\mathbf{R}$. The first possibility can be ruled out because it becomes boundlessly large when $\mathbf{R}$ increases in directions that are roughly parallel to $\mathbf{k}$. This would produce an eigenfunction that diverges and cannot be normalized. We therefore conclude that the second form applies, so Equation 7.10 can be written as

$$\psi(\mathbf{r}+\mathbf{R}) = \mathrm{e}^{\mathrm{i}\mathbf{k}\cdot\mathbf{R}}\psi(\mathbf{r}), \tag{Eqn 7.7}$$

and this completes the first part of the proof.

The proof is rounded off by showing that Equation 7.7 is equivalent to Bloch's theorem. We do this by writing the energy eigenfunction as

$$\psi(\mathbf{r}) = \mathrm{e}^{\mathrm{i}\mathbf{k}\cdot\mathbf{r}}u_{\mathbf{k}}(\mathbf{r}). \tag{Eqn 7.4}$$

This can always be done provided that we make no special assumptions about the function $u_{\mathbf{k}}(\mathbf{r})$. However, substituting Equation 7.4 into both sides of Equation 7.7, we obtain

$$\mathrm{e}^{\mathrm{i}\mathbf{k}\cdot(\mathbf{r}+\mathbf{R})}u_{\mathbf{k}}(\mathbf{r}+\mathbf{R}) = \mathrm{e}^{\mathrm{i}\mathbf{k}\cdot\mathbf{R}}\mathrm{e}^{\mathrm{i}\mathbf{k}\cdot\mathbf{r}}u_{\mathbf{k}}(\mathbf{r}),$$

Remember: $\mathrm{e}^{A+B} = \mathrm{e}^{A}\mathrm{e}^{B}$.

and so conclude that
$$u_\mathbf{k}(\mathbf{r}+\mathbf{R}) = u_\mathbf{k}(\mathbf{r}) \tag{Eqn 7.5}$$
for any lattice vector $\mathbf{R}$. This is precisely Bloch's theorem.

Exercise 7.4 Show that any lattice translation operator $\widehat{T}(\mathbf{R})$ commutes with a Hamiltonian operator $\widehat{H}$ that has the periodicity of the lattice. Also show that any two lattice translation operators $\widehat{T}(\mathbf{R}_1)$ and $\widehat{T}(\mathbf{R}_2)$ commute with one another.

Exercise 7.5 Show that Equation 7.7 is consistent with Equation 7.6. ∎

Labelling Bloch waves

The wave vector $\mathbf{k}$ serves as a label for a particular Bloch wave, and we generally include it as a subscript on $\psi(\mathbf{r})$ by writing
$$\psi_\mathbf{k}(\mathbf{r}) = e^{i\mathbf{k}\cdot\mathbf{r}} u_\mathbf{k}(\mathbf{r}).$$

It is therefore important to know what values $\mathbf{k}$ can have. As is often the case in quantum mechanics, quantization arises from the boundary conditions.

In solid state physics it is customary to use a special set of boundary conditions called **periodic boundary conditions**. Given a macroscopic cubic sample of material, with sides of length L, we require the eigenfunctions $\psi_\mathbf{k}(\mathbf{r})$ to have *identical values on opposite faces of the cube*. Although this condition may seem artificial, it produces no unphysical consequences in macroscopic systems, and the advantage is that we can imagine stacking identical cubes together to produce an infinite crystal in which the eigenfunctions vary continuously. This is a harmless mathematical trick which allows us to ignore any surface effects, yet still obtain quantization conditions appropriate for a finite sample.

To see the effect of periodic boundary conditions, let us suppose that the x-axis points along a line of atoms whose spacing is a. Taking L to be a large integer times a, we then impose periodic boundary conditions by requiring that
$$\psi_\mathbf{k}(\mathbf{r}+L\mathbf{e}_x) = \psi_\mathbf{k}(\mathbf{r}).$$

Using Bloch's theorem, together with the periodicity of $u_\mathbf{k}(\mathbf{r})$, we then have
$$e^{i\mathbf{k}\cdot(\mathbf{r}+L\mathbf{e}_x)} = e^{i\mathbf{k}\cdot\mathbf{r}},$$
which gives $e^{ik_x L} = 1$. This equation is solved by taking $k_x L$ to be any integer multiple of 2π. In a three-dimensional crystal, similar arguments apply to the other components of $\mathbf{k}$, so we conclude that the allowed values of k_x, k_y and k_z are
$$0, \pm\frac{2\pi}{L}, \pm\frac{4\pi}{L}, \ldots, \tag{7.12}$$
where L is the side length of our cubic sample. Hence periodic boundary conditions lead to a discrete set of allowed wave vectors $\mathbf{k}$, and a corresponding discrete set of energy eigenfunctions $\psi_\mathbf{k}(\mathbf{r})$.

It turns out to be sensible to restrict the values of $\mathbf{k}$ further, by considering only N different values centred around $\mathbf{k}=\mathbf{0}$, where N is the number of lattice sites in the crystal. To see the reason for this, it is helpful to recall that the wave vector $\mathbf{k}$

first appeared in our proof of Bloch's theorem via the factor $e^{i\mathbf{k}\cdot\mathbf{R}}$ in Equation 7.7. Now, it is possible to find pairs of wave vectors $\mathbf{k}$ and $\mathbf{k}'$ such that

$$e^{i\mathbf{k}\cdot\mathbf{R}} = e^{i\mathbf{k}'\cdot\mathbf{R}} \quad \text{for } all \text{ lattice vectors } \mathbf{R}.$$

This would allow both $\mathbf{k}$ and $\mathbf{k}'$ to label the same Bloch wave. We avoid this by restricting the values of $\mathbf{k}$ to ensure that each Bloch wave is labelled by a *unique* wave vector. The precise details depend on the lattice and do not matter for our purposes, but a specific example may illustrate the general idea: in the simple lattice shown in Figure 7.8, the components of $\mathbf{k}$ are usually restricted to lie in the ranges

$$-\frac{\pi}{a} < k_x \leq \frac{\pi}{a}, \quad -\frac{\pi}{a} < k_y \leq \frac{\pi}{a}, \quad -\frac{\pi}{a} < k_z \leq \frac{\pi}{a}, \quad (7.13)$$

where a is the distance between nearest neighbours. The benefit of making this restriction will emerge later (see Exercise 7.7 below).

We still have some freedom to choose the functions $u_\mathbf{k}(\mathbf{r})$, and there is no reason to suppose that a given value of $\mathbf{k}$ will correspond to just one function $u_\mathbf{k}(\mathbf{r})$. We can therefore add an extra index to the Bloch wave, writing it in full as

$$\psi_{n\mathbf{k}}(\mathbf{r}) = e^{i\mathbf{k}\cdot\mathbf{r}} u_{n\mathbf{k}}(\mathbf{r}). \quad (7.14)$$

The index n allows us to label different types of energy band deriving from different combinations of atomic orbital (e.g. 2s or 3d).

7.2.2 The tight-binding method

We shall now describe the **tight-binding method** for finding the states of electrons in crystals. In some ways, this is similar to the LCAO approach, but it is more satisfactory because it takes full account of Bloch's theorem. For simplicity, we shall consider a crystal in which each lattice site is occupied by a single atom.

The tight-binding method starts by taking a linear combination of atomic orbitals, just as in Equation 7.1. The sum now extends over all the atoms in the crystal, giving an expression with a vast number of terms, but one that is written down painlessly with a summation sign:

$$\psi(\mathbf{r}) = \sum_i c_i \, \phi(\mathbf{r} - \mathbf{R}_i). \quad (7.15)$$

Here, $\phi(\mathbf{r})$ is an atomic orbital of a given type (say a 2s orbital) centred on an atom at the origin, and $\phi(\mathbf{r} - \mathbf{R}_i)$ is a similar atomic orbital centred on an atom at lattice point $\mathbf{R}_i$.

Rather than determining the coefficients c_i by the variational method, as in the LCAO method, we now insist that Equation 7.15 should be consistent with Bloch's theorem. This can be achieved by choosing c_i to be proportional to $e^{i\mathbf{k}\cdot\mathbf{R}_i}$ and writing

$$\psi_\mathbf{k}(\mathbf{r}) = A_\mathbf{k} \sum_i e^{i\mathbf{k}\cdot\mathbf{R}_i} \phi(\mathbf{r} - \mathbf{R}_i), \quad (7.16)$$

where $A_\mathbf{k}$ is a normalization constant and $\mathbf{k}$ is the wave vector of the Bloch wave.

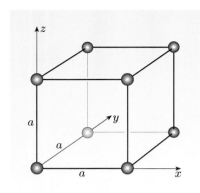

Figure 7.8 A simple lattice with nearest-neighbour spacing a. You should picture the structure as extending indefinitely in three dimensions. In this lattice, $e^{i\mathbf{k}\cdot\mathbf{R}} = e^{i\mathbf{k}'\cdot\mathbf{R}}$ for all lattice vectors $\mathbf{R}$ if $k_x - k'_x = 2\pi/a$, with similar results for the y- and z-components. The restriction imposed by Equation 7.13 allows us to avoid this double labelling problem.

To check that Equation 7.16 is consistent with Bloch's theorem, we can choose any lattice vector $\mathbf{R}_j$ and replace $\mathbf{r}$ by $\mathbf{r} + \mathbf{R}_j$ throughout Equation 7.16 to obtain

$$\psi_{\mathbf{k}}(\mathbf{r} + \mathbf{R}_j) = A_{\mathbf{k}} \sum_i e^{i\mathbf{k}\cdot\mathbf{R}_i} \phi(\mathbf{r} + \mathbf{R}_j - \mathbf{R}_i).$$

This can be rewritten as

$$\psi_{\mathbf{k}}(\mathbf{r} + \mathbf{R}_j) = e^{i\mathbf{k}\cdot\mathbf{R}_j} A_{\mathbf{k}} \sum_i e^{i\mathbf{k}\cdot(\mathbf{R}_i - \mathbf{R}_j)} \phi(\mathbf{r} - (\mathbf{R}_i - \mathbf{R}_j)).$$

Now, the sum on the right-hand side of this equation is just the same as the sum on the right-hand side of Equation 7.16; it uses a different lattice point from which to start the sum, but in an infinite crystal, this choice cannot matter. We therefore conclude that

$$\psi_{\mathbf{k}}(\mathbf{r} + \mathbf{R}_j) = e^{i\mathbf{k}\cdot\mathbf{R}_j} \psi_{\mathbf{k}}(\mathbf{r}),$$

and this agrees with Equation 7.7, which we have seen is equivalent to Bloch's theorem.

To find out more about the energy levels of an electron in a crystal, we shall make further approximations. Using an approach similar to the central-field model in atoms, we assume that it makes sense to treat each electron separately, subject to a single-particle Hamiltonian operator

$$\widehat{H} = -\frac{\hbar^2}{2m}\nabla^2 + \sum_j V_j(\mathbf{r}), \qquad (7.17)$$

where $V_j(\mathbf{r})$ is the effective potential energy contribution due to atom j, which includes electron–electron repulsion only in an averaged way, corresponding to screening of the positively-charged atomic nuclei.

We wish to find the possible energy eigenvalues, which we denote by $E(\mathbf{k})$. These are given by the time-independent Schrödinger equation

$$\widehat{H}\,\psi_{\mathbf{k}}(\mathbf{r}) = E(\mathbf{k})\,\psi_{\mathbf{k}}(\mathbf{r}),$$

or equivalently by

$$E(\mathbf{k}) = \frac{\langle\psi_{\mathbf{k}}|\widehat{H}|\psi_{\mathbf{k}}\rangle}{\langle\psi_{\mathbf{k}}|\psi_{\mathbf{k}}\rangle}. \qquad (7.18)$$

The energy eigenvalues depend on the wave vector $\mathbf{k}$ of the Bloch wave.

To simplify the notation, we shall represent the atomic orbital $\phi(\mathbf{r} - \mathbf{R}_i)$, centred on site i, by the ket vector $|\phi_i\rangle$, so that Equation 7.16 appears as

$$|\psi_{\mathbf{k}}\rangle = A_{\mathbf{k}} \sum_i e^{i\mathbf{k}\cdot\mathbf{R}_i} |\phi_i\rangle. \qquad (7.19)$$

Then the numerator of Equation 7.18 becomes

$$\langle\psi_{\mathbf{k}}|\widehat{H}|\psi_{\mathbf{k}}\rangle = |A_{\mathbf{k}}|^2 \sum_{ij} \langle\phi_i|\widehat{H}|\phi_j\rangle e^{-i\mathbf{k}\cdot(\mathbf{R}_i - \mathbf{R}_j)}, \qquad (7.20)$$

where i and j range over all the lattice sites. To simplify this expression we make a drastic, but not unreasonable, approximation. We retain matrix elements $\langle\phi_i|\widehat{H}|\phi_i\rangle$, where the two atomic orbitals refer to the *same site*, and we also retain matrix elements $\langle\phi_i|\widehat{H}|\phi_j\rangle$, where i and j are *nearest-neighbouring sites*.

However, we ignore all other matrix elements — those involving atomic orbitals centred on sites that are further apart than nearest neighbours.

Using this approximation in Equation 7.20, we obtain

$$\langle \psi_\mathbf{k} | \widehat{H} | \psi_\mathbf{k} \rangle = |A_\mathbf{k}|^2 \sum_i \left[\langle \phi_i | \widehat{H} | \phi_i \rangle + \sum_{j=\text{nn of }i} e^{-i\mathbf{k}\cdot(\mathbf{R}_i - \mathbf{R}_j)} \langle \phi_i | \widehat{H} | \phi_j \rangle \right],$$

where the sum over j is restricted to sites that are nearest neighbours of site i (denoted by nn of i). Because all lattice sites in an infinite crystal are equivalent, the sum in square brackets on the right-hand side cannot depend on i. It produces N identical terms, allowing us to write

$$\langle \psi_\mathbf{k} | \widehat{H} | \psi_\mathbf{k} \rangle = N |A_\mathbf{k}|^2 \left[\langle \phi_i | \widehat{H} | \phi_i \rangle + \sum_{j=\text{nn of }i} e^{-i\mathbf{k}\cdot(\mathbf{R}_i - \mathbf{R}_j)} \langle \phi_i | \widehat{H} | \phi_j \rangle \right],$$

where i is any lattice site whose nearest neighbours are labelled by j. A very similar calculation, carried out for $\langle \psi_\mathbf{k} | \psi_\mathbf{k} \rangle$, gives

$$\langle \psi_\mathbf{k} | \psi_\mathbf{k} \rangle = N |A_\mathbf{k}|^2 \left[1 + \sum_{j=\text{nn of }i} e^{-i\mathbf{k}\cdot(\mathbf{R}_i - \mathbf{R}_j)} \langle \phi_i | \phi_j \rangle \right],$$

and, substituting both these results into Equation 7.18, we conclude that

$$E(\mathbf{k}) = \frac{\langle \phi_i | \widehat{H} | \phi_i \rangle + \sum_{j=\text{nn of }i} e^{-i\mathbf{k}\cdot(\mathbf{R}_i - \mathbf{R}_j)} \langle \phi_i | \widehat{H} | \phi_j \rangle}{1 + \sum_{j=\text{nn of }i} e^{-i\mathbf{k}\cdot(\mathbf{R}_i - \mathbf{R}_j)} \langle \phi_i | \phi_j \rangle}. \tag{7.21}$$

If all the nearest neighbours of i are equivalent, we can introduce the notation

$$\langle \phi_i | \widehat{H} | \phi_i \rangle = E_0, \quad \langle \phi_i | \widehat{H} | \phi_j \rangle = B, \quad \text{and} \quad \langle \phi_i | \phi_j \rangle = S,$$

where E_0, B and S are independent of the lattice site i, and of any particular choice of its nearest neighbour, j. Equation 7.21 can then be written as

$$E(\mathbf{k}) = \frac{E_0 + B f(\mathbf{k})}{1 + S f(\mathbf{k})}, \tag{7.22}$$

where

$$f(\mathbf{k}) = \sum_{j=\text{nn of }i} e^{-i\mathbf{k}\cdot(\mathbf{R}_i - \mathbf{R}_j)}. \tag{7.23}$$

Finally, we shall assume that the overlap between neighbouring atomic orbitals is small, so that $|S| \ll 1$. Under these circumstances, we can expand the denominator of Equation 7.22 by the binomial theorem, and obtain the approximation

Normally, $\beta > 0$.

$$E(\mathbf{k}) = E_0 - \beta f(\mathbf{k}) \quad \text{where} \quad \beta = E_0 S - B. \tag{7.24}$$

The function $f(\mathbf{k})$ depends on the precise location of the nearest neighbours. To take a definite case, let us suppose that the atoms are arranged as in Figure 7.8, so that an atom at the origin has six nearest neighbours at $(\pm a, 0, 0)$, $(0, \pm a, 0)$ and $(0, 0, \pm a)$. In this case, the allowed energy levels are

$$E(\mathbf{k}) = E_0 - 2\beta \big[\cos(k_x a) + \cos(k_y a) + \cos(k_z a) \big]. \tag{7.25}$$

Exercise 7.6 For the arrangement of atoms in Figure 7.8, show that Equation 7.25 follows from Equations 7.23 and 7.24. ∎

Equation 7.25 is based on several approximations, and applies only to the crystal structure of Figure 7.8. Nevertheless, it incorporates many key ideas of solid state physics, and repays careful study.

The first point to notice is that the energy levels cover a finite range. The minimum and maximum energies are at

$$E_{\min} = E_0 - 6\beta \quad \text{and} \quad E_{\max} = E_0 + 6\beta, \tag{7.26}$$

so the energy eigenvalues are confined to a range of finite width 12β. We say that the energy levels form an *energy band* of width 12β, which is typically equal to several electronvolts for valence electrons. For a macroscopic crystal, there are very many different Bloch waves, all with energies within the energy band, and all very closely spaced. Provided that the crystal is macroscopic, the energy levels within a band are effectively indistinguishable from a continuum.

The width of the energy band depends on the quantity β in Equation 7.25. This is expected to increase as the overlap of orbitals centred on neighbouring atoms increases. Electrons in atomic orbitals close to the nucleus scarcely overlap, so are expected to have very narrow energy bands. However, electrons in outer orbitals will overlap more and produce much broader energy bands.

Consider how the energy bands change when we bring the atoms in a solid closer together. If we start with the atoms far apart, there will be almost no overlap of atomic orbitals based on different atoms, and the pattern of energy levels will be essentially the same as for isolated atoms. As the spacing between the atoms decreases, the overlap of atomic orbitals increases, and the energy bands broaden out and may overlap (Figure 7.9). The precise pattern of bands and gaps varies from material to material.

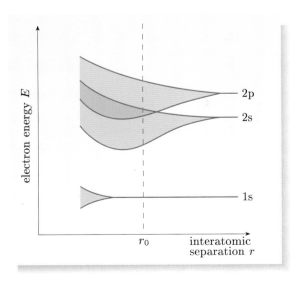

Figure 7.9 Energy bands in a solid as a function of interatomic separation. If the atoms are close enough, bands deriving from different atomic states may overlap.

Exercise 7.7 The energy $E(\mathbf{k})$ in Equation 7.25 is a periodic function of k_x, k_y and k_z, returning to the same value when any of these components changes by $2\pi/a$. What is the interpretation of this mathematical fact? ■

We can also investigate the behaviour of electrons near the bottom of the energy band, where k_x, k_y and k_z are small. In this case, we can approximate the cosine terms in Equation 7.25 by $\cos(k_x a) \simeq 1 - \frac{1}{2}(k_x a)^2$, etc. This leads to

$$E(\mathbf{k}) \simeq \text{constant} + \beta a^2 k^2, \quad \text{for} \quad k^2 = k_x^2 + k_y^2 + k_z^2 \ll 1/a^2.$$

Bearing in mind that the kinetic energy of a free particle in empty space is $\hbar^2 k^2/2m$, we can rewrite this equation in the form

$$E(\mathbf{k}) \simeq \text{constant} + \frac{\hbar^2 k^2}{2m_e^*}, \quad \text{where } m_e^* = \frac{\hbar^2}{2\beta a^2}. \tag{7.27}$$

m_e^* is the standard symbol used in solid state physics for the effective mass of an electron. The star does not imply complex conjugation!

The quantity m_e^* is called the **effective mass** for electrons near the bottom of the energy band. This quantity characterizes the way an electron near the bottom of the band will respond to forces, such as those due to an applied electric field. If the effective mass is high, for example, the electron will respond sluggishly. There is no reason to suppose that m_e^* will be the same as the mass of a free electron.

Exercise 7.8 Would you expect the effective mass of an electron at the bottom of a band to be larger in narrow bands or in wide bands? ■

7.2.3 Band structures in practice

The tight-binding model which we have just described is only a crude approximation. It tells us that a single atomic orbital in an isolated atom broadens out into an energy band in a solid. In practice, there are many different types of atomic orbital. Orbitals in the same shell (i.e. with the same n and l values) are degenerate, or at least have very closely spaced energies. So when the atoms are brought together in a solid, we can get energy bands that can hold several electrons per atom. For example, the d bands can hold up to 10 electrons per atom. The energy bands associated with different atomic shells may overlap. It is also possible for orbitals from *different* atomic shells to combine together in the same Bloch wave.

Detailed calculations take us far from the tight-binding model, and require considerable computing power. Nevertheless, it is possible to use the time-independent Schrödinger equation to find the dependence of energy on $\mathbf{k}$ in real solids. Figure 7.10 gives examples of this so-called **band structure** for copper and germanium.

In the case of copper, the 3d bands occupy a narrow range of energies, mainly between about -2 eV and -4 eV. The other energy bands shown for copper are 4s bands; these occupy a wider range of energies because the 4s atomic orbitals are more extended than the 3d atomic orbitals, and overlap more. The vertical arrow on the diagram indicates one of the dominant transitions that occur when copper absorbs visible light. This accounts for the reddish–brown colour of copper.

In the case of germanium, all of the energy bands shown are mixtures of 4s and 4p atomic orbitals. The interesting feature of this band structure is the fact that no energy levels are found in the narrow strip marked by dashed lines; this is a gap between two energy bands. You will see that it has a crucial role in determining the electrical properties of germanium.

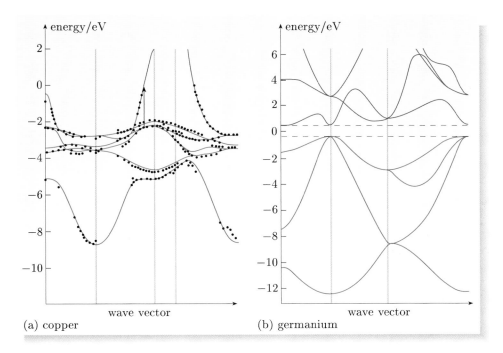

Figure 7.10 Band structure diagrams showing how the energies of electron states vary with wave vector **k** for (a) copper and (b) germanium. Because the wave vector is a vector, we can only show sample behaviour as **k** varies continuously in various directions; each grey vertical line marks a change in the direction of **k**. For copper, energy eigenvalues obtained from the time-independent Schrödinger equation (continuous curves) are in good agreement with experimental results obtained by ejecting electrons with photons (dots).

The next question is: how are the energy bands occupied by electrons? At absolute zero, the electrons occupy the states of lowest energy subject to the Pauli exclusion principle, which tells us that each Bloch wave can accommodate a maximum of two electrons — one with spin-up and the other with spin-down. In any given material, all the states below a characteristic energy E_F, called the **Fermi energy**, are occupied, and all the states above the Fermi energy are empty. In both parts of Figure 7.10, the energy zero has been placed at the Fermi energy, and the occupied states are shown in red while the unoccupied states are shown in blue.

Above absolute zero, we would expect electrons to be excited into states of higher energy. A typical thermal energy is of order kT, where k is Boltzmann's constant and T is the absolute temperature. At room temperature, $kT \simeq 0.025\,\text{eV}$, so we would expect some electrons to gain a fraction of an electronvolt compared to their energies at absolute zero. In quantum mechanics, they do this by moving into higher energy levels. However, the Pauli exclusion principle continues to exert a powerful influence. Figure 7.11 compares the way electrons occupy available states at absolute zero and at $300\,\text{K}$. Only a few states around the Fermi

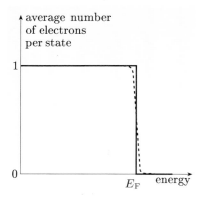

Figure 7.11 The average number of electrons of a given spin in a single quantum state at absolute zero (solid line) and at room temperature (dashed line). The Fermi energy is marked as E_F.

energy have their occupations changed. Nevertheless, these small changes have dramatic consequences, as you will see in the next section.

7.3 Conduction of electricity in solids

7.3.1 Metals, insulators and semiconductors

Different materials have vastly differing abilities to conduct electricity. For example, the electrical conductivity of silver is about 10^{19} times greater than that of pure diamond. This section will use the concept of an energy band to give a qualitative interpretation of electrical conduction in solids. In particular, we will explain the differences between insulators, metals and semiconductors. The following fact will be crucial:

> The electrons in a completely full band cannot carry an electric current, even if there is an electric field present.

The underlying reason is that the electron states in a band are arranged symmetrically, so for each state with wave vector $\mathbf{k}$, there is another state with wave vector $-\mathbf{k}$. The wave vector is proportional to a momentum, so there is no tendency for the electrons in a full band to flow in one direction rather than in the opposite direction: there is no current. In the absence of an electric field, this is not surprising, but the important point is that an electric field makes no difference. This is because the Pauli exclusion principle blocks any transfer of electrons into states that are already full. Moreover, the electrons cannot gain enough energy from an applied electric field to cross the gap between energy bands, so a full band remains full and is unable to contribute to an electric current even in the presence of an electric field.

By contrast, electrons in a partly-full band *can* produce electric currents. The energy levels within a band are extremely closely spaced, practically forming a continuum, so there is no difficulty in an electron moving from an occupied state to an unoccupied state within the same band. Electrons are negatively-charged, so they tend to move in the opposite direction to an applied electric field. In quantum-mechanical terms, an electric field pointing in the z-direction has the effect of causing more electrons to occupy states with $k_z < 0$ than states with $k_z > 0$, and it is this *asymmetry* of occupied states that constitutes an electric current.

We can now understand the major difference between a conductor and an insulator. First, let us consider the situation at absolute zero. If the highest occupied band is not completely full (Figure 7.12a), the material can conduct electricity with ease, and it is classified as a **conductor**. If all the electrons are in full bands (Figure 7.12b), the material is unable to conduct electricity, and is classified as an **insulator**.

The distinction between an insulator and a conductor is less clear-cut at non-zero temperatures. If even a small fraction of the electrons in the highest full band are thermally excited into the lowest empty band, then neither of these bands will be completely full or empty, and the material will be able to conduct electricity. In practice, the number of electrons transferred between bands is negligible if the band gap is greater than about $80kT$, where k is Boltzmann's constant and T is the absolute temperature. However, if the band gap is smaller than $80kT$, significant conduction can take place, and the material is classified as a **semiconductor**. At room temperature, semiconductors have band gaps that are smaller than about $2\,\text{eV}$ (Figure 7.12c).

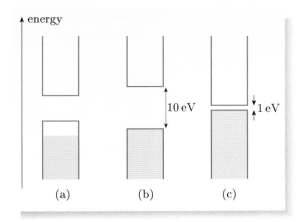

In Figure 7.12 (only) energy levels that are empty at $T = 0$ are left unshaded

Figure 7.12 (a) A conductor has a partly-full energy band; (b) an insulator has a large band gap between a band that is completely full and a band that is completely empty; (c) a semiconductor has a band gap that is small enough for electrons to be thermally excited across it.

In some cases it is easy to predict that a material will be a conductor. For example, sodium atoms have the ground-state configuration $1s^2\,2s^2\,2p^6\,3s$, with a single 3s electron in their valence shell. When sodium atoms come together to form a solid, this 3s shell broadens out to form a band that is capable of holding two electrons per atom (the factor of two arises because electrons can be spin-up or spin-down). Hence the 3s band in sodium is half full, and is able to conduct electricity. Since most atoms have valence electrons in partly-full atomic shells, it is not surprising that most elements in the Periodic Table are conductors.

Sometimes it is harder to make predictions. Magnesium atoms have the ground-state configuration $1s^2\,2s^2\,2p^6\,3s^2$, so one might suppose that the 3s band in the solid would be completely full, and that magnesium would be an insulator. In practice, the 3s band in magnesium overlaps with the 3p band, giving a combined 3s–3p band that can accommodate 8 electrons; this band is only partly full, so magnesium is a conductor.

The case of diamond is particularly tricky to predict. Carbon atoms have the ground-state configuration $1s^2\,2s^2\,2p^2$, but 2s and 2p states mix together in diamond, producing two separate s–p bands, each capable of holding four electrons per atom at the equilibrium interatomic separation, r_0 (Figure 7.13). In diamond, the lower of these bands is completely full, and is separated from an empty band by a band gap of $5.5\,\text{eV}$, so diamond is an insulator at room temperature. Similar results apply to silicon and germanium but with much smaller band gaps, making these materials semiconductors at room temperature (they are the only semiconducting elements).

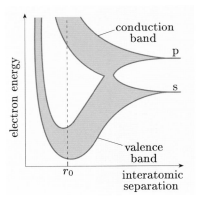

Figure 7.13 A schematic diagram showing how the energy bands in diamond, silicon and germanium split when the atoms are brought together. The equilibrium interatomic separation, r_0, is marked. Two s–p bands (called the valence band and the conduction band) are produced, each capable of holding four electrons per atom.

Electrical conductivity in metals

The broad division of materials into conductors, insulators and semiconductors does not explain why some metals such as silver and copper are better conductors than others such as titanium or steel. We will now give a very brief survey of the factors that affect electrical conductivity in metals.

Long before quantum mechanics, Paul Drude developed a classical model of conduction based on the idea that some electrons in a metal become detached from their parent atoms and are free to flow. According to Drude, the free electrons are driven onwards by an applied electric field, but their progress is impeded by collisions with the positive ions in the metal. This led to the following formula for electrical conductivity:

$$\sigma = \frac{ne^2\tau}{m_e}, \qquad (7.28)$$

The electrical conductivity σ is a measure of the ease with which a material conducts electricity: a wire of length l and cross-sectional area A has electrical resistance $R = l/A\sigma$.

where n is the number density of free electrons (the number per unit volume), e/m_e is the magnitude of the charge-to-mass ratio of an electron, and τ is a characteristic **relaxation time** (the average time an electron spends between collisions).

Remarkably enough, quantum mechanics leads to a similar formula, but with some significant differences in interpretation. The electrons whose number density is counted are those in a partly-full band, and the mass m_e is replaced by an effective mass m_e^* (which may not be the same as m_e, as you saw in Equation 7.27). However, the main point we wish to emphasize here is the nature of the collisions that impede the flow of electrons and give rise to the relaxation time τ.

In a perfect crystal, there is no reason for a particular Bloch wave, or a wave packet made up of different Bloch waves, to be scattered. This was understood by Heisenberg as soon as he learned about Bloch's theorem. What scatters electrons in a crystal is any *deviation* from a perfect periodic lattice. A crystal may contain defects of various kinds — empty lattice sites, foreign atoms or boundaries between different crystalline grains, for example. Also, the atomic cores vibrate around their equilibrium sites, so an electron passing through the crystal does not see a perfectly periodic potential energy function. These vibrations become increasingly vigorous at higher temperatures, so the conductivity falls with increasing temperature. For example, the conductivity of copper drops by a factor of 0.2 as the temperature increases from 100 K to 300 K.

7.3.2 Semiconductors

A semiconductor is a material with an electrical conductivity that is intermediate between metals and insulators; its electrical conductivity generally increases with temperature and is extremely sensitive to the presence of impurities.

Semiconducting behaviour was observed by Faraday in 1833, and early radio sets used natural semiconducting crystals as rectifiers, but the sensitivity of semiconductors to impurities plagued experimental investigations, and Pauli counselled against studying them at all! It was not until the 1940s that techniques of crystal growth and purification became good enough to allow semiconductors

7.3 Conduction of electricity in solids

to be properly characterized. The semiconductor age dawned on 23 December 1947, when the first semiconducting transistor was constructed.

Pure semiconductors

We begin by considering semiconductors that are so pure that the effects of impurities can be neglected. The essential feature of a pure semiconductor is that at absolute zero it has an energy band that is full, separated by a small gap from an energy band that is empty. The full band is called the **valence band**, and the empty band above it is called the **conduction band**. Although there are many other bands, the **band gap** of the material invariably refers to the gap between the valence and conduction bands (Table 7.1).

Table 7.1 Band gaps ΔE_{Si} and ΔE_{Ge} of silicon and germanium at 0 K and 300 K.

T	ΔE_{Si}	ΔE_{Ge}
0 K	1.17 eV	0.75 eV
300 K	1.12 eV	0.67 eV

Exercise 7.9 Table 7.1 shows that the band gaps of silicon and germanium decrease with temperature. Can you suggest why? ∎

Above absolute zero, some electrons from the valence band are thermally excited into the conduction band, and electrical conduction takes place in both bands. At room temperature, the number density of electrons in the conduction bands of pure silicon and pure germanium are 1.5×10^{16} m^{-3} and 2.4×10^{19} m^{-3}, respectively, well below typical values for metals (which are around 10^{29} m^{-3}). Referring to Equation 7.28, we can say that the main reason for the relatively low conductivity of semiconductors is that they have low number densities of charge carriers. However, these number densities increase rapidly with temperature, as more electrons are thermally excited across the band gap, and this is why the electrical conductivity of a pure semiconductor increases with temperature, in spite of the increasingly disruptive effect of vibrating atomic cores.

In a pure semiconductor, *both* the conduction band and the valence band contribute to the conduction of electricity. The flow of current can be conveniently measured by the **current density**, which is a vector quantity **J** pointing in the direction of current flow, with a magnitude equal to the current per unit cross-sectional area perpendicular to the current flow. A classical argument (whose details are not needed here) shows that the current density is given by

$$\mathbf{J} = \frac{-e}{V} \sum_i \mathbf{v}_i, \qquad (7.29)$$

where $\mathbf{v}_i$ is the velocity of the ith electron, and the sum is over electrons in a volume V. The minus sign appears because electrons have charge $-e$.

Now let us suppose that the valence band is completely full, except for electron j, which has been thermally excited into the conduction band. Then the remaining electrons in the valence band contribute a current

$$\mathbf{J} = \frac{-e}{V} \sum_{i \neq j} \mathbf{v}_i = \frac{-e}{V} \sum_{\text{all } i} \mathbf{v}_i + \frac{e}{V} \mathbf{v}_j.$$

However, the first term on the right-hand side is the current density due to a full band, which we have already seen is equal to zero. We therefore conclude that:

> An otherwise full band in which one electron is missing, behaves just like an empty band containing a particle of charge $+e$; this 'particle' is called a **hole**.

This may seem an odd way of looking at things, but it is very convenient to replace a description involving a large number of electrons in the valence band by one that involves a small number of holes. Electrical conduction in a pure semiconductor is due to a combination of positively-charged holes in the valence band, flowing in the direction of an applied electric field, and negatively-charged electrons in the conduction band, flowing in the opposite direction.

Exercise 7.10 In pure germanium at $300\,\text{K}$, the number density of electrons in the conduction band is $2.4 \times 10^{19}\,\text{m}^{-3}$. What is the number density of holes in the valence band? ■

Taking the energy zero to be in the middle of the band gap, as in Figure 7.14, we define the energy E_h of a hole in the valence band to be the *magnitude* of the energy difference between this energy zero and the vacated electron state. The energy E_e of an electron in the conduction band is the magnitude of the energy difference between the energy level occupied by the electron and the energy zero. Thus, the total energy required to promote the electron shown in Figure 7.14 from the valence band to the conduction band is $E_h + E_e$. Note that hole states that are deeper down in the valence band have higher energies than those at the top of the band. This may look strange on diagrams, but makes good sense because it costs more energy to remove an electron from a state deep within the valence band than from the top.

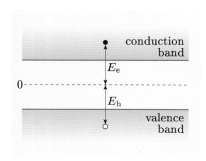

Figure 7.14 The energies E_e and E_h of electron and hole states relative to an energy zero in the middle of the band gap.

It is possible for an electron in the conduction band to drop back into an empty state in the valence band with the emission of light. When this happens, the energy of the emitted photon is the sum of the energies of the electron and the hole. Also, just as an electron and a positron can form positronium, so an electron and a hole in a semiconductor can form a bound hydrogen-like system called an **exciton**. For reasons that will emerge shortly, excitons are very weakly bound, and they are observed only at low temperatures.

Doped semiconductors

The number density of electrons in the conduction band of pure silicon is too small for practical semiconductor devices. Fortunately, it is possible to add impurities to silicon that have the effect of increasing the number of electrons in the conduction band or increasing the number of holes in the valence band. This process is called **doping**. Because germanium has a smaller band gap than silicon, it has a greater density of thermally-excited conduction electrons and can be used in essentially pure form. Nevertheless, it is common to dope germanium as well, as this gives greater control over its electrical properties.

To understand how doping works, it is important to note that semiconductors such as silicon or germanium are in Group 14 of the Periodic Table, and this means that they have four valence electrons which completely fill the valence band. What happens if we add an impurity such as arsenic, which has five valence electrons, to a silicon crystal? The arsenic atom substitutes for one of the silicon atoms, but it has an extra electron which goes into the conduction band of the crystal. Thus, by adding arsenic atoms, we can boost the number of electrons in the conduction band. With sufficient arsenic impurities, the electrical behaviour of the doped silicon is dominated by the electrons in the conduction band; the thermally-excited holes in the valence band play a negligible role. Because its charge carriers are negatively-charged, the material is then said to be an **n-type semiconductor**. Atoms, such as arsenic, which donate an electron to the conduction band are called **donor atoms**.

By contrast, an impurity atom (such as boron) with three valence electrons has a missing electron that can be drawn from the valence band of the silicon crystal, leaving a hole behind. Atoms, such as boron, which accept an electron from the valence band are called **acceptor atoms**. With a sufficient density of acceptor atoms, the electrical behaviour of the doped silicon is dominated by holes in the valence band, and the material is called a **p-type semiconductor**.

Even diamond can be a semiconductor if it contains enough impurities. Naturally-occurring blue diamonds are p-type semiconductors because they contain boron.

For a more quantitative understanding, we can picture a donor atom as a single outer electron attracted to an atomic core. Because of shielding by inner electrons, the outer electron sees an effective nuclear charge that is close to $+e$, so the system can be modelled as a hydrogen-like atom. The escape of the outer electron into the conduction band of the semiconductor can then be regarded as a process of ionization.

Ignoring small effects associated with reduced mass, the energy levels of an isolated hydrogen atom are given by

$$E_n = -\left(\frac{e^2}{4\pi\varepsilon_0}\right)^2 \frac{m_e}{2\hbar^2} \frac{1}{n^2} = -\frac{13.6\,\text{eV}}{n^2}.$$

We ignore the small difference between the reduced mass μ and the electron mass m_e.

Two adjustments are needed to obtain the energy levels for a donor atom in a semiconductor. Firstly, we must replace the free electron mass m_e by the *effective mass* m_e^* of conduction electrons in the semiconductor. Secondly, we must also allow for the fact that the Coulomb attraction felt by charges immersed in a medium is smaller than it would be in free space; this is accounted for by replacing ε_0 by $\varepsilon\varepsilon_0$, where ε is a constant known as the *relative permittivity* of the medium, and has a value greater than 1. For germanium, $m_e^* = 0.12 m_e$ and $\varepsilon = 15.8$, so the energy levels for a donor atom in germanium are

$$E_n^{\text{Ge}} = \frac{0.12}{15.8^2} \times -\frac{13.6\,\text{eV}}{n^2} = -\frac{6.5\,\text{meV}}{n^2}.$$

In this context, ionization corresponds to the release of an electron into the conduction band, so donor energy levels lie just below the bottom of the conduction band (Figure 7.15a). A similar description applies to acceptor atoms. An acceptor atomic core, surrounded by the same number of electrons as in silicon, but with one less proton, is negatively charged and can weakly bind with a hole in the valence band. Because they refer to the energies of holes, acceptor energy levels lie just above the top of the valence band (Figure 7.15b).

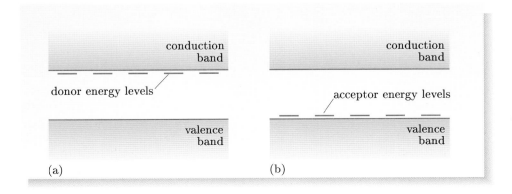

Figure 7.15 (a) Donor atoms produce electron energy levels just below the bottom of the conduction band. (b) Acceptor atoms produce hole energy levels just above the top of the valence band.

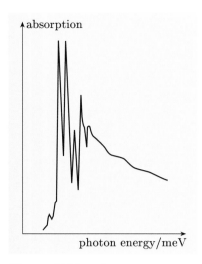

Figure 7.16 A low-temperature absorption spectrum of germanium doped with antimony donor atoms. Peaks correspond to transitions between bound states associated with the donor atoms.

Because the ionization energies of donors and acceptors are so small, thermal energy causes nearly all of them to release their electrons and holes at room temperature. As the temperature is reduced, however, the electrons and holes start to be recaptured, and the conductivity of a doped semiconductor starts to fall. At sufficiently low temperatures, it is possible to get spectroscopic evidence for the different energy levels of electrons or holes captured by donors or acceptors (Figure 7.16).

Exercise 7.11 Calculate the scaled Bohr radius for an electron bound to a donor impurity in germanium.

Exercise 7.12 Assuming that a hole in the valence band of germanium has the same effective mass as an electron in the conduction band, $0.12m_e$, calculate the binding energy of an exciton (see page 190) in germanium. ■

Doped semiconducting materials play a central role in the electronics and communications industries. Nowadays, it is possible to deposit atoms layer by layer, so that a solid is built up according to a predetermined recipe. This means that we can prepare new materials, which do not occur naturally, but which are tailored to have specific properties.

In a **quantum dot**, for example, a material with a narrow band gap is surrounded by a material with a wider band gap. In Book 1, Chapter 3, we described quantum dots in terms of their confining effect on electrons. A more complete picture is shown in Figure 7.17. Because of the very small size of the dot, there are discrete electron states above the band gap, and discrete hole states below the band gap. Electrons and holes in these states are confined to the region of the quantum dot because no states are available to them in the wide band gap of the surrounding material. And, because the quantum dot is tiny, the energy levels are clearly separated. Light is emitted when electrons above the band gap make radiative transitions into hole states below the band gap.

The discrete energy levels in a single quantum dot may broaden into energy bands if a material contains a high density of quantum dots, and the energy eigenfunctions in one quantum dot overlap with those from neighbouring quantum dots.

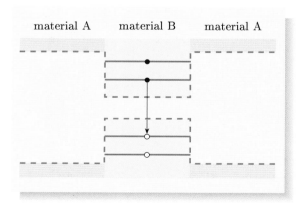

Figure 7.17 An energy-level diagram of a quantum dot. The dashed red lines show the positions of the top of the valence band and the bottom of the conduction band in the bulk materials.

Exercise 7.13 Some materials, including graphite and tin, are classified as *semimetals*. Such materials have valence and conduction bands that overlap very slightly, giving a tiny number of electrons in the conduction band, and leading to conductivities intermediate between good conductors and insulators. How would you expect semimetals to differ from semiconductors? ∎

Summary of Chapter 7

Section 7.1 A crystal is a solid structure in which a particular arrangement of atoms is repeated over a regular lattice. In ionic bonding, one type of atom loses electrons and another type of atom gains them; the resulting positive and negative ions bind together through electrostatic forces. In covalent bonding, valence electrons occupy bonding orbitals between nearest neighbours, with a high electron probability density between atoms. In metallic bonding, atoms beyond nearest neighbours contribute to bonding and the valence electrons spread out over the whole crystal.

The LCAO method correctly predicts that electron states are organized in energy bands, but ignores the requirement that the properties of a crystal should remain unchanged by any translation through a lattice vector.

Section 7.2 Bloch's theorem states that the energy eigenfunctions of electrons in an infinite crystal take the Bloch wave form

$$\psi_{\mathbf{k}}(\mathbf{r}) = e^{i\mathbf{k}\cdot\mathbf{r}} u_{\mathbf{k}}(\mathbf{r}),$$

where $\mathbf{k}$ is the wave vector of the Bloch wave, and $u_{\mathbf{k}}(\mathbf{r})$ is a function with the periodicity of the lattice (i.e. $u_{\mathbf{k}}(\mathbf{r} + \mathbf{R}) = u_{\mathbf{k}}(\mathbf{r})$ for any lattice vector $\mathbf{R}$). The values of $\mathbf{k}$ are restricted to a discrete set by periodic boundary conditions, and are further restricted to ensure that each Bloch wave has a unique wave vector.

The tight-binding method takes the electron energy eigenfunctions to be

$$\psi_{\mathbf{k}}(\mathbf{r}) = A_{\mathbf{k}} \sum_i e^{i\mathbf{k}\cdot\mathbf{R}_i} \phi(\mathbf{r} - \mathbf{R}_i),$$

where $\phi(\mathbf{r} - \mathbf{R}_i)$ is an atomic orbital of a given type, centred on lattice site $\mathbf{R}_i$, and the sum is over all the lattice sites in the crystal. These eigenfunctions are consistent with Bloch's theorem. Interactions between valence electrons are ignored, and terms like $\langle \phi_i | \widehat{H} | \phi_j \rangle$ are dropped, where i and j refer to different, non-adjacent sites. The method predicts that the energy levels are practically continuous within finite energy bands, and that these bands are separated by band gaps. The width of an energy band increases as atoms are brought closer together, so some energy bands may overlap. At absolute zero, all the energy levels up to the Fermi energy are occupied, while higher energy levels are empty. At room temperature, only a few states around the Fermi energy have their occupations modified from those at absolute zero.

Section 7.3 A completely full energy band cannot conduct electricity. At absolute zero, a material with a partly-full band is a conductor, while a material whose bands are all completely full or completely empty is an insulator. The electrical conductivity of a metal is proportional to the number density of electrons in the partly-full band. It is reduced by imperfections in the crystal associated with vacant lattice sites, foreign atoms or boundaries between crystalline grains; it is also reduced by vibrations of the atomic cores about their equilibrium sites, and therefore decreases with increasing temperature.

At a non-zero temperature T, a pure material with a band gap that is less than $80kT$ is classified as a semiconductor; conduction occurs because electrons in the valence band are thermally excited into the conduction band. Each excited electron leaves behind a positive hole, and conduction takes place through flows of electrons in the conduction band and holes in the valence band. The electrical conductivity increases with temperature as the number of charge carriers increases.

Semiconductors are often doped with specific impurity atoms. Donor atoms such as arsenic have an extra valence electron which can be donated to the conduction band. Acceptor atoms such as boron lack a valence electron which can be drawn from the valence band, leaving a hole behind. With sufficient concentrations of donor or acceptor atoms, it is possible to produce an n-type semiconductor in which conduction is dominated by electrons, or a p-type semiconductor in which conduction is dominated by holes.

Achievements from Chapter 7

After studying this chapter, you should be able to:

7.1 Explain the meanings of the newly defined (emboldened) terms and symbols, and use them appropriately.

7.2 Characterize the type of translational symmetry found in crystalline solids.

7.3 Distinguish between ionic, covalent and metallic bonding.

7.4 Describe how the LCAO method is extended to crystalline solids, and explain why this method does not give a correct description of the energy eigenfunctions of electrons.

7.5 State Bloch's theorem and apply it in simple cases.

7.6 Describe the principles of the tight-binding method, and show that it is consistent with Bloch's theorem.

7.7 Describe how the width of an energy band varies with interatomic spacing.

7.8 Relate the differences between conductors, insulators and semiconductors to band structures and the way electrons fill energy levels.

7.9 Describe the principal causes of electrical resistance in metals.

7.10 Describe how both pure and doped semiconductors conduct electricity, and calculate binding energies and scaled Bohr radii for excitons and electrons or holes bound to donor or acceptor atoms.

Chapter 8 Light and matter

Introduction

To complete this book, and the whole course, we shall consider how matter interacts with light. The simple idea that atoms exist in discrete energy levels, and that they can jump between these levels by absorbing or emitting photons, has been with us since the beginning of Book 1. Absorption and emission spectra carry 'fingerprint' signatures that identify individual atoms and molecules even in the depths of space. Indeed, the bulk of our knowledge of the Universe, beyond our own planet, has been obtained by analyzing the spectra of radiation collected by telescopes of one kind or another.

In this chapter, light is taken to mean electromagnetic radiation of any wavelength.

The idea of atoms emitting light when making transitions between discrete energy levels was deeply mysterious when Niels Bohr first proposed it, and the full explanation had to await a deeper understanding of the way light interacts with matter. This understanding, once achieved, explained many things, such as the fact that some spectral lines are brighter than others, and the fact that transitions between some pairs of energy levels appear to be forbidden. It also led to the invention of the laser, now ubiquitous wherever CDs and DVDs are to be found.

A complete account of the interaction of matter with light is based on a quantum theory of the electromagnetic field. Dirac initiated this theory in 1927, about a year and a half after Heisenberg's invention of quantum mechanics. Many people contributed to it over the next 20 years, until it was completed by Feynman, Schwinger and others. This theory is called *quantum electrodynamics* (*QED*) and is well beyond the scope of this course. Fortunately, very accurate predictions can be made by treating atoms according to quantum physics, and the electromagnetic field *mainly* according to classical physics, and that is what we shall do here.

The letters of 'laser' stand for light amplification by the stimulated emission of radiation.

When an atom is exposed to light, its Hamiltonian operator is no longer that for an isolated atom, but includes terms that describe the energy of interaction between the atom and the electric and magnetic fields in the light, which we treat classically. Because the electric and magnetic fields oscillate as the light propagates, the new interaction terms are time-dependent. Therefore, for the first time in this course, we shall consider a Hamiltonian operator that depends on time. In general, the interaction term coupling the atom to the light is small and can be treated as a perturbation — albeit one that depends on time. In order to cope with this situation, we shall develop a useful approximation method called *time-dependent perturbation theory*.

The chapter is organized as follows. Section 8.1 introduces the three different processes that arise in the interaction between electromagnetic radiation and matter; *absorption*, *stimulated emission* and *spontaneous emission*. It then introduces the *electric dipole approximation*, which provides a good description of the interaction between matter and radiation in many situations. Section 8.2 introduces time-dependent perturbation theory, which is the key tool needed to analyze the interaction between light and matter. This allows us to estimate the probability that an atom, initially in state ψ_i at $t = 0$, will absorb light of a given frequency and be found in some other state ψ_f at time $t > 0$. Section 8.3 discusses cases where the probability of making such a radiative transition is estimated to be zero. This leads to *selection rules*, which indicate which pairs of quantum states produce spectral lines that are readily observed. Section 8.4 discusses the absorption and stimulated emission of light by atoms, both for light of a single frequency and for the more usual situation where atoms are illuminated by light covering a range of different frequencies. Finally, Section 8.5 revisits the spontaneous emission of light by excited atoms. The full quantum electrodynamic theory is side-stepped using a beautiful argument due to Einstein; this allows us to relate the rate of spontaneous emission of light to the rate of absorption calculated in Section 8.4.

8.1 The interaction of light with matter

Electromagnetic radiation interacts with matter in its many different forms — solids, plasmas, molecules, atomic nuclei, and so on. In this chapter, we focus on the interaction of light with the electrons in atoms; however, the methods we use, and the conclusions we shall reach, apply far more generally. We shall sometimes indicate this by referring to other systems, such as vibrating or rotating molecules.

8.1.1 Three types of radiative transition

In previous chapters you studied the hydrogen atom, and other atoms, and have seen how the time-independent Schrödinger equation leads to quantized energy levels for the bound electrons. Left to itself, an atom will eventually settle down in the state of lowest energy — the ground state. However, atoms can interact with external electromagnetic fields, and become excited from an initial state, often the ground state, to states of higher energy. In quantum-mechanical terms, we say that an atom *absorbs* a photon. The photon then ceases to exist, but its energy is transferred to the atom as it jumps to a state of higher energy.

If the initial state of the atom had energy E_1, the final state will have energy $E_2 = E_1 + hf$, where f is the frequency associated with the photon. If hf is large enough, the atom will become ionized and have an energy that forms part of a continuum, but at low energies the atom jumps into one of a series of discrete bound states. We only consider bound states here. This process of the **absorption** of light by an atom is illustrated in Figure 8.1a.

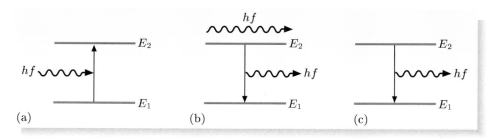

Figure 8.1 Three types of transition in an atom between energy levels E_1 and E_2, with $E_2 > E_1$, involving a photon of energy $hf = E_2 - E_1$: (a) absorption; (b) stimulated emission; (c) spontaneous emission.

An atom in an excited state can jump to a state of lower energy by emitting a photon. The energy hf of the emitted photon is equal to the energy difference between the initial and final states of the atom, respecting the law of conservation of energy. It is a remarkable fact that the emission of a photon by an excited atom can be encouraged by shining light on the atom. An atom initially in a state of energy E_2 can be *stimulated* to descend to a state of lower energy E_1 by light whose photons have energy $hf = E_2 - E_1$. When it jumps to the lower energy level, the atom emits a photon of energy $E_2 - E_1$. The emitted photon shares all the characteristics of the incident photon; it has the same energy, state of polarization and direction of propagation, and the emitted light has the same wavelength and phase as the incident light (it is said to be *coherent* with it). In effect, the photons in the incident beam encourage the atom to emit a similar photon that comes to join them. This process is called **stimulated emission**, and is at the heart of the operation of lasers; it is illustrated in Figure 8.1b.

Finally, an isolated atom in an excited state can *spontaneously* emit a photon, making a transition to a state of lower energy even when there is no external light. Exactly the same process occurs when an excited atomic nucleus emits a gamma-ray photon. This process is called **spontaneous emission**, and is illustrated in Figure 8.1c. Spontaneous emission produces the emission spectrum of a gas that has been heated in a flame. This is the process that so perplexed Bohr and his contemporaries: given an isolated atom in an excited stationary state, how and why does it jump to the state of lower energy? The measurable properties of a stationary state do not depend on time, so why should the atom, free from external influences, suddenly decide to change from one stationary state to another?

The key to answering this question is found by examining the phrase 'free from external influences'. Quantum electrodynamics (QED) tells us that an isolated atom in a vacuum is never truly free from external influences. The vacuum is not empty but is a seething soup of 'virtual particles', which appear from nowhere and then disappear again over very short timescales. These fleeting virtual particles, in effect, shake atoms out of excited stationary states. We have no need to consider the details here because of a discovery made by Einstein in 1916, before the development of quantum mechanics. He showed that the rate of spontaneous emission of light can be related to the rate of absorption, or to the rate of stimulated emission, in the presence of an external light beam. We can therefore use the rates of the processes illustrated in Figures 8.1a or 8.1b to predict the rate of the process in Figure 8.1c. Moreover, the absorption and stimulated emission rates can be predicted by treating the external light as a classical time-dependent field, neglecting the quantum fluctuations in the light.

The three processes in Figure 8.1 are known as **radiative transitions** because they involve the absorption or emission of electromagnetic radiation. There are also *non-radiative* processes that can cause atoms to make energy jumps, for example collisions between atoms or molecules in a gas. The details of non-radiative processes are not the subject of this chapter, but they do provide a source of atoms in excited states that subsequently undergo radiative transitions. For example, passing an electric current through a partially-ionized gas produces excited atoms that spontaneously emit electromagnetic radiation; this is the principle behind fluorescent lighting tubes.

Most of this chapter concerns absorption and stimulated emission; we will come back to spontaneous emission in Section 8.5.

Chapter 8 Light and matter

When an electric current is passed through hydrogen gas in the laboratory, some free hydrogen atoms will be produced in excited states, and they produce a characteristic spontaneous emission spectrum — a fingerprint for the hydrogen atom. The part of this spectrum that lies in the visible range is shown in Figure 8.2. Quantum mechanics is the same *everywhere*, so hydrogen atoms in distant galaxies have precisely the same spectrum, a fact with profound implications for the study of cosmology. By the time the light reaches us from distant galaxies, the spectrum is bodily shifted towards the red, a signal that the Universe is expanding.

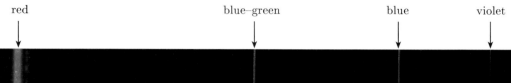

Figure 8.2 Spectral lines emitted by hydrogen atoms in the visible part of the electromagnetic spectrum.

It is not just the positions of spectral lines that we need to explain. Some lines are much stronger than others; for example, the red line in Figure 8.2 is stronger than the violet. Indeed, there are many pairs of energy levels that do not correspond to spectral lines at all. A case in point is the absorption spectrum from the rotational states of HCl, as discussed in Chapter 2 of Book 2 and shown in Figure 8.3.

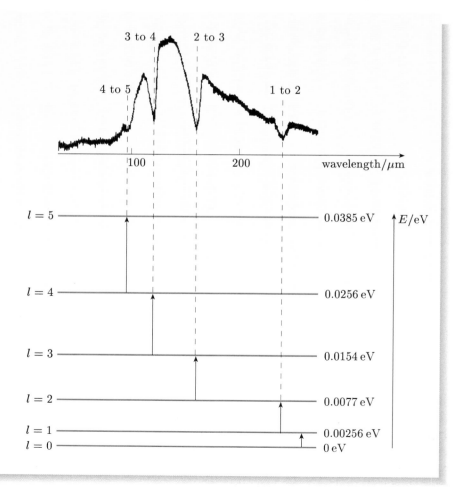

Figure 8.3 The absorption spectrum for the transmission of infrared radiation through HCl gas. Note that only certain radiative transitions are allowed. Here, the quantum number l refers to the orbital angular momentum of the whole molecule. This serves as a reminder that radiative processes are not restricted to electronic states; they also occur between rotational states and vibrational states in molecules.

Notice the absent lines: why, for example, is there no line corresponding to absorption from the state marked $l = 1$ to the state marked $l = 3$? A similar situation arose in Chapter 5 of Book 1, which restricted radiative transitions between the vibrational states of HCl to those between states adjacent to one another in the energy ordering. In order to understand the origin of such *selection rules*, we need to consider more closely how light interacts with matter.

8.1.2 The electric dipole approximation

To take a definite case, we consider an atom that is exposed to a light wave propagating in the y-direction. We shall suppose that the light wave is monochromatic, with frequency f (and angular frequency $\omega = 2\pi f$).

From a classical perspective, light is a transverse electromagnetic wave, with its electric and magnetic fields oscillating perpendicular to the direction of propagation of the wave. We assume that the wave is plane-polarized, so that its electric field vector oscillates in a fixed direction, taken to be the z-direction (Figure 8.4). The magnetic field then oscillates along the x-direction, perpendicular to both the electric field and the direction of propagation. The electric and magnetic fields oscillate in phase with one another.

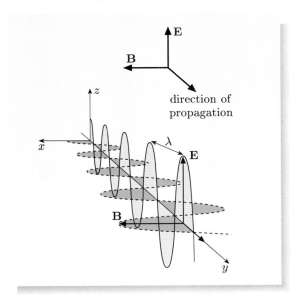

Figure 8.4 A linearly polarized electromagnetic wave propagating in the y-direction. The electric field oscillates in the z-direction.

Electromagnetic radiation is classified into various types according to its frequency f and wavelength λ. In a vacuum, these are related by

$$c = f\lambda,$$

where $c = 3.00 \times 10^8$ m s^{-1} is the speed of light in a vacuum. The entire spectrum, ranging from gamma rays to radio waves, is shown in Figure 8.5. For our purposes, the most important parts are the ultraviolet and visible regions (associated with electronic transitions in atoms and molecules), the infrared region (associated with vibrations in molecules) and the microwave region (associated with rotations in molecules).

Chapter 8 Light and matter

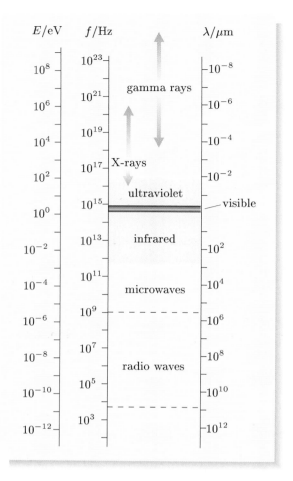

Figure 8.5 The electromagnetic spectrum, from gamma rays to radio waves.

We assume that the electromagnetic wave can be treated classically, as a source of time-dependent electric and magnetic fields that interact with the atom. This is valid provided that the light wave is made up of a vast number of photons.

Exercise 8.1 A light source emits 100 W of electromagnetic radiation at a wavelength of 5.0×10^{-7} m. How many photons does it emit per second? ■

How does an atom interact with an electromagnetic wave? We are treating the electric and magnetic fields of the wave classically, but the states of the atom must be treated quantum-mechanically. To do this, we begin with a classical picture of the interaction, and then translate it into quantum-mechanical terms. This follows the well-trodden path of obtaining a Hamiltonian operator by first writing down the corresponding classical Hamiltonian function.

There are many different effects to consider. Both the electrons and the nucleus in an atom are charged particles, which also behave as tiny magnetic dipoles. So both the electrons and the nucleus respond to the oscillating electric and magnetic field in an electromagnetic wave. However, we shall make some reasonable approximations.

1. We neglect the response of the nucleus. This is reasonable because nuclei have much smaller charge-to-mass ratios than electrons. Their magnetic

dipole moments are also about 1000 times smaller than those of electrons. We shall therefore treat the nucleus of an atom as being fixed and undisturbed by the electromagnetic wave.

2. We ignore the interactions between the electrons and the magnetic field part of the wave. This is reasonable because the magnetic interactions are much weaker than the interactions with the electric field. In a typical case, the rate of transitions caused by the magnetic field is about 10^4 times smaller than the rate of transitions caused by the electric field.

3. We assume that the wavelength of the electromagnetic wave is much larger than the size of an atom. This is a good assumption for the types of radiation we shall consider; visible light, for example, has a wavelength that is about 10^4 times greater than the Bohr radius.

With these assumptions, we need only consider the energy of interaction between the electrons in an atom and the oscillating *electric* field of the electromagnetic radiation. According to Point 3, this field is effectively *uniform* in the vicinity of the atom at each instant.

In a uniform electric field $\mathcal{E}$, an electron of charge $-e$ experiences a force $\mathbf{F} = -e\mathcal{E}$. This force is *minus* the gradient of a potential energy function. If the electron has zero potential energy when placed at the origin, it has potential energy

$$-\mathbf{F} \cdot \mathbf{r} = e\mathcal{E} \cdot \mathbf{r}$$

at position $\mathbf{r}$. So the total potential energy of all the electrons in an atom due to their interaction with the electric field in an electromagnetic wave is

$$V(t) = e \sum_i \mathcal{E}(t) \cdot \mathbf{r}_i, \tag{8.1}$$

where $\mathcal{E}(t)$ is the common electric field experienced by all the electrons in the atom due to the electromagnetic wave, $\mathbf{r}_i$ is the position of the ith electron, and the sum is over all the electrons. We can now state our key approximation, called the **electric dipole approximation**: we assume that the time-dependent potential energy of interaction between the atom and the electromagnetic wave is given by Equation 8.1. With the origin fixed at the nucleus, we call $-e\sum_i \mathbf{r}_i$ the **electric dipole moment** of the atom and $V(t)$ is the electric dipole potential energy.

You might wonder whether the slight distortion of the electron cloud caused by the applied electric field will modify the kinetic energies and electron–nuclear potential energies within the atom. However, such effects turn out to be negligible, giving a very simple result: the total Hamiltonian operator of the atom in the presence of an electromagnetic wave takes the form

$$\widehat{H}(t) = \widehat{H}^{(0)} + V(t), \tag{8.2}$$

where $\widehat{H}^{(0)}$ is the Hamiltonian operator in the absence of external fields, and $V(t)$ is the electric dipole potential energy of the electrons, given by Equation 8.1.

This is the first time in the course that you have seen a time-dependent Hamiltonian operator. Just as in the time-independent case, we can write down Schrödinger's equation

$$i\hbar \frac{\partial \Psi}{\partial t} = \widehat{H}(t)\Psi, \tag{8.3}$$

so the Hamiltonian operator determines the time-development of the wave function. In the time-dependent case, we cannot separate Schrödinger's equation into position-dependent and time-dependent parts, and so we cannot write down a time-independent Schrödinger equation. However, this will not prevent us from using Schrödinger's equation directly to estimate the probability that a light wave induces a radiative transition between two given states.

Equation 8.3 is, of course, harder to solve that the corresponding equation for an isolated atom. We therefore need to make further approximations. We shall assume that the second term in Equation 8.2 is small compared to the first, so that we can treat $\widehat{H}^{(0)}$ as our zeroth-order approximation, and $V(t)$ as a perturbation. We are therefore led to consider the use of perturbation theory.

8.2 Time-dependent perturbation theory

Chapter 3 introduced perturbation theory in the context of the time-independent Schrödinger equation, telling us how much a time-independent perturbation shifts the energy eigenvalues. This was **time-independent perturbation theory**.

We now introduce **time-dependent perturbation theory**, which applies to Schrödinger's (time-dependent!) equation and determines how a time-dependent perturbation influences the wave function of a system. This approximation scheme was devised by Paul Dirac in 1926.

> The following subsection presents first-order time-dependent perturbation theory, and hence provides the tools needed to analyze the effect of a light wave on an atom. The key results are given on page 205 in the box that includes Equation 8.20. You should read the argument that leads up to this result, as this will help you understand its meaning and significance, but you will not be asked to reproduce steps in the derivation.

8.2.1 Solving Schrödinger's equation

We consider a system described by a Hamiltonian operator like that in Equation 8.2:

$$\widehat{H}(t) = \widehat{H}^{(0)} + \widehat{V}(t),$$

where $\widehat{H}^{(0)}$ is the unperturbed Hamiltonian, and $\widehat{V}(t) = V(t)$ is a time-dependent perturbation. Note that the unperturbed Hamiltonian is independent of time, while the perturbation is time-dependent. Following the convention adopted for time-independent perturbation theory in Chapter 3, quantities that do not depend on the perturbation are indicated by a (0) superscript.

The exact form of Schrödinger's equation is

$$i\hbar \frac{\partial \Psi(x,t)}{\partial t} = \widehat{H}(t)\, \Psi(x,t), \tag{8.4}$$

where, for simplicity, we have written the wave function as depending on only one spatial coordinate, x. It will be straightforward to extend our results to more complicated cases later.

If the perturbation were absent, Schrödinger's equation would reduce to

$$i\hbar \frac{\partial \Psi^{(0)}(x,t)}{\partial t} = \widehat{H}^{(0)} \Psi^{(0)}(x,t), \qquad (8.5)$$

which is simpler because the Hamiltonian operator does not depend on time. The stationary-state solutions of this unperturbed Schrödinger equation are

$$\Psi_k^{(0)}(x,t) = \psi_k(x)\,e^{-iE_k t/\hbar}, \qquad (8.6)$$

where E_k is an energy eigenvalue and $\psi_k(x)$ is the corresponding energy eigenfunction, a solution of the time-independent Schrödinger equation

$$\widehat{H}^{(0)} \psi_k(x) = E_k\, \psi_k(x). \qquad (8.7)$$

We assume that the energy eigenfunctions $\psi_k(x)$ form a complete orthonormal set, so that any function can be expressed as a linear combination of them. In particular, the exact wave function of the system can be written as

$$\Psi(x,t) = \sum_k c_k(t)\, \psi_k(x),$$

where the coefficients $c_k(t)$ are functions of time. In fact, it will turn out to be convenient to write these coefficients as

$$c_k(t) = a_k(t)\, e^{-iE_k t/\hbar},$$

so that the exact wave function is expressed as

$$\Psi(x,t) = \sum_k a_k(t)\, \psi_k(x)\, e^{-iE_k t/\hbar}. \qquad (8.8)$$

- How does this wave function differ from all those presented in the course so far?
- The coefficients $a_k(t)$ depend on time.

At each instant t, the wave function $\Psi(x,t)$ in Equation 8.8 is a linear combination of the unperturbed stationary-state wave functions, with coefficients $a_k(t)$. If there were no time-dependent perturbation, these coefficients would be independent of time; their time-dependence arises because the perturbation depends on time. Assuming that the unperturbed energy eigenfunctions and eigenvalues are known, the coefficients $a_k(t)$ *provide the missing information about how the state of the system evolves in time*; they are the functions we need to find.

To obtain an equation for the $a_k(t)$ functions, we substitute Equation 8.8 into both sides of the exact Schrödinger equation (Equation 8.4). The left-hand side of Schrödinger's equation gives

$$\begin{aligned}
i\hbar \frac{\partial \Psi(x,t)}{\partial t} &= i\hbar \frac{\partial}{\partial t} \sum_k a_k(t)\, \psi_k(x)\, e^{-iE_k t/\hbar} \\
&= i\hbar \sum_k \left(\frac{da_k(t)}{dt} - \frac{iE_k}{\hbar} a_k(t) \right) \psi_k(x)\, e^{-iE_k t/\hbar} \\
&= \sum_k \left(i\hbar \frac{da_k(t)}{dt} + E_k\, a_k(t) \right) \psi_k(x)\, e^{-iE_k t/\hbar}. \qquad (8.9)
\end{aligned}$$

Similarly, substituting Equation 8.8 into the right-hand side of Schrödinger's equation gives

$$\widehat{H}\,\Psi(x,t) = \widehat{H}\sum_k a_k(t)\,\psi_k(x)\,e^{-iE_kt/\hbar}$$
$$= \sum_k a_k(t)\left(\widehat{H}^{(0)} + \widehat{V}(t)\right)\psi_k(x)\,e^{-iE_kt/\hbar}$$
$$= \sum_k a_k(t)\left(E_k + \widehat{V}(t)\right)\psi_k(x)\,e^{-iE_kt/\hbar}, \tag{8.10}$$

where the last step follows because $\psi_k(x)$ is an eigenfunction of the unperturbed Hamiltonian (Equation 8.7).

Now we can equate the two sides of Schrödinger's equation. The final expressions in both Equations 8.9 and 8.10 involve identical sums involving $E_k\,a_k(t)$ which cancel out, so we are left with

$$\sum_k i\hbar\frac{da_k(t)}{dt}\psi_k(x)\,e^{-iE_kt/\hbar} = \sum_k a_k(t)\,\widehat{V}(t)\,\psi_k(x)\,e^{-iE_kt/\hbar}. \tag{8.11}$$

The unperturbed energy eigenfunctions are orthonormal:

$$\int_{-\infty}^{\infty}\psi_n^*(x)\,\psi_k(x)\,dx = \delta_{nk},$$

so multiplying both sides of Equation 8.11 by $\psi_n^*(x)$ and integrating over all x gives

$$\sum_k i\hbar\frac{da_k(t)}{dt}\delta_{nk}\,e^{-iE_kt/\hbar} = i\hbar\frac{da_n(t)}{dt}\,e^{-iE_nt/\hbar}$$
$$= \sum_k a_k(t)\,e^{-iE_kt/\hbar}\int_{-\infty}^{\infty}\psi_n^*(x)\,\widehat{V}(t)\,\psi_k(x)\,dx.$$

Using the notation

$$V_{nk}(t) = \langle\psi_n|\,\widehat{V}(t)\,|\psi_k\rangle = \int_{-\infty}^{\infty}\psi_n^*(x)\,\widehat{V}(t)\,\psi_k(x)\,dx, \tag{8.12}$$

and defining $\omega_{nk} = (E_n - E_k)/\hbar$, our final result can be expressed as

$$i\hbar\frac{da_n(t)}{dt} = \sum_k e^{i\omega_{nk}t}\,V_{nk}(t)\,a_k(t), \tag{8.13}$$

where $V_{nk}(t)$ is called the nk-**matrix element of the perturbation**.

Now, let us suppose that the perturbation is introduced at time $t = 0$ and that at earlier times the system is in the ith unperturbed stationary state of energy E_i. Then the initial wave function, just as the perturbation is switched on, is

$$\Psi(x,0) = \psi_i(x)\,e^{-iE_it/\hbar}. \tag{8.14}$$

- Express Equation 8.14 in words.

○ The system at time $t = 0$, just as the perturbation is turned on, is in the ith stationary state of the unperturbed system.

Comparing with Equation 8.8, we obtain the initial condition

$$a_k(0) = \delta_{ki}. \tag{8.15}$$

Solving Equation 8.13 subject to this initial condition would give us an exact solution of the full Schrödinger equation, but would be a difficult task. In perturbation theory, we are looking for approximate solutions starting from the assumption that the perturbation is *small*. With this in mind, we write the coefficients as

$$a_k(t) = \delta_{ki} + a_k^{(1)}(t) + \cdots, \tag{8.16}$$

where $a_k^{(1)}(t)$ is a first-order correction, and the dots signify the small higher-order corrections that could be included in principle. The first term, δ_{ki}, represents the coefficient before the perturbation is switched on (Equation 8.15).

Substituting Equation 8.16 into Equation 8.13, we obtain

$$i\hbar \frac{da_n^{(1)}(t)}{dt} = \sum_k e^{i\omega_{nk}t} V_{nk}(t)\,\delta_{ki} + \sum_k e^{i\omega_{nk}t} V_{nk}(t)\, a_k^{(1)}(t) + \cdots. \tag{8.17}$$

The second sum on the right-hand side involves one small quantity, $V_{nk}(t)$, multiplied by another, $a_k^{(1)}(t)$: small × small = very small. To obtain a consistent 'first-order' approximation, we drop this sum. Noting that the delta function kills off all terms in the remaining sum except for the term with $k = i$, we obtain

$$i\hbar \frac{da_n^{(1)}(t)}{dt} = e^{i\omega_{ni}t} V_{ni}(t), \tag{8.18}$$

which can be integrated to give

$$a_n^{(1)}(t) - a_n^{(1)}(0) = \frac{1}{i\hbar} \int_0^t e^{i\omega_{ni}t'} V_{ni}(t')\,dt'. \tag{8.19}$$

Here, we have put a prime on the dummy variable of integration to distinguish it from the time t at which we want to evaluate the coefficients, which appears in the upper limit of integration. The lower limit of integration is the time when the perturbation is switched on. We can now collect everything together.

First-order approximation for the wave function

If the system is initially in the ith unperturbed stationary state, and a small perturbation $V(t)$ is switched on at time $t = 0$, then the wave function of the system at a later time t is

$$\Psi(x,t) = \sum_k a_k(t)\,\psi_k(x)\,e^{-iE_k t/\hbar}, \tag{Eqn 8.8}$$

where, in our first-order approximation,

$$a_k(t) \simeq \delta_{ki} + \frac{1}{i\hbar} \int_0^t e^{i\omega_{ki}t'} V_{ki}(t')\,dt', \tag{8.20}$$

Do not confuse the index i of the initial state with $i = \sqrt{-1}$.

with $V_{ki}(t') = \langle \psi_k | \widehat{V}(t') | \psi_i \rangle$. This approximation is valid so long as the first-order corrections $a_k^{(1)}(t)$ all remain small (that is, $\ll 1$).

Exercise 8.2 Suppose that $V_{ki}(t) \neq 0$ in general, but $V_{ji}(t) = 0$ for a specific matrix element with $j \neq i$. Will $a_j^{(1)}(t)$ become non-zero according to first-order perturbation theory? ∎

8.2.2 The probability of making a transition

Time-dependent perturbation theory tells us how the wave function of a system responds to a time-dependent perturbation. Using this theory, we can estimate the probability that an atom exposed to a light wave will make a radiative transition to a state of higher energy (absorption) or to a state of lower energy (stimulated emission).

Strictly speaking, the presence of a time-dependent perturbation prevents us from defining precise energy levels for an atom. The mathematical reason is that the perturbation renders the Schrödinger equation non-separable, so we are unable to derive a time-independent Schrödinger equation. In more physical terms, the perturbation makes it impossible to distinguish the energy of the atom from the energy of the light wave that it interacts with. If the perturbation is very small, we may *approximate* the possible energies of the atom by the eigenvalues of the unperturbed Hamiltonian. More generally, we will suppose that the light source is switched off at time t, and that the energy of the atom is measured immediately afterwards. Although the atom is then unperturbed, its history of being exposed to light in the time interval between 0 and t has caused its wave function to evolve into the linear combination of stationary states given by Equations 8.8 and 8.20.

We use the symbol $P_{i \to f}(t)$ to denote the probability that an atom, initially in state i at $t = 0$, will be found to be in the state f at time t, immediately after the perturbation has been switched off. This probability is given by the general overlap rule of quantum mechanics. Applying this rule to the wave function $\Psi(x, t)$ obtained from first-order time-dependent perturbation theory, we see that

$$P_{i \to f}(t) = |\langle \psi_f | \Psi \rangle|^2 = |a_f(t)|^2. \tag{8.21}$$

Combining this with Equation 8.20, we reach the following conclusion:

> **First-order probability of a transition**
>
> If a system is initially in state i at time $t = 0$, the probability that it will be found to be in state $f \neq i$ at time t is
>
> $$P_{i \to f}(t) = \frac{1}{\hbar^2} \left| \int_0^t e^{i\omega_{fi} t'} V_{fi}(t') \, dt' \right|^2, \tag{8.22}$$
>
> where $V(t)$ is the time-dependent perturbation that drives the transition, $V_{fi}(t') = \langle \psi_f | \widehat{V}(t') | \psi_i \rangle$ and $\omega_{fi} = (E_f - E_i)/\hbar$.

Exercise 8.3 A system is initially in the state i. A perturbation is switched on at $t = 0$ and has the constant matrix element $V_{fi}(t) = U$ until time t, when it is switched off. Show that the probability of the system being in state f at time t is

$$P_{i \to f}(t) = \frac{U^2}{\hbar^2} \left[\frac{\sin(\omega_{fi} t/2)}{(\omega_{fi}/2)} \right]^2.$$

Hint: You may assume that $P_{i\to f} \ll 1$. It will be useful in this exercise, and also later in the chapter, to note that

$$e^{iat} - 1 = e^{iat/2}(e^{iat/2} - e^{-iat/2}) = 2ie^{iat/2}\sin(at/2). \tag{8.23}$$

for any a. ∎

8.2.3 Transitions induced by monochromatic light

We now turn to the case of major interest: transitions between atomic states that are driven by an electromagnetic wave. In this case, the time-dependent perturbation is given by the electric dipole approximation of Section 8.1.2:

$$V(t) = e\sum_i \boldsymbol{\mathcal{E}}(t) \cdot \mathbf{r}_i. \tag{Eqn 8.1}$$

For simplicity, we consider a case where only one electron matters. This may be because the atom has only one electron, as in hydrogen or He^+, or because it has a single valence electron, as in sodium or potassium. The perturbation then becomes

$$V(t) = e\boldsymbol{\mathcal{E}}(t) \cdot \mathbf{r}, \tag{8.24}$$

where $\mathbf{r}$ is the position vector of the electron.

The electric field $\boldsymbol{\mathcal{E}}(t)$ depends on the nature of the light encountered by the atom. The simplest case is monochromatic linearly polarized light, which corresponds to an electric field vector oscillating sinusoidally at a single frequency in a fixed direction. Taking the direction of polarization to be the z-direction, we then have

$$V(t) = e\,\mathcal{E}_0\cos(\omega t)\,z. \tag{8.25}$$

In this expression, $\mathcal{E}_0$ and ω are the amplitude and angular frequency of the light, and z is the z-component of the electron's position relative to an origin at the nucleus. Since $V(t)$ is a quantum-mechanical operator (it is part of the Hamiltonian operator), we can also write

$$\widehat{V}(t) = e\,\mathcal{E}_0\cos(\omega t)\,\widehat{z}. \tag{8.26}$$

The operator $\widehat{z}$ just tells us to multiply by z, so the hat could be omitted. However, including it makes it clear that $\widehat{V}(t)$ is just some numerical factors times the simple quantum-mechanical operator $\widehat{z}$.

Now, let us see what effect this perturbation has on an atom initially in the state i. Combining Equations 8.22 and 8.26, we see that the **transition probability** from state i to state f is

$$P_{i\to f}(t) = \frac{e^2\mathcal{E}_0^2}{\hbar^2}\left|\langle\psi_f|\widehat{z}|\psi_i\rangle\right|^2 \left|\int_0^t e^{i\omega_{fi}t'}\cos(\omega t')\,\mathrm{d}t'\right|^2. \tag{8.27}$$

We shall now consider separately the consequences of two important terms in this equation. First, there is the matrix element $\langle\psi_f|\widehat{z}|\psi_i\rangle$, which affects the magnitude of the transition probability, but is independent of angular frequency or time. Second, there is the integral, which contains the time t in its upper limit, and therefore tells us how the probability of the atom being in state f depends upon time. We shall consider the matrix element first and then come back to consider the time-dependence.

Chapter 8 Light and matter

8.3 Selection rules for radiative transitions

When we examine the spectra of quantum systems, whether they be atoms, molecules or nuclei, we generally see a complex pattern of spectral lines. Although the number of spectral lines can sometimes be vast, the number of lines observed is much smaller than the number we would expect if transitions between all energy levels were possible. In general, there are many pairs of energy levels for which there are no discernible spectral lines. This is because **selection rules** exist, which restrict radiative transitions to special pairs of quantum states. We can now understand the origin of these rules. Looking back to Equation 8.27, we see that:

> The first-order probability for a transition between states i and f will vanish if the matrix element $\langle \psi_f | \hat{z} | \psi_i \rangle$ is equal to zero. This is true for both absorption and stimulated emission, driven by monochromatic light, linearly polarized in the z-direction.

The following worked example illustrates how this principle works for a transition between two particular states in a hydrogen atom.

Essential skill

Determining whether a transition between two states is allowed by the electric dipole approximation and first-order perturbation theory

The factor $1/\sqrt{4\pi}$ in ψ_{1s} and ψ_{2s} is the Y_{00} spherical harmonic.

Worked Example 8.1

Show that weak, linearly polarized light cannot excite a hydrogen atom in its ψ_{1s} ground state to a ψ_{2s} state.

Solution

With no loss of generality, we can take the light to be polarized in the z-direction, and we then need to calculate the matrix element $\langle \psi_{2s} | \hat{z} | \psi_{1s} \rangle$. The hydrogen atom wave functions are given in Tables 1.1 and 2.1 as

$$\psi_{1s} = \psi_{1,0,0} = 2\left(\frac{1}{a_0}\right)^{3/2} e^{-r/a_0} \times \frac{1}{\sqrt{4\pi}},$$

$$\psi_{2s} = \psi_{2,0,0} = 2\left(\frac{1}{2a_0}\right)^{3/2} \left(1 - \frac{r}{2a_0}\right) e^{-r/2a_0} \times \frac{1}{\sqrt{4\pi}}.$$

Using spherical coordinates to write $z = r\cos\theta$, the required matrix element is

$$\langle \psi_{2s} | \hat{z} | \psi_{1s} \rangle = \frac{1}{\pi} \left(\frac{1}{2a_0^2}\right)^{3/2} I,$$

where

$$I = \int_0^{2\pi} \int_0^{\pi} \int_0^{\infty} \left(1 - \frac{r}{2a_0}\right) e^{-r/2a_0} \, r\cos\theta \, e^{-r/a_0} \, r^2 \sin\theta \, dr \, d\theta \, d\phi.$$

Separating the integrals over different variables, we get

$$I = \int_0^{\infty} r^3 \left(1 - \frac{r}{2a_0}\right) e^{-r/2a_0} e^{-r/a_0} \, dr \int_0^{\pi} \cos\theta \sin\theta \, d\theta \int_0^{2\pi} d\phi.$$

> If any one of these integrals is zero, the transition will not be allowed. Let's look at the integral over θ. Using a trigonometric identity, we have
>
> $$\int_0^\pi \cos\theta \sin\theta \, d\theta = \frac{1}{2}\int_0^\pi \sin(2\theta) \, d\theta = -\frac{1}{4}\Big[\cos(2\theta)\Big]_0^\pi = 0.$$
>
> It therefore follows that the matrix element $\langle \psi_{2s}|\widehat{z}|\psi_{1s}\rangle = 0$. This means that the transition will not take place, at least within the approximations assumed here (the electric dipole approximation and first-order perturbation theory).

Remember:
$\sin(2\theta) = 2\sin\theta\cos\theta$.

Often, when integrals vanish, there are shortcuts that involve the use of symmetry arguments. This is the case here. The 1s and 2s hydrogen atom eigenfunctions depend only on $r = \sqrt{x^2 + y^2 + z^2}$, so are unchanged by the operation $z \to -z$, which reverses the sign of z. However, the perturbation is proportional to z and does change sign under this operation. The integral over all space of three functions, two of which do not change sign when $z \to -z$ and one of which (z itself) does change sign, must be zero and hence $\langle \psi_{2s}|\widehat{z}|\psi_{1s}\rangle = 0$.

The above calculation was made using an approximation that includes only the effects that are normally dominant. We found that a particular transition does not take place within this framework. This means that the transition is either totally forbidden, or is very weak, driven by effects we have neglected. For example, it is possible for transitions to occur via the magnetic field in light or via higher-order effects in perturbation theory. When this happens, the associated spectral lines are generally very weak (perhaps 10 000 times weaker than normal lines) and may pass unnoticed. To avoid distracting caveats, we shall say that a radiative transition is **forbidden** if it is not allowed by the electric dipole approximation and first-order perturbation theory. This does not imply that it is completely impossible.

The calculation in Worked Example 8.1 can be generalized to deal with radiative transitions between any pair of energy states in a one-electron atom. We know that the general form of such a state is

$$\psi_{nlm}(r,\theta,\phi) = R_{nl}(r)\, Y_{lm}(\theta,\phi).$$

So, for a transition from an initial state

$$\psi_i = \psi_{n_i,l_i,m_i}(r,\theta,\phi)$$

to a final state

$$\psi_f = \psi_{n_f,l_f,m_f}(r,\theta,\phi),$$

the appropriate matrix element of $z = r\cos\theta$ is

$$\langle \psi_f|\widehat{z}|\psi_i\rangle = \int_0^\infty R^*_{n_f,l_f}(r)\, R_{n_i,l_i}(r)\, r^3\, dr$$
$$\times \int_0^{2\pi}\int_0^\pi Y^*_{l_f,m_f}(\theta,\phi)\, \cos\theta\, Y_{l_i,m_i}(\theta,\phi)\, \sin\theta\, d\theta\, d\phi.$$

(8.28)

Just as in the worked example, the selection rules arise from the angular integral. Now, it can be shown (though we shall not do so) that spherical harmonics Y_{lm} have the following property:

$$\cos\theta\, Y_{lm}(\theta,\phi) = C_{lm} Y_{l+1,m}(\theta,\phi) + D_{lm} Y_{l-1,m}(\theta,\phi), \quad (8.29)$$

where C_{lm} and D_{lm} depend on the indices l and m, but are not functions of θ and ϕ. Substituting this relationship into the angular integral in Equation 8.28, we get two integrals of the form

$$C_{l_i,m_i} \int_0^{2\pi} \int_0^\pi Y^*_{l_f,m_f}(\theta,\phi)\, Y_{l_i+1,m_i}(\theta,\phi) \sin\theta\, \mathrm{d}\theta\, \mathrm{d}\phi$$

and

$$D_{l_i,m_i} \int_0^{2\pi} \int_0^\pi Y^*_{l_f,m_f}(\theta,\phi)\, Y_{l_i-1,m_i}(\theta,\phi) \sin\theta\, \mathrm{d}\theta\, \mathrm{d}\phi.$$

However, we know that different spherical harmonics are orthogonal to one another, so it follows that the angular integral in Equation 8.28 will vanish unless the following conditions are met:

$$l_f = l_i \pm 1 \quad \text{and} \quad m_f = m_i. \quad (8.30)$$

These are the selection rules for light that is linearly polarized in the z-direction (and propagates perpendicular to the z-direction). It turns out that the rule $l_f = l_i \pm 1$ is completely general, and applies for all directions of polarization, but the rule $m_f = m_i$ applies specifically to linear polarization in the z-direction.

Exercise 8.4 Derive the selection rule for the magnetic quantum number m in Equation 8.30 without using Equation 8.29, but using the fact that $Y_{lm}(\theta,\phi)$ is a function of θ times $\mathrm{e}^{\mathrm{i}m\phi}$. ■

What happens if the light is linearly polarized in some other direction? To take a definite case, let us suppose it is linearly polarized in the x-direction. In this case, the relevant matrix element is

$$\langle\psi_f|\widehat{\mathrm{x}}|\psi_i\rangle = \langle\psi_f|r\sin\theta\cos\phi|\psi_i\rangle$$
$$= \frac{1}{2}\left[\langle\psi_f|r\sin\theta\, \mathrm{e}^{\mathrm{i}\phi}|\psi_i\rangle + \langle\psi_f|r\sin\theta\, \mathrm{e}^{-\mathrm{i}\phi}|\psi_i\rangle\right]. \quad (8.31)$$

Remember:
$\mathrm{e}^{\mathrm{i}\phi} = \cos\phi + \mathrm{i}\sin\phi.$

A detailed calculation, similar to that given for linear polarization in the z-direction, then leads to the selection rules:

$$l_f = l_i \pm 1 \quad \text{and} \quad m_f = m_i \pm 1, \quad (8.32)$$

and of course, the same results apply for linear polarization along the y-direction.

Exercise 8.5 Use Equation 8.31 to prove the selection rule for the magnetic quantum number m for light that is linearly polarized in the x-direction. ■

You may recall from Chapter 6 of Book 2 that circularly polarized photons can be regarded as being in a superposition of two linearly polarized states with a relative phase factor of $\pi/2$. From a classical perspective, circularly polarized light propagating in the z-direction is a superposition of waves that are linearly polarized in the x and y-directions. It therefore obeys the selection rules of Equation 8.32, rather than Equation 8.30.

We can understand this result as follows. A circularly polarized photon propagating in the z-direction carries a z-component of spin angular momentum that is equal to $+\hbar$ for left-handed circular polarization, and is equal to $-\hbar$ for right-handed circular polarization. Hence, when an atom absorbs a circularly polarized photon travelling along the z-axis, the law of conservation of angular momentum requires the z-component of the angular momentum of the atom to change by $\pm\hbar$. This is achieved by an electronic transition in which m changes by ± 1. The magnetic quantum number m may increase or decrease, depending on the handedness of photon, and whether it propagates in the positive or negative z-direction.

We can now summarize the selection rules that apply to one-electron atoms for light of any polarization or direction of propagation. These selection rules assume that the electric dipole approximation and first-order perturbation theory are both valid.

Selection rules for a one-electron atom

$$l_f = l_i \pm 1 \quad \text{and} \quad m_f = m_i - 1, m_i \text{ or } m_i + 1. \tag{8.33}$$

The 'or' is inclusive; for example circularly polarized light propagating in the x-direction produces all three of the possible changes in m.

Our discussion has centred so far on absorption and stimulated emission, but we shall see that these rules also apply to spontaneous emission spectra.

Exercise 8.6 (a) Would you expect linearly polarized light to excite a helium atom from a singlet state to a triplet state?

(b) Given your answer to part (a), how might triplet states be produced (starting from a helium atom in its ground state)?

In connection with Exercise 8.6 we note that:

1. It is a striking experimental fact that the spectrum of helium appears to contains two apparently independent sets of spectral lines: one associated with transitions between singlet states, and another associated with transitions between triplet states. There are no spectral lines corresponding to transitions between singlet states and triplet states. This fact provided many important clues about the existence of spin and the properties of identical particles in the early days of quantum mechanics.

2. The helium spectrum normally seen in a laboratory experiment arises from the spontaneous emission of photons from helium atoms that have been excited by an electrical discharge; the selection rules for spontaneous emission are the same as for absorption and stimulated emission.

8.4 Absorption and stimulated emission

Let's return to our expression for the probability of a transition from state i to state f in the presence of monochromatic light of angular frequency ω, linearly polarized in the z-direction:

$$P_{i \to f}(t) = \frac{e^2 \mathcal{E}_0^2}{\hbar^2} |\langle \psi_f | \widehat{z} | \psi_i \rangle|^2 \left| \int_0^t e^{i\omega_{fi} t'} \cos(\omega t') \, dt' \right|^2. \tag{Eqn 8.27}$$

This expression is based on the electric dipole approximation and first-order time-dependent perturbation theory. So far, we have looked at the matrix element $\langle\psi_f|\widehat{z}|\psi_i\rangle$ and have seen that this can vanish, leading to selection rules. Now we shall look more closely at the rest of the expression, especially the integral, which will tell us how the transition probability depends on angular frequency and on time.

We can evaluate the integral by noting that

$$e^{i\alpha t}\cos(\beta t) = e^{i\alpha t}\left(\frac{e^{i\beta t}+e^{-i\beta t}}{2}\right) = \frac{e^{i(\alpha+\beta)t}+e^{i(\alpha-\beta)t}}{2}.$$

Hence

$$\int_0^t e^{i\omega_{fi}t'}\cos(\omega t')\,\mathrm{d}t' = \frac{1}{2}\int_0^t \left(e^{i(\omega_{fi}+\omega)t'} + e^{i(\omega_{fi}-\omega)t'}\right)\mathrm{d}t'$$

$$= \frac{1}{2i}\left[\frac{e^{i(\omega_{fi}+\omega)t'}}{\omega_{fi}+\omega} + \frac{e^{i(\omega_{fi}-\omega)t'}}{\omega_{fi}-\omega}\right]_0^t$$

$$= \frac{1}{2i}\left[\frac{e^{i(\omega_{fi}+\omega)t}-1}{\omega_{fi}+\omega} + \frac{e^{i(\omega_{fi}-\omega)t}-1}{\omega_{fi}-\omega}\right].$$

Writing $z_{fi} = \langle\psi_f|\widehat{z}|\psi_i\rangle$, we conclude that the transition probability is

$$P_{i\to f}(t) = \frac{e^2\mathcal{E}_0^2|z_{fi}|^2}{4\hbar^2}\left|\frac{e^{i(\omega_{fi}+\omega)t}-1}{\omega_{fi}+\omega} + \frac{e^{i(\omega_{fi}-\omega)t}-1}{\omega_{fi}-\omega}\right|^2. \quad (8.34)$$

This formula shows that the transition probability is a function of the angular frequency ω of the light, and of the time t. This is not surprising: we know that we have to choose the frequency of the light very carefully to induce a particular transition, and we would expect the probability of finding an atom or molecule in a given final state to depend on how long it has been exposed to the light.

Let us fix the time t and see how the transition probability depends on the angular frequency ω of the light (which we take to be a positive quantity: $\omega > 0$). Large effects may be observed when one of the denominators inside the modulus in Equation 8.34 becomes very small. There are two cases to consider, illustrated in Figure 8.6.

1. **Absorption**   If the energy of the final state is greater than that of the initial state, we have $E_f > E_i$ and $\omega_{fi} > 0$ (Figure 8.6a). In this case, the second denominator in Equation 8.34 becomes small when $\omega_{fi} - \omega \simeq 0$. The second term inside the modulus sign is then much greater than the first, which can be ignored. Since the energy of the atom is increased, this process corresponds to the *absorption* of light.

2. **Stimulated emission**   If the energy of the final state is less than that of the initial state, we have $E_f < E_i$ and $\omega_{fi} < 0$ (Figure 8.6b). In this case, the first denominator in Equation 8.34 becomes small when $\omega_{fi} + \omega \simeq 0$. The first term inside the modulus sign is then much greater than the second, which can be ignored. Since the energy of the atom is decreased, this process corresponds to *stimulated emission* of light.

We shall focus on the case of absorption in the regime where $\omega \simeq \omega_{fi}$. Under these circumstances, the atom is said to be in *resonance* with the electromagnetic

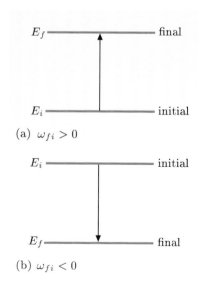

Figure 8.6   Transitions from an initial state i to a final state f: (a) absorption ($E_f > E_i$); (b) stimulated emission ($E_f < E_i$).

wave, and the first term inside the modulus sign of Equation 8.34 can be neglected. Multiplying both the denominator and numerator by t^2, $P_{i \to f}(t)$ then becomes

$$P_{i \to f}(t) = \frac{e^2 \mathcal{E}_0^2 |z_{fi}|^2 t^2}{4\hbar^2} \left| \frac{e^{i(\omega_{fi}-\omega)t} - 1}{(\omega_{fi}-\omega)t} \right|^2. \tag{8.35}$$

This can be tidied up a bit using Equation 8.23:

$$\left| e^{iat} - 1 \right|^2 = \left| 2i\, e^{iat/2} \sin(at/2) \right|^2$$
$$= 4 \left| e^{iat/2} \right|^2 \left| \sin(at/2) \right|^2 = 4\sin^2(at/2).$$

Hence

$$P_{i \to f}(t) = \frac{e^2 \mathcal{E}_0^2 |z_{fi}|^2 t^2}{4\hbar^2} \left[\frac{\sin[(\omega_{fi}-\omega)t/2]}{(\omega_{fi}-\omega)t/2} \right]^2. \tag{8.36}$$

The resulting transition probability is plotted against ω in Figure 8.7a, and Figure 8.7b shows how this function narrows and becomes more peaked around $\omega = \omega_{fi}$ as time progresses. Notice the following:

- The transition probability has a maximum at $\omega = \omega_{fi}$. The peak value of the transition probability increases with time, being proportional to t^2.

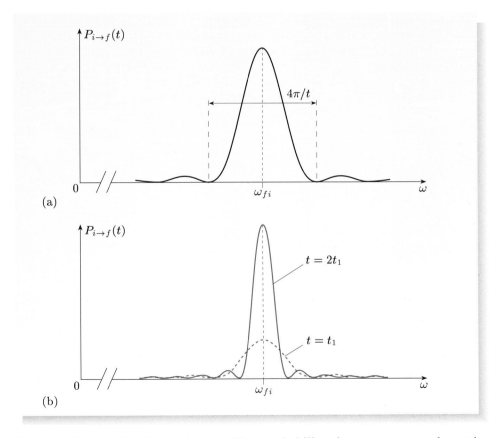

Figure 8.7 (a) The first-order transition probability close to resonant absorption plotted against the angular frequency ω of the radiation. (b) Comparison of $P_{i \to f}$ at two different times, $t = t_1$ and $t = 2t_1$.

- It is not essential for the photon energy $\hbar\omega$ to match the energy difference between the levels $\hbar\omega_{fi}$ exactly. The frequency range of the radiation that contributes significantly to absorption is of order $4\pi/t$, where t is the time since the perturbation has been switched on. This range is inversely proportional to the time.

Exercise 8.7 A two-level system is perturbed by interaction with electromagnetic waves. What range of angular frequencies can drive a transition between the two levels if the wave is switched on for (a) $t = 3 \times 10^{-14}$ s and (b) $t = 3 \times 10^{-3}$ s? ■

Let us fix the angular frequency ω and see how the transition probability depends on time. First, consider the peak value, which occurs when ω is exactly equal to ω_{fi}. At this angular frequency, the function inside the large square brackets of Equation 8.36 takes the limiting value of 1, so the transition probability is proportional to t^2. However, this behaviour is exceptional. More typically, ω is close to, but not quite equal to ω_{fi}. In this case, the transition probability will be zero whenever $\sin[(\omega_{fi} - \omega)t/2] = 0$, i.e. when $(\omega_{fi} - \omega)t/2$ is a multiple of π. As a result, $P_{i \to f}(t)$ oscillates as shown in Figure 8.8. It reaches its maximum value at $t = \pi/|\omega_{fi} - \omega|$, and then starts to decrease, reaching zero at $t = 2\pi/|\omega_{fi} - \omega|$.

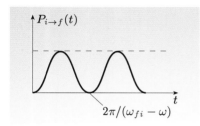

Figure 8.8 The first-order transition probability close to resonant absorption plotted against time at a fixed angular frequency close to, but not equal to, ω_{fi}.

Exercise 8.8 A two-level system with $\omega_{21} = 2.0 \times 10^{10}$ s^{-1} interacts with a monochromatic light wave of angular frequency $\omega = 1.9 \times 10^{10}$ s^{-1}. What is the minimum interaction time needed to maximize the transition from the ground state to the excited state? ■

No doubt this behaviour seems rather strange, but effects like these are actually observed when monochromatic light from a laser source is used. More usually, the light source is not monochromatic, but has a range of different frequencies and something rather different is seen. Ordinary light, such as that emitted by an electric light bulb, covers a range of different wavelengths. To find the transition probability induced by such light, we must sum the expression in Equation 8.36 over all relevant angular frequencies, ω, while keeping ω_{fi} fixed. However, any light source will be more intense at some angular frequencies than others, and this must be taken into account. The details are rather intricate, and are not assessable, but are sketched below for completeness.

1. In classical electromagnetic theory, monochromatic light with an electric field of amplitude $\mathcal{E}_0$ has a time-averaged energy density $U = \tfrac{1}{2}\varepsilon_0 \mathcal{E}_0^2$, so $\mathcal{E}_0^2$ in Equation 8.36 can be expressed as $2U/\varepsilon_0$.

2. To describe non-monochromatic light, we introduce the **spectral energy density function**, $u(\omega)$. This is defined in such a way that $u(\omega)\,d\omega$ is the contribution to the energy density from a small range of angular frequencies, centred on ω, of width $\delta\omega$.

3. Using Equation 8.36, and bearing Points 1 and 2 in mind, the transition probability induced by light with a range of different angular frequencies is written as

$$P_{i \to f}(t) = \frac{2e^2 |z_{fi}|^2}{\varepsilon_0 \hbar^2} \int_0^\infty \left[\frac{\sin\left[(\omega_{fi} - \omega)t/2\right]}{\omega_{fi} - \omega} \right]^2 u(\omega)\,d\omega.$$

Energy density is energy per unit volume.

4. As t becomes large, the part of the integrand that does not involve $u(\omega)$ becomes very strongly-peaked around $\omega = \omega_{fi}$ In this limit, $u(\omega)$ can be replaced by the constant value $u(\omega_{fi})$ and taken outside the integral sign. The remaining integral is not easy, but turns out to be well-approximated by $\pi t/2$, so we have

$$P_{i \to f}(t) = \frac{\pi e^2 |z_{fi}|^2}{\varepsilon_0 \hbar^2} u(\omega_{fi}) \, t.$$

5. Finally, we note that the calculation so far assumes that the light is linearly polarized in the z-direction. Most ordinary light sources emit unpolarized light. Taking the atoms to be bathed in isotropic, unpolarized light, it is appropriate to make the replacement $|z_{fi}|^2 \to (|x_{fi}|^2 + |y_{fi}|^2 + |z_{fi}|^2)/3$. Hence we conclude that

$$P_{i \to f}(t) = \frac{\pi e^2}{3\varepsilon_0 \hbar^2} \big(|x_{fi}|^2 + |y_{fi}|^2 + |z_{fi}|^2\big) u(\omega_{fi}) \, t. \tag{8.37}$$

This is the result we have been seeking for the transition probability between two given states i and f, driven by electromagnetic radiation covering a continuous range of frequencies. The most important point is that the transition probability is proportional to time. This is just what we would expect intuitively, since it implies a constant probability *per unit time* for the transition. As always in quantum mechanics, we cannot say when the transition will occur for an individual atom, but we can talk about its probability per unit time.

Because we started from Equation 8.36, our calculation applies specifically to absorption, but a parallel calculation, carried out for stimulated emission gives exactly the same formula. And, because $|x_{ab}|^2 = |x_{ba}|^2$, etc., the transition probabilities per unit time for absorption and stimulated emission between any pair of states are identical. The reason that light tends to be absorbed rather than emitted, when it is shone on a sample of matter, is simply that most atoms are initially in their ground states, and are unable to jump to states of lower energy. This situation is reversed inside a laser, where materials temporarily have an excess of atoms in excited states (a so-called *population inversion*); stimulated emission then becomes the dominant process and produces intense, coherent laser light.

8.5 Spontaneous emission and the Einstein coefficients

So far we have concentrated on two processes that take place in the presence of external radiation: absorption and stimulated emission. We have said very little about the third type of radiative process, spontaneous emission, apart from mentioning some of the difficulties in dealing with it. But it is, after all, a critically important phenomenon, being the process whereby excited atoms emit photons to yield their familiar emission spectra, such as the Balmer series of hydrogen, or the yellow sodium lines. It is also the process by which an atomic nucleus de-excites by emitting a gamma-ray photon. But in the quantum mechanics of the course so far, a stationary state of an isolated atom is just that: stationary. So what prompts a hydrogen atom in a 3d stationary state to emit a photon of red light as it drops to the 2p stationary state? The key point is that no atom in the universe is truly isolated, because the vacuum itself is active. We had some hints of its activity in Chapter 4 where we saw how, according to quantum electrodynamics (QED), quantum fluctuations of the vacuum induce the Lamb shift in the hydrogen spectrum and slightly modify the electron magnetic dipole moment. The same fluctuations are responsible for prompting an atom in an excited stationary state to emit a photon.

We cannot give here an account of quantum electrodynamics but we can give a general argument, due to Einstein in 1916, that relates the rate of spontaneous emission to the rates of stimulated emission and absorption.

We start by considering, for simplicity, a collection of many atoms, each with two states, 1 and 2, of energies E_1 and E_2, with $E_2 > E_1$. Suppose that N_1 atoms are in state 1, and N_2 are in state 2. We also suppose that external radiation is present, and that it is characterized by a spectral energy density function $u(\omega)$. This radiation will be absorbed by atoms and also stimulate them to emit radiation, but Einstein realized that spontaneous emission would be involved too. For the system to come to equilibrium, as it must, spontaneous emission must be related to the other two processes. In considering how this equilibrium comes about, Einstein discovered the link that allows us to calculate the probability for spontaneous emission in terms of the probabilities for absorption and stimulated emission.

Einstein's first step was to write down plausible equations describing the rates of the three types of radiative process. He assumed that:

1. **Absorption**   The rate of transitions from state 1 to state 2 due to absorption is proportional to the number N_1 in state 1, and is also proportional to the spectral energy density of the radiation at the appropriate angular frequency:

$$\left(\frac{dN_1}{dt}\right)_{abs} = -B_{12} N_1 \, u(\omega_{21}), \qquad (8.38)$$

The minus sign appears because absorption causes the number of atoms in state 1 to decrease.

where B_{12} is some unknown coefficient and $\omega_{21} = (E_2 - E_1)/\hbar$.

2. **Stimulated emission**   The rate at which atoms leave state 2 due to stimulated emission is proportional to the number N_2 in that state and to $u(\omega_{21})$:

$$\left(\frac{dN_2}{dt}\right)_{stim} = -B_{21} N_2 \, u(\omega_{21}), \qquad (8.39)$$

where B_{21} is another unknown coefficient.

3. **Spontaneous emission**   The rate of spontaneous emission is proportional to the number of atoms in the excited state, but is independent of the intensity of the external radiation (this is what it means to say that the process is *spontaneous*):

*B_{12} and B_{21} are called **Einstein B-coefficients** and A_{21} is called the **Einstein A-coefficient**.*

$$\left(\frac{dN_2}{dt}\right)_{spon} = -A_{21} N_2, \qquad (8.40)$$

where A_{21} is a third unknown coefficient.

Einstein's argument relating the three coefficients B_{12}, B_{21} and A_{21} to one another goes as follows. Let us suppose that the atoms are all in a box whose walls are at temperature T. The box also contains radiation which is in equilibrium with

8.5 Spontaneous emission and the Einstein coefficients

the atoms and with the walls of the box. The radiation will induce absorption from atoms in the lower state and stimulated emission from atoms in the upper state. There will also be spontaneous emission from the upper state. However, eventually an equilibrium will be reached in which the numbers of atoms in the two states will remain constant. In this state of equilibrium, the absorption from state 1 must be balanced by the two types of emission from state 2. So we have

$$\left(\frac{dN_1}{dt}\right)_{abs} = \left(\frac{dN_2}{dt}\right)_{stim} + \left(\frac{dN_2}{dt}\right)_{spon}$$

or, using Equations 8.38–8.40,

$$B_{12}\, u(\omega_{21})\, N_1 = B_{21}\, u(\omega_{21})\, N_2 + A_{21} N_2. \tag{8.41}$$

The rest of the argument requires two pieces of knowledge from thermal physics that may not be familiar to you. We shall just quote the results that are needed. First, Ludwig Boltzmann showed in 1877 that, in thermal equilibrium at absolute temperature T, the average numbers of atoms in states 1 and 2 are related by

$$\frac{N_2}{N_1} = \frac{e^{-E_2/kT}}{e^{-E_1/kT}} = e^{-(E_2-E_1)/kT} = e^{-\hbar\omega_{21}/kT}, \tag{8.42}$$

where we have written $E_2 - E_1 = \hbar\omega_{21}$. This is the **Boltzmann distribution law**, and the constant k is Boltzmann's constant.

Secondly, Max Planck in 1900 showed that the spectral energy density of radiation that is in equilibrium with matter at temperature T is given by

$$u(\omega) = \frac{\hbar\omega^3}{\pi^2 c^3} \frac{1}{e^{\hbar\omega/kT} - 1}. \tag{8.43}$$

This is the famous **Planck distribution law** — the first equation in physics to involve Planck's constant.

Now let's put all these results together. We rearrange Equation 8.41 to get an expression for $u(\omega_{21})$, and use Equation 8.42 to get

$$u(\omega_{21}) = \frac{A_{21}}{B_{12}\, e^{\hbar\omega_{21}/kT} - B_{21}}. \tag{8.44}$$

We can set $\omega = \omega_{21}$ in Equation 8.43 and set the two expressions for $u(\omega_{21})$ equal to each other:

$$\frac{\hbar\omega_{21}^3}{\pi^2 c^3} \frac{1}{e^{\hbar\omega_{21}/kT} - 1} = \frac{A_{21}}{B_{12}\, e^{\hbar\omega_{21}/kT} - B_{21}}.$$

In order for this equation to be true at all temperatures, we must have

$$B_{21} = B_{12} \quad \text{and} \quad A_{21} = \frac{\hbar\omega_{21}^3}{\pi^2 c^3} B_{12}. \tag{8.45}$$

The fact that $B_{21} = B_{12}$ agrees with our previous finding that the transition probabilities for absorption and stimulated emission are the same. However, the second relationship, between 'Einstein's A-coefficient' and 'Einstein's B-coefficient' is new, and is what makes Einstein's deduction so important. This relation tells us how to calculate the rate of spontaneous emission if we know the rate of absorption.

Comparing Equations 8.38 and 8.37, we see that the Einstein B-coefficient is given by

$$B_{12} = \frac{\pi e^2}{3\varepsilon_0 \hbar^2}\left(|x_{12}|^2 + |y_{12}|^2 + |z_{12}|^2\right), \tag{8.46}$$

and it follows from Equation 8.45 that the A-coefficient is

$$A_{21} = \frac{e^2 \omega_{21}^3}{3\pi\varepsilon_0 \hbar c^3}\left(|x_{12}|^2 + |y_{12}|^2 + |z_{12}|^2\right). \tag{8.47}$$

The physical significance of the Einstein A-coefficient is that, if spontaneous emission from state 2 to state 1 is the only mechanism removing atoms from this excited state, then Equation 8.40 gives

This is just like the discussion of radioactive decay in Chapter 1 of Book 1.

$$N_2(t) = N_2(0)\,\mathrm{e}^{-A_{21}t},$$

so A_{21} is a decay constant for spontaneous decay, inversely proportional to the average lifetime of an excited state. Equation 8.47 then shows that a state that is separated by only a very small energy from states with lower energy will be relatively long-lived. This is a very general result, and applies equally to the emission of light by excited states of atoms and the emission of gamma-ray photons by nuclei.

Exercise 8.9 If the energy difference between two states of a nucleus were to be increased by a factor of ten, with everything else the same, by how much would the average lifetime of the state be changed?

Exercise 8.10 Do the same selection rules apply for spontaneous emission as for stimulated emission? ∎

Final remarks: Einstein, lasers, God playing dice, and all that

We cannot omit mention of the fact that the 1916 paper by Einstein, in which he introduced the A- and B-coefficients that we have just discussed, had a richness and profundity that we cannot do justice to here. It also introduced the concept of stimulated emission for the first time; Einstein saw it as essential for the establishment of an equilibrium between matter and radiation. The consequences are with you every time you play a CD or DVD, since stimulated emission is the underlying principle of the laser. This paper was also the first to make *probability* an important part of our understanding of the quantum world; this was ironical from the man commonly identified as the one who could not stand the idea of 'God playing dice'.

At the end of the quantum journey

> We shall not cease from exploration
> And the end of all our exploring
> Will be to arrive where we started
> And know the place for the first time.
> T. S. Eliot, *Little Gidding*

8.5 Spontaneous emission and the Einstein coefficients

This course began with spectra, the fingerprints of atoms that allow us to find the composition of the most distant visible galaxies. Spectra have again been centre-stage in this last chapter. Over the span of the three books, we have, step by step, filled in the gaps to provide explanations for these remarkable patterns of emission and absorption by quantum systems.

The first step of the journey, taken in Book 1, was to show how discrete energy levels arise from the time-independent Schrödinger equation. We concentrated mainly on one-dimensional situations and, in the context of a harmonic oscillator, you saw the first example of a selection rule. Book 1 also distinguished between stationary states and wave packets, and showed how the overlap rule allows us to predict the probabilities of possible experimental outcomes, a key ingredient used to calculate transition probabilities in this last chapter.

Book 2 started by introducing a new language for quantum mechanics, replacing wave functions by vectors in an abstract vector space. The compact notation $\langle\psi|\phi\rangle$ has saved a lot of clutter ever since. Book 2 went on to introduce several key concepts needed in applications of quantum mechanics. Orbital angular momentum is needed to describe three-dimensional systems such as atoms, which have spherically-symmetric Hamiltonians. Both the magnitude and the individual components of orbital angular momentum are quantized but it is impossible to find states with well-defined non-zero values of two *different* components of angular momentum. In general, our ability to label energy eigenfunctions with the values of a given set of observables depends on whether the Hamiltonian of the system and the quantum-mechanical operators representing the observables form a mutually commuting set.

We also saw that particles such as electrons have spin angular momentum, which behaves in many respects like orbital angular momentum, but is described using Dirac notation or matrices rather than wave functions. Identical particles have a special status in quantum mechanics, and a collection of electrons must be described by an antisymmetric total wave function. This led to the Pauli exclusion principle, which has profound implications for the electronic structure of atoms and molecules.

Book 3 has demonstrated how quantum mechanics is used to model real systems, from the hydrogen atom to many-electron atoms, diatomic molecules and solids. It is possible to find exact energy eigenfunctions and eigenvalues for the Coulomb model of the hydrogen atom, but additional small effects, not included in the Coulomb model, place exact solutions for the hydrogen atom beyond our grasp. In more complicated systems, it is never possible to obtain exact solutions, and this is why approximation methods are so important. Time-independent perturbation theory gives the approximate energy levels of a system in the presence of a small time-independent perturbation. We concentrated on the first-order estimate of the energy levels, but perturbation theory can also provide a series of higher-order corrections that systematically improve our estimates. Alternatively, we can use the variational method; this is especially valuable for ground states, but it is also used for excited states, with less guarantee of success. The ability of quantum mechanics to provide good approximations in complex situations is well-illustrated by the band-structure diagram of copper (Figure 7.10) — here, there is excellent agreement between theoretical predictions and experimental data, in spite of the fearsome complications of electron–electron interactions.

Finally, arriving at this last chapter, we have taken it for granted that an atom will have a series of quantum states whose approximate energies are known. We then let the atom interact with electromagnetic radiation. To deal with this situation, a new approximation method was introduced — time-dependent perturbation theory. Using this method, together with the electric dipole approximation, we predicted the transition probabilities associated with the absorption and stimulated emission of light. In particular, we derived some selection rules that govern these processes. The remaining puzzle was the process of spontaneous emission. While this process is ultimately explained by quantum electrodynamics, a consistency argument due to Einstein connects spontaneous emission to absorption and stimulated emission, and allows us to predict the rate of a given spontaneous transition. In particular, the selection rules for spontaneous emission are the same as those for absorption or stimulated emission, allowing us to explain the overall pattern of spectral lines emitted by excited atoms.

Of course, the story of spectra is only one facet of quantum mechanics. You have also seen how scattering is described in quantum mechanics, how tunnelling explains the working of the scanning tunnelling microscope, how atoms form Bose–Einstein condensates and how silicon conducts electricity. Not least are the new developments described at the end of Book 2; developments such as quantum cryptography, quantum teleportation and quantum computing, which deeply challenge our intuitions, but seem likely to be of increasing technological importance in the quantum world.

Summary of Chapter 8

Section 8.1 There are three types of radiative process: absorption of light, stimulated emission of light, and spontaneous emission of light. The first two of these can be analyzed in terms of the interaction between an atom (treated quantum-mechanically) and external radiation (treated classically). This is done in the electric dipole approximation, which represents the atom–light interaction by a time-dependent potential energy term

$$V(t) = e \sum_i \boldsymbol{\mathcal{E}}(t) \cdot \mathbf{r}_i,$$

where $\boldsymbol{\mathcal{E}}(t)$ is the electric field at the position of the atom and $\mathbf{r}_i$ is the position of the ith electron.

Section 8.2 If the electric dipole interaction is small, its effect can be found using time-dependent perturbation theory. This allows us to obtain an approximation for the wave function in a system that has been exposed to a small time-dependent perturbation. The overlap rule allows us to calculate the transition probability from an initial state to a final state. For monochromatic light, polarized in the z-direction and switched on at time 0, and off at time t, the probability of a transition from state i of energy E_i to state f of energy E_f is

$$P_{i \to f}(t) = \frac{e^2 \mathcal{E}_0^2}{\hbar^2} |\langle \psi_f | \widehat{z} | \psi_i \rangle|^2 \left| \int_0^t e^{i\omega_{fi} t'} \cos(\omega t') \, dt' \right|^2,$$

where $\mathcal{E}_0$ is the electric field amplitude and $\omega_{fi} = (E_f - E_i)/\hbar$.

Section 8.3 If $\langle \psi_f | \widehat{z} | \psi_i \rangle = 0$, light polarized in the z-direction does not induce first-order radiative transitions between states i and f, and the transition is said to be forbidden. The general rules restricting transitions between quantum states are called selection rules. For a one-electron atom, the selection rules are: $l_f = l_i \pm 1$ and $m_f = m_i - 1, m_i$ or $m_i + 1$. For some directions of polarization, the changes in m may be further restricted.

Section 8.4 The probability that monochromatic radiation induces a radiative transition (either absorption or stimulated emission) depends on the angular frequency of the radiation, and on time. For monochromatic radiation, first-order perturbation theory predicts a transition probability that oscillates in time, but ordinary (non-monochromatic) light produces a transition probability that is proportional to time.

Section 8.5 The rates of radiative processes can be expressed in terms of various coefficients: B-coefficients for absorption and stimulated emission, and A-coefficients for spontaneous emission. Assuming a state of equilibrium, and using the Boltzmann and Planck distribution laws, Einstein related these coefficients to each other. This provides a way of calculating the rate of spontaneous emission in terms of the rate of absorption (or of stimulated emission).

Achievements from Chapter 8

After studying this chapter, you should be able to:

8.1 Explain the meanings of the newly defined (emboldened) terms and symbols, and use them appropriately.

8.2 Describe the three processes that occur when light interacts with matter. Recall the difference between radiation produced by spontaneous emission and stimulated emission.

8.3 Give an account of the electric dipole approximation and the assumptions underlying it.

8.4 Give a qualitative description of time-dependent perturbation theory.

8.5 Apply time-dependent perturbation theory to simple problems in which the effects of monochromatic light are treated as a perturbation.

8.6 Explain how the electric dipole approximation leads to selection rules for linearly and circularly polarized light.

8.7 Apply selection rules to determine whether specific transitions are allowed or forbidden.

8.8 Give an account of the transition between states for stimulated emission and absorption for both monochromatic radiation and radiation covering a range of frequencies.

8.9 Give an account of the Einstein coefficients. Explain the basis for the relation between these coefficients that makes it possible to calculate the rate of spontaneous emission.

Acknowledgements

Grateful acknowledgement is made to the Science Photo Library for permission to use the image on the cover of this book. We apologize unreservedly for the omission of this acknowledgement from the companion volumes *Wave mechanics* and *Quantum mechanics and its interpretation* and promise to rectify this omission at the earliest opportunity.

Grateful acknowledgement is made to the following sources for permission to include material within this book:

Figure 1.5: Courtesy of Raymond Mackintosh;

Figures 2.3, 2.4, 2.6, 2.7, 2.11 and 2.16: French, A. P and Taylor, E. F. (1978), *An Introduction to Quantum Physics*, The Massachusetts Institute of Technology;

Figure 2.15: McCarthy, I. E., and Weigold, E., (1983), 'A real "thought" experiment for the hydrogen atom', *American Journal of Physics*, Copyright © 2007 All rights reserved.

Figure 4.2: Picture provided by 'RIKEN-RAL';

Figure 4.5: Photograph taken from http://www.nrao.edu/whatisra/hst_ewenpurcell.shtml

Figure 4.6: Goronwy Tudor Jones, University of Birmingham;

Table 6.2: Levine, I. N., (2000), 'Electronic structure of diatomic molecules', *Quantum Chemistry*, 5th edition, 2000, Prentice Hall, Inc.;

Figure 6.11: Levine I. N., (2000) 'Electronic structure of diatomic molecules', *Quantum Chemistry*, 5th edition, 2000, Prentice Hall, Inc.

Every effort has been made to contact copyright holders. If any have been inadvertently overlooked the publishers will be pleased to make the necessary arrangements at the first opportunity.

Solutions to exercises

Ex 1.1 (a) If you turn to Equation 1.20, the right-hand side is unity, dimensionless, expressing the fact that the total probability of finding the particle somewhere or other is 1. In the integral, $|\psi|^2$ is multiplied by a volume element with dimensions of (length)^3, so $|\psi|^2$ must have dimensions of (length)$^{-3}$. It follows that $\psi(r,\theta,\phi)$ (or, indeed, $\psi(x,y,z)$) must have dimensions of (length)$^{-3/2}$.

(b) The Laplacian is a second derivative with respect to length, and must have dimensions of (length)$^{-2}$. Equation 1.17 tells us that $\widehat{X}$ must have the same dimensions.

Ex 1.2 The operator $\widehat{L}_z$ involves differentiation with respect to ϕ. It commutes with any operator that does not include a *function* of ϕ (derivatives with respect to ϕ are harmless). Thus,

$$\frac{\partial}{\partial \phi}\left[\frac{1}{\sin\theta}\frac{\partial}{\partial \theta}\left(\sin\theta\frac{\partial}{\partial \theta}\right) + \frac{1}{\sin^2\theta}\frac{\partial^2}{\partial \phi^2}\right]$$
$$= \left[\frac{1}{\sin\theta}\frac{\partial}{\partial \theta}\left(\sin\theta\frac{\partial}{\partial \theta}\right) + \frac{1}{\sin^2\theta}\frac{\partial^2}{\partial \phi^2}\right]\frac{\partial}{\partial \phi},$$

and so $\widehat{L}_z\widehat{L}^2 = \widehat{L}^2\widehat{L}_z$.

With a spherically-symmetric potential energy function, the Hamiltonian operator $\widehat{H}$ involves only terms in r and $\widehat{L}^2$. All of these terms are unaffected by differentiation with respect to ϕ, so $\widehat{L}_z$ also commutes with $\widehat{H}$.

Finally, $\widehat{L}^2$ involves differentiation with respect to θ and ϕ. It therefore commutes with all r-dependent terms and (like any operator) it also commutes with itself. Thus, $\widehat{L}^2$ commutes with $\widehat{H}$.

Ex 1.3 Since $20\hbar^2$ corresponds to $l=4$, we have, writing combinations as (l,m):
$(0,0)$, $(1,-1)$, $(1,0)$, $(1,1)$, $(2,-2)$, $(2,-1)$, $(2,0)$, $(2,1)$, $(2,2)$, $(3,-3)$, $(3,-2)$, $(3,-1)$, $(3,0)$, $(3,1)$, $(3,2)$, $(3,3)$, $(4,-4)$, $(4,-3)$, $(4,-2)$, $(4,-1)$, $(4,0)$, $(4,1)$, $(4,2)$, $(4,3)$, $(4,4)$.

Ex 1.4 Substituting $\cos\theta$ for $\Theta(\theta)$ in Equation 1.34, differentiating and multiplying through by $\sin^2\theta$, we get

$$2\sin^2\theta\cos\theta + m^2\cos\theta = l(l+1)\cos\theta\sin^2\theta.$$

This can be true for all θ if and only if $m=0$ and $l=1$, so, apart from a normalizing factor, $\cos\theta$ is $\Theta_{1,0}$, the solution of Equation 1.34 for $l=1$ and $m=0$.

Ex 1.5 (a) l takes the values $0,1,2,3$, and in each case m takes the values of the positive and negative integers, including zero, with $|m|\leq l$.

(b) In order of appearance in the text, $Y_{0,0}$, $Y_{1,-1}$ and $Y_{1,1}$, and, in Exercise 1.4, $Y_{1,0}$.

(c) This has $l=2$ and $m=-1$ and so is

$$Y_{2,-1}(\theta,\phi) = \sqrt{\tfrac{15}{8\pi}}\cos\theta\sin\theta\, e^{-i\phi}.$$

Ex 1.6 (a) $Y_{0,0} = 1/\sqrt{4\pi}$, so the required integral of $|Y_{0,0}|^2$ is

$$\int_0^{2\pi}\int_0^{\pi}\frac{1}{4\pi}\sin\theta\,d\theta\,d\phi = \frac{1}{4\pi}\int_0^{2\pi}d\phi\int_0^{\pi}\sin\theta\,d\theta$$
$$= \frac{1}{4\pi}\times 2\pi\times 2 = 1,$$

where we have used

$$\int_0^{2\pi} d\phi = 2\pi \quad \text{and} \quad \int_0^{\pi}\sin\theta\,d\theta = 2.$$

To verify the normalization for $l=1$, $m=1$, note from Table 1.1 that $Y_{1,1} = -\sqrt{\tfrac{3}{8\pi}}\sin\theta\,e^{+i\phi}$, so $|Y_{1,1}|^2 = \tfrac{3}{8\pi}\sin^2\theta$, which is independent of ϕ. As a result, the normalization integral can be written as a product of a trivial ϕ integral and a less trivial θ integral:

$$\frac{3}{8\pi}\int_0^{2\pi}d\phi\times\int_0^{\pi}\sin^3\theta\,d\theta = \frac{3}{8\pi}\times 2\pi\times\int_0^{\pi}\sin^3\theta\,d\theta.$$

From the list of integrals given inside the back cover, we have

$$\int_0^{\pi}\sin^3 x\,dx = \frac{4}{3}.$$

So, putting the factors together, the normalization integral becomes

$$\frac{3}{8\pi}\times 2\pi\times\frac{4}{3} = 1,$$

verifying that $Y_{1,1}$ is normalized.

(b) To show that $Y_{0,0}$ and $Y_{1,1}$ are orthogonal, we must show that

$$\int_0^{2\pi}\int_0^{\pi}Y_{0,0}^*Y_{1,1}\sin\theta\,d\theta\,d\phi = 0.$$

The θ integral cannot be zero since the $\sin\theta$ factor from $Y_{1,1}$ implies that the integrand is a constant times $\sin^2\theta$,

Solutions to exercises

which is always positive. We therefore consider the ϕ integral:

$$\int_0^{2\pi} e^{i\phi}\, d\phi = \frac{1}{i}[e^{2\pi i} - e^0] = \frac{1}{i}[1-1] = 0.$$

This shows that $Y_{0,0}$ and $Y_{1,1}$ are orthogonal.

Ex 1.7 We have

$$\cos(\pi - \theta) = -\cos\theta$$
$$\sin(\pi - \theta) = \sin\theta$$
$$e^{i(\phi+\pi)} = -e^{i\phi}$$
$$e^{2i(\phi+\pi)} = e^{2i\phi}.$$

Referring to Table 1.1, we see that, for $m = 0, \pm 2$, the ϕ-dependent term does not change sign under the transformation $\phi \to \phi + \pi$, and neither does the θ-dependent term under $\theta \to \pi - \theta$. For $m = \pm 1$, the ϕ-term does change sign under the transformation $\phi \to \phi + \pi$, and the θ term changes sign under the transformation $\theta \to \pi - \theta$. Hence, for all $l = 2$ spherical harmonics, the overall sign is unchanged under the parity transformation.

Ex 1.8 All spherical harmonics $Y_{lm}(\theta, \phi)$ have a factor $e^{im\phi}$. The other factor is independent of ϕ. The complex conjugate Y_{lm}^* therefore has a factor $e^{-im\phi}$, and the product $Y_{lm}^* Y_{lm}$ has a factor $e^{-im\phi} e^{+im\phi} = 1$, so $|Y_{lm}|^2$ is independent of ϕ for all l and m.

Ex 1.9 (a) The z-axis (see Exercise 1.8).

(b) They have a common factor of $\sin\theta$, which is zero for $\theta = 0$ or π, so are zero at the poles. Since $\sin\theta$ is greatest where $\theta = \pi/2$, these spherical harmonics have their largest magnitudes in the equatorial plane.

(c) From Table 1.1, all of these have a factor of $\sin^l\theta$, and so all are zero for $\theta = 0$ or $\theta = \pi$, i.e. 'at the poles'.

Ex 1.10 Yes, because $\widehat{J}_z(Y_{2,0}|\!\uparrow\rangle + Y_{2,1}|\!\downarrow\rangle)$ is a number times $(Y_{2,0}|\!\uparrow\rangle + Y_{2,1}|\!\downarrow\rangle)$, as seen from Equation 1.46. The eigenvalue is $\frac{1}{2}\hbar$.

Ex 1.11

$$\widehat{J}_x\widehat{J}_y - \widehat{J}_y\widehat{J}_x = (\widehat{L}_x + \widehat{S}_x)(\widehat{L}_y + \widehat{S}_y)$$
$$- (\widehat{L}_y + \widehat{S}_y)(\widehat{L}_x + \widehat{S}_x)$$
$$= \widehat{L}_x\widehat{L}_y - \widehat{L}_y\widehat{L}_x + \widehat{S}_x\widehat{S}_y - \widehat{S}_y\widehat{S}_x$$
$$+ \widehat{L}_x\widehat{S}_y - \widehat{S}_y\widehat{L}_x + \widehat{S}_x\widehat{L}_y - \widehat{L}_y\widehat{S}_x$$
$$= i\hbar \widehat{L}_z + i\hbar \widehat{S}_z + 0 + 0$$
$$= i\hbar \widehat{J}_z.$$

The two zeros appear because orbital and spin operators commute with each other, so $\widehat{S}_x\widehat{L}_y - \widehat{L}_y\widehat{S}_x = 0$, for example.

Ex 1.12 There are two ways of answering this question. The direct way is to write

$$\left[\widehat{J}_z, \widehat{\mathbf{L}}\cdot\widehat{\mathbf{S}}\right] = \left[\widehat{L}_z + \widehat{S}_z, \widehat{L}_x\widehat{S}_x + \widehat{L}_y\widehat{S}_y + \widehat{L}_z\widehat{S}_z\right],$$

and then use the commutation relations for orbital angular momentum (Equation 1.5) and the analogous relations for spin. This gives

$$\left[\widehat{J}_z, \widehat{\mathbf{L}}\cdot\widehat{\mathbf{S}}\right] = \left[\widehat{L}_z, \widehat{L}_x\right]\widehat{S}_x + \left[\widehat{L}_z, \widehat{L}_y\right]\widehat{S}_y$$
$$+ \widehat{L}_x\left[\widehat{S}_z, \widehat{S}_x\right] + \widehat{L}_y\left[\widehat{S}_z, \widehat{S}_y\right]$$
$$= i\hbar\left(\widehat{L}_y\widehat{S}_x - \widehat{L}_x\widehat{S}_y + \widehat{L}_x\widehat{S}_y - \widehat{L}_y\widehat{S}_x\right)$$
$$= 0.$$

The alternative method uses Equation 1.49 to write

$$\widehat{\mathbf{L}}\cdot\widehat{\mathbf{S}} = \tfrac{1}{2}\!\left(\widehat{\mathbf{J}}^2 - \widehat{\mathbf{L}}^2 - \widehat{\mathbf{S}}^2\right).$$

From Equation 1.48, $\widehat{J}_z$ commutes with $\widehat{\mathbf{J}}^2$. To show that $\widehat{J}_z$ commutes with $\widehat{\mathbf{L}}^2$, we note that $\widehat{J}_z = \widehat{L}_z + \widehat{S}_z$, and both $\widehat{L}_z$ and $\widehat{S}_z$ commute with $\widehat{\mathbf{L}}^2$. A similar argument shows that $\widehat{J}_z = \widehat{L}_z + \widehat{S}_z$ commutes with $\widehat{\mathbf{S}}^2$, so we conclude that $\widehat{J}_z$ commutes with $\widehat{\mathbf{L}}\cdot\widehat{\mathbf{S}}$.

Either way, since $\widehat{J}_z$ involves no differentiation with respect to r, it commutes with all the r-dependent terms in $\widehat{H}_{so}$, as well as with $\widehat{\mathbf{L}}\cdot\widehat{\mathbf{S}}$, and so commutes with $\widehat{H}_{so}$ itself.

Ex 1.13 The possible values of j are $5/2$ and $7/2$.

For $j = 7/2$, there are 8 possible values of m_j:

$$7/2, 5/2, 3/2, 1/2, -1/2, -3/2, -5/2 \text{ and } -7/2.$$

For $j = 5/2$, there are 6 possible values of m_j:

$$5/2, 3/2, 1/2, -1/2, -3/2 \text{ and } -5/2.$$

So there are 14 possible states in all.

Ex 1.14 Taking $|1, \tfrac{1}{2}, \tfrac{1}{2}\rangle$ from Equation 1.51 and using $\widehat{J}_z = \widehat{L}_z + \widehat{S}_z$, we have

$$\widehat{J}_z |1, \tfrac{1}{2}, \tfrac{1}{2}\rangle = (\widehat{L}_z + \widehat{S}_z)|1, \tfrac{1}{2}, \tfrac{1}{2}\rangle$$
$$= (\widehat{L}_z + \widehat{S}_z)(\sqrt{\tfrac{1}{3}} Y_{1,0}|\uparrow\rangle - \sqrt{\tfrac{2}{3}} Y_{1,1}|\downarrow\rangle)$$
$$= \sqrt{\tfrac{1}{3}}(\widehat{L}_z Y_{1,0})|\uparrow\rangle - \sqrt{\tfrac{2}{3}}(\widehat{L}_z Y_{1,1})|\downarrow\rangle$$
$$+ \sqrt{\tfrac{1}{3}} Y_{1,0}(\widehat{S}_z|\uparrow\rangle) - \sqrt{\tfrac{2}{3}} Y_{1,1}(\widehat{S}_z|\downarrow\rangle).$$

But $\widehat{L}_z Y_{1,m} = m\hbar Y_{1,m}$ and $\widehat{S}_z|\uparrow\rangle = \tfrac{1}{2}\hbar|\uparrow\rangle$, etc., so we obtain

$$\widehat{J}_z |1, \tfrac{1}{2}, \tfrac{1}{2}\rangle$$
$$= \hbar\left[0 - \sqrt{\tfrac{2}{3}} Y_{1,1}|\downarrow\rangle + \sqrt{\tfrac{1}{3}} Y_{1,0} \tfrac{1}{2}|\uparrow\rangle + \sqrt{\tfrac{2}{3}} Y_{1,1} \tfrac{1}{2}|\downarrow\rangle\right]$$
$$= \tfrac{1}{2}\hbar\left[\sqrt{\tfrac{1}{3}} Y_{1,0}|\uparrow\rangle - \sqrt{\tfrac{2}{3}} Y_{1,1}|\downarrow\rangle\right]$$
$$= \tfrac{1}{2}\hbar |1, \tfrac{1}{2}, \tfrac{1}{2}\rangle,$$

as required.

Ex 1.15 To show that $|1, \tfrac{1}{2}, \tfrac{1}{2}\rangle$ in Equation 1.51 is normalized we must verify that $\langle 1, \tfrac{1}{2}, \tfrac{1}{2}|1, \tfrac{1}{2}, \tfrac{1}{2}\rangle = 1$, and the key ingredients are the orthonormality of the spinors: $\langle\uparrow|\downarrow\rangle = 0$, $\langle\uparrow|\uparrow\rangle = \langle\downarrow|\downarrow\rangle = 1$, and of the spherical harmonics (Equation 1.39). We then proceed term by term, noting that a factor $\tfrac{1}{3}$ can conveniently be taken out as a common factor:

$$\langle 1, \tfrac{1}{2}, \tfrac{1}{2}|1, \tfrac{1}{2}, \tfrac{1}{2}\rangle$$
$$= \tfrac{1}{3}\Big[\langle\uparrow|\uparrow\rangle \int_0^{2\pi}\!\!\int_0^{\pi} Y_{1,0}^* Y_{1,0} \sin\theta\, d\theta\, d\phi$$
$$- \sqrt{2}\langle\uparrow|\downarrow\rangle \int_0^{2\pi}\!\!\int_0^{\pi} Y_{1,0}^* Y_{1,1} \sin\theta\, d\theta\, d\phi$$
$$- \sqrt{2}\langle\downarrow|\uparrow\rangle \int_0^{2\pi}\!\!\int_0^{\pi} Y_{1,1}^* Y_{1,0} \sin\theta\, d\theta\, d\phi$$
$$+ 2\langle\downarrow|\downarrow\rangle \int_0^{2\pi}\!\!\int_0^{\pi} Y_{1,1}^* Y_{1,1} \sin\theta\, d\theta\, d\phi\Big]$$
$$= \tfrac{1}{3}[1 - 0 - 0 + 2] = 1,$$

as required.

The two zero terms are in fact zero twice over:
(i) because $\langle\uparrow|\downarrow\rangle = \langle\downarrow|\uparrow\rangle = 0$, and (ii) because $Y_{1,0}$ and $Y_{1,1}$ are orthogonal.

To show that $|1, \tfrac{1}{2}, \tfrac{1}{2}\rangle$ is orthogonal to $|1, \tfrac{1}{2}, -\tfrac{1}{2}\rangle$, we must verify that $\langle 1, \tfrac{1}{2}, \tfrac{1}{2}|1, \tfrac{1}{2}, -\tfrac{1}{2}\rangle = 0$.

Using the shorthand notation $|l, m\rangle$ for $Y_{lm}(\theta, \phi)$ and taking results from Table 1.2, we have

$$\langle 1, \tfrac{1}{2}, \tfrac{1}{2}|1, \tfrac{1}{2}, -\tfrac{1}{2}\rangle = \tfrac{1}{3}\big[\sqrt{2}\langle\uparrow|\uparrow\rangle\langle 1, 0|1, -1\rangle$$
$$- \langle\uparrow|\downarrow\rangle\langle 1, 0|1, 0\rangle$$
$$- 2\langle\downarrow|\uparrow\rangle\langle 1, 1|1, -1\rangle$$
$$+ \sqrt{2}\langle\downarrow|\downarrow\rangle\langle 1, 1|1, 0\rangle\big]$$
$$= 0,$$

because each of the four terms in the sum vanishes either because the spherical harmonics are orthogonal or because the spinors are orthogonal.

Ex 2.1 In the Bohr model, the magnitude of the orbital angular momentum is quantized. An electron has mass $m = m_e$, so in the orbit with quantum number n, we have

$$m_e v_n r_n = m_e v_n n^2 a_0 = n\hbar,$$

so

$$v_n = \frac{\hbar}{n m_e a_0}.$$

For the $n = 1$ orbit, this gives

$$v_1 = \frac{1.06 \times 10^{-34}\,\text{J s}}{9.11 \times 10^{-31}\,\text{kg} \times 5.29 \times 10^{-11}\,\text{m}}$$
$$= 2.2 \times 10^6\,\text{m s}^{-1}.$$

This is less than 1% of the speed of light, c. The effects of special relativity are generally of order v^2/c^2, so the fractional error in using classical physics will be of order 10^{-4}.

Ex 2.2 Substituting $u(\rho) = \rho^k$ into Equation 2.17, we obtain

$$-\frac{d^2 \rho^k}{d\rho^2} + \frac{l(l+1)}{\rho^2}\rho^k$$
$$= -k(k-1)\rho^{k-2} + l(l+1)\rho^{k-2}$$
$$= 0,$$

which is satisfied provided that $k(k-1) = l(l+1)$. This quadratic equation has two solutions, which are easily seen to be $k = -l$ and $k = l+1$.

Ex 2.3 We have

$$\frac{d}{d\rho}\rho^{l+1}e^{-\beta\rho} = (l+1)\rho^l e^{-\beta\rho} - \beta\rho^{l+1} e^{-\beta\rho},$$
$$= [(l+1)\rho^l - \beta\rho^{l+1}]e^{-\beta\rho}.$$

Solutions to exercises

$$\frac{d^2}{d\rho^2}\rho^{l+1}e^{-\beta\rho}$$
$$= [l(l+1)\rho^{l-1} - \beta(l+1)\rho^l]e^{-\beta\rho}$$
$$\quad - \beta[(l+1)\rho^l - \beta\rho^{l+1}]e^{-\beta\rho}$$
$$= \left(\frac{l(l+1)}{\rho^2} - \frac{2\beta(l+1)}{\rho} + \beta^2\right)\rho^{l+1}e^{-\beta\rho}.$$

So, substituting $u(\rho) = C\rho^{l+1}e^{-\beta\rho}$ into Equation 2.15 and rearranging gives

$$-\left(\frac{l(l+1)}{\rho^2} - \frac{2\beta(l+1)}{\rho} + \beta^2\right)C\rho^{l+1}e^{-\beta\rho}$$
$$+ \left(\frac{l(l+1)}{\rho^2} - \frac{2}{\rho} + \beta^2\right)C\rho^{l+1}e^{-\beta\rho} = 0.$$

The $l(l+1)/\rho^2$ terms cancel, and so do the terms in β^2. We therefore see that the trial solution satisfies Equation 2.15 provided that $2\beta(l+1) = 2$. Hence $\beta = 1/(l+1)$.

Ex 2.4 We require that

$$1 = \int_0^\infty |R(r)|^2 r^2 \, dr = \frac{|B|^2}{a_0^2} \int_0^\infty r^4 e^{-r/a_0} \, dr.$$

Using a standard integral from inside the back cover, we obtain

$$1 = \frac{|B|^2}{a_0^2} 4! \, a_0^5 = 24 a_0^3 |B|^2.$$

Choosing B to be real and positive then gives $B = 1/(\sqrt{24}\, a_0^{3/2})$.

Ex 2.5 The numbers of nodes are given in Table S2.1. (Remember that zeros at $r = 0$ are not counted as nodes.)

Table S2.1 Nodes for various radial functions.

n	l	Number of nodes
1	0	0
2	1	0
2	0	1
3	2	0
3	1	1
3	0	2

The table verifies that the number of nodes is given by $n - l - 1$.

Ex 2.6 According to Table 1.1, the spherical harmonic $Y_{2,1}(\theta, \phi)$ is proportional to $\cos\theta \sin\theta$. It will therefore vanish where $\cos\theta = 0$, that is, where $\theta = 90°$ (the equatorial plane), and where $\sin\theta = 0$, that is, where $\theta = 0$ or $180°$ (the north and south poles). The $3d_1$ radial function has $n - l - 1 = 3 - 2 - 1 = 0$ nodes, so there are no spherical nodal surfaces.

Ex 2.7 The radial function for a 3d state is $R_{3,2}(r)$ so

$$\langle r \rangle = \int_0^\infty r^3 R_{3,2}^2(r) \, dr.$$

Substituting for $R_{3,2}(r)$ from Table 2.1, we have

$$\langle r \rangle = \int_0^\infty r^3 \left(\frac{8}{(3a_0)^3 \times 27^2 \times 5}\right)\left(\frac{r^4}{a_0^4}\right) e^{-2r/3a_0} \, dr.$$

Using a standard integral from inside the back cover, $\int_0^\infty r^7 e^{-2r/3a_0} \, dr = 7!\,(3a_0/2)^8$. So

$$\langle r \rangle = \left(\frac{8}{(3a_0)^3 \times 27^2 \times 5}\right)\left(\frac{1}{a_0^4}\right) \times \frac{7!\,(3a_0)^8}{2^8} = \frac{21 a_0}{2}.$$

Ex 2.8 The Coulomb potential energy is given by $E_{\text{Coul}} = -e^2/(4\pi\varepsilon_0 r)$. In the ground state of a hydrogen atom, this has the expectation value

$$\langle E_{\text{Coul}} \rangle = -\frac{e^2}{4\pi\varepsilon_0}\left\langle \frac{1}{r} \right\rangle$$
$$= -\frac{e^2}{4\pi\varepsilon_0} \int_0^\infty \frac{1}{r} R_{1,0}^2(r)\, r^2 \, dr.$$

Substituting $R_{1,0}(r) = (2/a_0^{3/2})\,e^{-r/a_0}$ from Table 2.1, we obtain

$$\langle E_{\text{Coul}} \rangle = -\frac{e^2}{4\pi\varepsilon_0} \times \frac{4}{a_0^3} \int_0^\infty r\, e^{-2r/a_0}\, dr.$$

Using a standard integral from inside the back cover,

$$\int_0^\infty r\, e^{-2r/a_0}\, dr = 1!\left(\frac{a_0}{2}\right)^2.$$

Hence,

$$\langle E_{\text{Coul}} \rangle = -\frac{e^2}{4\pi\varepsilon_0} \times \frac{4}{a_0^3} \times \frac{a_0^2}{4} = -\frac{e^2}{4\pi\varepsilon_0 a_0}.$$

However, Equations 2.3 and 2.4 give

$$E_{\text{R}} = \frac{\hbar^2}{2\mu a_0^2} = \frac{1}{2a_0^2}\frac{e^2 a_0}{4\pi\varepsilon_0} = \frac{e^2}{8\pi\varepsilon_0 a_0},$$

so we conclude that $\langle E_{\text{Coul}} \rangle = -2E_{\text{R}}$ in the ground state. Since the ground-state energy is $-E_{\text{R}}$, this negative potential energy must be partially offset by a positive kinetic energy of expectation value E_{R}.

Ex 2.9 (a) The volume element should be centred on the point where the probability density is largest. In the ground state of a hydrogen atom, the probability density at a point **r** is

$$|\psi_{1,0,0}(\mathbf{r})|^2 = |R_{1,0}(r)|^2 |Y_{0,0}(\theta,\phi)|^2 = \frac{1}{4\pi}|R_{1,0}(r)|^2.$$

Referring to Figure 2.7, we see that this has its maximum value at $r = 0$, so the volume element should be located at the origin, corresponding to zero electron–proton separation.

(b) The most likely electron–proton distance is given by the position of the maximum radial probability density. Referring to Figure 2.11 and the discussion in the paragraphs below it, we see that the most likely electron–proton separation is a_0. This differs from the answer to part (a) because the radial probability density includes a factor r^2 which takes account of the fact that all points on the surface of a sphere are the same distance from the origin.

(c) Worked Example 2.3 showed that the expectation value of the electron–proton distance is $\frac{3}{2}a_0$. This differs from the answer to part (b) because the radial probability density is not symmetrical about its maximum value.

Ex 2.10 The energy levels are given by $-E_R/n^2$. For $n = 100$, $E_{100} = -13.6\,\text{eV}/100^2 = -1.36\,\text{meV}$. The energy required to ionize the atom from this state is $1.36\,\text{meV}$. The energy required to excite the atom to this state is $(13.6 - 13.6 \times 10^{-4})\,\text{eV} \approx 13.6\,\text{eV}$.

Ex 3.1 To evaluate Equation 3.6, we simply evaluate the numerator and denominator separately, and then recombine the results. For the numerator,

$$\int_{-L/2}^{L/2} \left(\frac{L^2}{4} - x^2\right)\left(-\frac{\hbar^2}{2m}\frac{\mathrm{d}^2}{\mathrm{d}x^2}\right)\left(\frac{L^2}{4} - x^2\right)\mathrm{d}x$$

$$= \int_{-L/2}^{L/2}\left(\frac{L^2}{4} - x^2\right)\left(\frac{\hbar^2}{m}\right)\mathrm{d}x$$

$$= \frac{\hbar^2}{m}\left[\frac{L^2 x}{4} - \frac{x^3}{3}\right]_{-L/2}^{L/2} = \frac{\hbar^2 L^3}{6m}.$$

For the denominator,

$$\int_{-L/2}^{L/2}\left(\frac{L^2}{4} - x^2\right)^2 \mathrm{d}x$$

$$= \int_{-L/2}^{L/2}\left(\frac{L^4}{16} - \frac{L^2 x^2}{2} + x^4\right)\mathrm{d}x$$

$$= \left[\frac{L^4 x}{16} - \frac{L^2 x^3}{6} + \frac{x^5}{5}\right]_{-L/2}^{L/2} = \frac{L^5}{30}.$$

Substituting the numerator and denominator back into the expression for $E_{1,t}$ gives

$$E_{1,t} = \frac{\hbar^2 L^3}{6m} \Big/ \frac{L^5}{30} = \frac{5\hbar^2}{mL^2},$$

as required.

Ex 3.2 An estimate for the ground-state energy is found in exactly the same way as in Worked Example 3.1, using the potential energy function $\frac{1}{2}Dx^4$ in place of $\frac{1}{2}Cx^2$, but using the same trial function as before.

We can immediately write down the expression for the expectation value of the energy:

$$E_{0,t} = \frac{\int_{-\infty}^{\infty} e^{-b^2 x^2}\left(-\frac{\hbar^2}{2m}\frac{\mathrm{d}^2}{\mathrm{d}x^2} + \frac{1}{2}Dx^4\right)e^{-b^2 x^2}\mathrm{d}x}{\int_{-\infty}^{\infty} e^{-2b^2 x^2}\mathrm{d}x}.$$

The kinetic energy term will be identical to that in Worked Example 3.1, which contributed $\hbar^2 b^2/2m$ to the energy associated with the trial function.

The potential energy term requires the evaluation of an integral with an x^4 term in place of the x^2 term, and this term is

$$\frac{\frac{1}{2}D\int_{-\infty}^{\infty} x^4 e^{-2b^2 x^2}\mathrm{d}x}{\int_{-\infty}^{\infty} e^{-2b^2 x^2}\mathrm{d}x}.$$

To evaluate the integral in the numerator, we make the substitution $x = y/(\sqrt{2}b)$, so

$$\int_{-\infty}^{\infty} x^4 e^{-2b^2 x^2}\mathrm{d}x = \int_{-\infty}^{\infty}\frac{y^4}{(\sqrt{2}b)^4}e^{-y^2}\frac{\mathrm{d}y}{\sqrt{2}b}$$

$$= \frac{1}{(\sqrt{2}b)^5}\int_{-\infty}^{\infty} y^4 e^{-y^2}\mathrm{d}y$$

$$= \frac{1}{(\sqrt{2}b)^5}\frac{3\sqrt{\pi}}{4},$$

where we have used the standard integral $\int_{-\infty}^{\infty} y^4 e^{-y^2}\, dy = 3\sqrt{\pi}/4$ from inside the back cover. So the numerator in the potential energy term becomes

$$\tfrac{1}{2}D \frac{1}{(\sqrt{2}b)^5} \frac{3\sqrt{\pi}}{4} = \frac{3}{32}\frac{D}{b^5}\sqrt{\frac{\pi}{2}}.$$

The integral in the denominator is equal to $\sqrt{\pi}/(\sqrt{2}b)$, as shown in Worked Example 3.1. Combining the results for the numerator and denominator, the potential energy term is $3D/32b^4$ when $V = \tfrac{1}{2}Dx^4$, compared with $C/8b^2$ when $V = \tfrac{1}{2}Cx^2$. We therefore conclude that

$$E_{0,\text{t}} = \frac{\hbar^2 b^2}{2m} + \frac{3D}{32b^4}.$$

To minimize this, we differentiate with respect to b and set the result to zero:

$$\frac{dE_{0,\text{t}}}{db} = \frac{\hbar^2 b}{m} - \frac{3D}{8b^5} = 0,$$

which gives

$$b = \left(\frac{3mD}{8\hbar^2}\right)^{1/6}.$$

Substituting this back into the expression for $E_{0,\text{t}}$ we have

$$E_{0,\text{t}} = \frac{\hbar^2}{2m}\left(\frac{3mD}{8\hbar^2}\right)^{1/3} + \frac{3D}{32}\left(\frac{8\hbar^2}{3mD}\right)^{2/3}$$

$$= \frac{1}{4}\left(\frac{3\hbar^4 D}{m^2}\right)^{1/3} + \frac{1}{8}\left(\frac{3\hbar^4 D}{m^2}\right)^{1/3}$$

$$= \frac{3}{8}\left(\frac{3\hbar^4 D}{m^2}\right)^{1/3}.$$

Ex 3.3 (a) We have

$$\widehat{H} = \begin{cases} \infty & \text{for } x < -L/2 \text{ and } x > L/2, \\ -\dfrac{\hbar^2}{2m}\dfrac{d^2}{dx^2} + v(x) & \text{for } |x - x_0| < w/2, \\ -\dfrac{\hbar^2}{2m}\dfrac{d^2}{dx^2} & \text{elsewhere.} \end{cases}$$

(b) The solutions for the eigenvalues and eigenfunctions of an infinite square well without a crater were discussed in Book 1, Chapter 3. So a suitable choice of unperturbed Hamiltonian is

$$\widehat{H}^{(0)} = \begin{cases} \infty & \text{for } x < -L/2 \text{ and } x > L/2, \\ -\dfrac{\hbar^2}{2m}\dfrac{d^2}{dx^2} & \text{for } -L/2 \le x \le L/2. \end{cases}$$

(c) The perturbation in this case is

$$\delta\widehat{H} = v(x) \quad \text{for } |x - x_0| < w/2.$$

In Section 3.2.3 you will see how a first-order correction for the energy levels can be calculated from $\delta\widehat{H}$ and the eigenfunctions for the unperturbed well.

Ex 3.4 The Hamiltonian operator for the oscillator is

$$\widehat{H} = -\frac{\hbar^2}{2m}\frac{d^2}{dx^2} + \tfrac{1}{2}Cx^2 + \tfrac{1}{2}Dx^4.$$

This is the same as the Hamiltonian operator for a harmonic oscillator, except for the additional potential energy term $\tfrac{1}{2}Dx^4$. We take

$$\widehat{H}^{(0)} = -\frac{\hbar^2}{2m}\frac{d^2}{dx^2} + \tfrac{1}{2}Cx^2,$$

and

$$\delta\widehat{H} = \tfrac{1}{2}Dx^4.$$

We are given $Da^2/C \ll 1$, so $\tfrac{1}{2}Da^4 \ll \tfrac{1}{2}Ca^2$; since a is a typical value of x, we can regard the perturbation $\delta\widehat{H} = \tfrac{1}{2}Dx^4$ as being small compared to $\widehat{H}^{(0)}$.

The zeroth-order approximation is the energy eigenvalue of the first excited state of the *unperturbed* oscillator:

$$E_1^{(0)} = \tfrac{3}{2}\hbar\left(\frac{C}{m}\right)^{1/2}.$$

The first-order correction is found by substituting the energy eigenfunction for the first excited state of the unperturbed oscillator, and the perturbation of the Hamiltonian, into Equation 3.22 to obtain:

$$E_1^{(1)} \simeq \langle \psi_1^{(0)} | \delta\widehat{H} | \psi_1^{(0)} \rangle$$

$$= \int_{-\infty}^{\infty} \tfrac{1}{2}Dx^4 \left[\left(\frac{1}{2\sqrt{\pi}a}\right)^{1/2} \frac{2x}{a} e^{-x^2/2a^2}\right]^2 dx$$

$$= \frac{1}{2\sqrt{\pi}a}\frac{4}{a^2}\frac{D}{2}\int_{-\infty}^{\infty} x^6 e^{-x^2/a^2}\, dx.$$

We can evaluate the integral by making the substitution $y = x/a$, so that

$$\int_{-\infty}^{\infty} x^6 e^{-x^2/a^2}\, dx = a^7 \int_{-\infty}^{\infty} y^6 e^{-y^2}\, dy$$

$$= a^7 \times \frac{15\sqrt{\pi}}{8},$$

where we have used a standard integral from inside the back cover. Hence

$$E_1^{(1)} = \frac{1}{2\sqrt{\pi}a}\frac{4}{a^2}\frac{D}{2} \times a^7 \times \frac{15\sqrt{\pi}}{8}$$

$$= \frac{15Da^4}{8},$$

or, substituting $a = \hbar^{1/2}/(mC)^{1/4}$,

$$E_1^{(1)} = \frac{15\hbar^2 D}{8mC}.$$

Adding the zeroth-order approximation and the first-order correction we conclude that

$$E_1 \simeq E_1^{(0)} + E_1^{(1)}$$

$$= \frac{3}{2}\hbar\left(\frac{C}{m}\right)^{1/2} + \frac{15\hbar^2 D}{8mC}.$$

Ex 3.5 (a) The Hamiltonian operator is

$$\widehat{H} = -\frac{\hbar^2}{2\mu}\nabla^2 + V(r),$$

where the potential energy term $V(r)$ has the form given in the question.

In Chapter 2 we discussed the solutions for the Coulomb model of the hydrogen atom, for which the Hamiltonian is

$$\widehat{H}^{(0)} = -\frac{\hbar^2}{2\mu}\nabla^2 - \frac{e^2}{4\pi\varepsilon_0 r},$$

and this is a good choice for the unperturbed Hamiltonian. The perturbation is the difference between the perturbed and unperturbed Hamiltonians, namely

$$\delta\widehat{H} = \begin{cases} -\dfrac{e^2}{4\pi\varepsilon_0}\left(\dfrac{3}{2R} - \dfrac{r^2}{2R^3} - \dfrac{1}{r}\right) & \text{for } r \leq R, \\ 0 & \text{for } r \geq R. \end{cases}$$

(b) The first-order approximation for the ground-state energy when this perturbation is included can be determined using Equation 3.23. The unperturbed ground-state energy is given by $E_1^{(0)} = -E_R$, where E_R is the Rydberg energy. The first-order correction is given by

$$E_1^{(1)} = \langle \psi_{1,0,0}^{(0)} | \delta\widehat{H} | \psi_{1,0,0}^{(0)} \rangle,$$

with

$$\psi_{1,0,0}^{(0)}(\mathbf{r}) = R_{1,0}(r)Y_{0,0}(\theta,\phi),$$

where

$$R_{1,0}(r) = \frac{2}{a_0^{3/2}}e^{-r/a_0}$$

and

$$Y_{0,0}(\theta,\phi) = \frac{1}{\sqrt{4\pi}}$$

(see Tables 2.1 and 1.1).

Now, the perturbation $\delta\widehat{H}$ is independent of θ and ϕ, and the spherical harmonics are normalized, so carrying out the sandwich integral for $E_1^{(1)}$ in spherical coordinates gives

$$E_1^{(1)} = \int_0^\infty R_{1,0}^*(r)\,\delta\widehat{H}\,R_{1,0}(r)\,r^2\,dr$$

$$= -\frac{e^2}{4\pi\varepsilon_0}\frac{4}{a_0^3}\int_0^R \left(\frac{3r^2}{2R} - \frac{r^4}{2R^3} - r\right)e^{-2r/a_0}\,dr$$

$$= -\frac{e^2}{\pi\varepsilon_0 a_0^3}\int_0^R \left(\frac{3r^2}{2R} - \frac{r^4}{2R^3} - r\right)e^{-2r/a_0}\,dr,$$

as required.

(c) Approximating e^{-2r/a_0} by 1 inside the integral, we obtain

$$E_1^{(1)} \simeq -\frac{e^2}{\pi\varepsilon_0 a_0^3}\int_0^R \left(\frac{3r^2}{2R} - \frac{r^4}{2R^3} - r\right)dr$$

$$= -\frac{e^2}{\pi\varepsilon_0 a_0^3}\left[\frac{r^3}{2R} - \frac{r^5}{10R^3} - \frac{r^2}{2}\right]_0^R$$

$$= -\frac{e^2}{\pi\varepsilon_0 a_0^3} \times \left(-\frac{R^2}{10}\right)$$

$$= \frac{2}{5}\left(\frac{R}{a_0}\right)^2 \frac{e^2}{4\pi\varepsilon_0 a_0}.$$

This is the first-order energy shift of the ground state. It is positive because, for $r < R$, the electron feels a smaller attraction towards the centre of the proton than it would for a point-like proton.

Comment: In hydrogen, the energy shift is tiny (about $2 \times 10^{-10}E_R$), but for heavier atoms, such as mercury, the effect is much larger, and provides a useful way of comparing the sizes of different nuclei.

Ex 4.1 The radial functions for the deuterium atom differ from those for hydrogen because the reduced masses of the systems are different. Denoting by a_D the

Solutions to exercises

value of the scaled Bohr radius in deuterium, we have from Equation 4.4,

$$\frac{\langle r \rangle_D}{\langle r \rangle_H} = \frac{a_D}{a_0} = \frac{\mu_H}{\mu_D} = \frac{1}{1.000\,272} = 0.999\,728.$$

Ex 4.2 To the level of approximation of Worked Example 4.1, we can ignore the change in reduced mass and simply use $Z = 3$ to get

$$E_n = -9 \times 13.6\,\text{eV}\,\frac{1}{n^2} = -\frac{122.4}{n^2}\,\text{eV}.$$

The energies for $n = 1, 2, 3$ are therefore $-122.4\,\text{eV}$, $-30.6\,\text{eV}$ and $-13.6\,\text{eV}$, respectively.

Similarly, the values of $\langle r \rangle$ for the three states are approximately one-third of the corresponding values in a hydrogen atom. Using data given in Worked Example 4.1, this gives $a_0/2$, $5a_0/3$ and $7a_0/2$, for the 1s, 2p and 3d states respectively, where $a_0 = 5.29 \times 10^{-11}\,\text{m}$.

Ex 4.3 (a) This is an application of Equations 4.1 and 4.2, with the reduced mass a factor of 186 larger than for the hydrogen atom. For muonic hydrogen, therefore, the energy of the nth state is $E_n = -186 \times 13.6\,\text{eV}/n^2 = -2.53\,\text{keV}/n^2$. For a transition from $n = 2$ to $n = 1$, the emitted photon has energy

$$\left(\frac{1}{1^2} - \frac{1}{2^2}\right) \times 2.53\,\text{keV} = 1.90\,\text{keV}.$$

(b) To calculate $\langle r \rangle$ for the ground state, we apply Equation 4.7 with $\mu_\mu/\mu_H = 186$. Noting that $\langle r \rangle$ for the hydrogen ground state is $\tfrac{3}{2}a_0$, for the ground state of muonic hydrogen we have

$$\langle r \rangle = \tfrac{3}{2} \times (5.29 \times 10^{-11}\,\text{m}/186)$$
$$= 4.27 \times 10^{-13}\,\text{m}.$$

Ex 4.4 The reduced mass of a muon in an orbit about a nucleus containing A nucleons is

$$\mu = \frac{1840 A m_e \times 207 m_e}{1840 A m_e + 207 m_e}$$
$$= \frac{1840 A}{1840 A + 207} \times 207 m_e.$$

For $A = 1$, $\mu = 186 m_e$, as we found before for muonic hydrogen, but as A becomes larger, around 120 for tin, the second term in the denominator becomes negligible compared to the first, and μ closely approaches $207 m_e$.

With this in mind, Equation 4.7 implies that $\langle r \rangle$ in muonic tin will be some 207 times smaller than in ordinary tin (rather than 186 times smaller, as in muonic hydrogen).

Ex 4.5 An antiproton, like a proton, has a mass of about $1840 m_e$ compared to the muon's mass of $207 m_e$. Since the mass of a heavy nucleus is many times larger than the mass of either a muon or an antiproton, the ratio of the reduced masses can be taken to be the ratio of the masses of the antiproton and muon. From Equation 4.7 we deduce that measures of radial extent such as $\langle r \rangle$ will all be smaller for antiprotons than for muons by a factor of $207/1840 = 0.113$.

Ex 4.6 We have $c = f\lambda$ and $E_{\text{photon}} = hf$, so $\lambda = hc/E_{\text{photon}}$, where $E_{\text{photon}} = \tfrac{3}{4} \times (Z-1)^2 \times 13.6\,\text{eV}$. The wavelength of the K_α copper line is

$$\lambda_{\text{Cu}} = \frac{6.63 \times 10^{-34}\,\text{J s} \times 3.00 \times 10^8\,\text{m s}^{-1} \times 4}{3 \times 28^2 \times 13.6 \times 1.60 \times 10^{-19}\,\text{J}}$$
$$= 1.55 \times 10^{-10}\,\text{m}.$$

For tungsten, we have

$$\lambda_W = \frac{6.63 \times 10^{-34}\,\text{J s} \times 3.00 \times 10^8\,\text{m s}^{-1} \times 4}{3 \times 73^2 \times 13.6 \times 1.60 \times 10^{-19}\,\text{J}}$$
$$= 2.29 \times 10^{-11}\,\text{m}.$$

The tungsten K_α line has a much shorter wavelength than the copper K_α line because tungsten has a higher atomic number Z than copper, and therefore a higher $(Z-1)^2$ factor in the expression for the photon energy.

Ex 4.7 We need to evaluate

$$\tfrac{1}{2}\left[j(j+1) - l(l+1) - \tfrac{3}{4}\right]\hbar^2$$

for $j = l - \tfrac{1}{2}$ and for $j = l + \tfrac{1}{2}$. For $j = l - \tfrac{1}{2}$, we find

$$\tfrac{1}{2}\left[(l - \tfrac{1}{2})(l + \tfrac{1}{2}) - l^2 - l - \tfrac{3}{4}\right]\hbar^2 = -(l+1)\frac{\hbar^2}{2},$$

and for $j = l + \tfrac{1}{2}$, we find

$$\tfrac{1}{2}\left[(l + \tfrac{1}{2})(l + \tfrac{3}{2}) - l^2 - l - \tfrac{3}{4}\right]\hbar^2 = l\,\frac{\hbar^2}{2}.$$

For the case $l = 0$, the orbital angular momentum is zero, so there can be no spin–orbit interaction. In this case we have $j = l + \tfrac{1}{2} = \tfrac{1}{2}$ only, so that $\left[j(j+1) - l(l+1) - \tfrac{3}{4}\right]\hbar^2 = 0$ as expected.

Ex 4.8 (a) The ground state has $n = 1$ and $j = \tfrac{1}{2}$, so the second term multiplying $-E_R/n^2$ is $\alpha^2(1 - \tfrac{3}{4})$, which is positive. So the negative energy is sightly increased in magnitude, i.e. the energy is lowered.

(b) There is a factor α^2/n in the perturbation correction in Equation 4.27, so the level splitting scales roughly as $1/n$.

(c) For $n = 3$, j can be $\frac{1}{2}$, $\frac{3}{2}$ or $\frac{5}{2}$. For these three cases, the term in parentheses in Equation 4.27 is, respectively, $\frac{3}{4}$, $\frac{1}{4}$, $\frac{1}{12}$. The difference between the first two terms is three times larger than the difference between the second and last terms, which shows that the splitting between the $j = \frac{1}{2}$ and $j = \frac{3}{2}$ levels is greater than the splitting between the $j = \frac{3}{2}$ and $j = \frac{5}{2}$ levels.

(d) The perturbation is very small because of the α^2 term (we are considering 'fine structure' after all), and would be invisible without being magnified by a large factor. A factor of $1/\alpha^2 \sim 137^2$ is a natural factor to use.

(e) There is no l in Equation 4.27.

Ex 4.9 According to Equations 4.1 and 4.2, the energy of the ground state ($n = 1$) is proportional to the reduced mass of the system. In this case $\mu = \frac{1}{2}m_e$, half of the reduced mass of a hydrogen atom (to one part in 1840), so the energy of the ground state is $-13.6\,\text{eV}/2 = -6.8\,\text{eV}$. Therefore at least $6.8\,\text{eV}$ must be supplied in order for positronium in its ground state to be split into a free electron and a positron.

Ex 5.1 We have

$$\sum_{i=1}^{4}\sum_{j>i}^{4} A_{ij} = (A_{12} + A_{13} + A_{14}) \\ + (A_{23} + A_{24}) + A_{34},$$

so each pair of distinct indices appears just once.

Ex 5.2 We have

$$\left(\widehat{h}_1 + \widehat{h}_2\right) \phi_r(\mathbf{r}_1) \phi_s(\mathbf{r}_2) \\ = \left[\widehat{h}_1 \phi_r(\mathbf{r}_1)\right]\phi_s(\mathbf{r}_2) + \phi_r(\mathbf{r}_1)\left[\widehat{h}_2 \phi_s(\mathbf{r}_2)\right] \\ = \left[E_r \phi_r(\mathbf{r}_1)\right]\phi_s(\mathbf{r}_2) + \phi_r(\mathbf{r}_1)\left[E_s \phi_s(\mathbf{r}_2)\right] \\ = (E_r + E_s) \phi_r(\mathbf{r}_1) \phi_s(\mathbf{r}_2).$$

Similarly,

$$\left(\widehat{h}_1 + \widehat{h}_2\right) \phi_s(\mathbf{r}_1) \phi_r(\mathbf{r}_2) \\ = \left[\widehat{h}_1 \phi_s(\mathbf{r}_1)\right]\phi_r(\mathbf{r}_2) + \phi_s(\mathbf{r}_1)\left[\widehat{h}_2 \phi_r(\mathbf{r}_2)\right] \\ = \left[E_s \phi_s(\mathbf{r}_1)\right]\phi_r(\mathbf{r}_2) + \phi_s(\mathbf{r}_1)\left[E_r \phi_r(\mathbf{r}_2)\right] \\ = (E_r + E_s) \phi_s(\mathbf{r}_1) \phi_r(\mathbf{r}_2).$$

Adding these equations and multiplying by $1/\sqrt{2}$ establishes the result.

Ex 5.3 The ground state of helium is described by a spatial wave function in which both electrons occupy the same orbital. This is symmetric under exchange of electron labels, so the spin ket must be antisymmetric in order to produce an antisymmetric total wave function. The spin ket is therefore the singlet state $|0,0\rangle$.

Comment: Historically, this worked the other way around: the experimental fact that the ground state of helium is a singlet, rather than a triplet, provided the primary evidence that led Pauli to his exclusion principle.

Ex 5.4 We know that s shells can hold at most 2 electrons, p shells can hold at most 6 electrons, and d shells can hold at most 10 electrons. Using the ordering of Figure 5.4, the required ground-state configurations are as follows.

O: $1s^2\,2s^2\,2p^4$
Ne: $1s^2\,2s^2\,2p^6$
Na: $1s^2\,2s^2\,2p^6\,3s$
Ar: $1s^2\,2s^2\,2p^6\,3s^2\,3p^6$
Fe: $1s^2\,2s^2\,2p^6\,3s^2\,3p^6\,4s^2\,3d^6$

The last configuration arises because iron follows the general rule that the 4s shell is lower in energy than the 3d shell (see Figure 5.4).

Ex 5.5 The maximum number of electrons with quantum number n is

$$\sum_{l=0}^{n-1} 2(2l + 1) = 4\sum_{l=0}^{n-1} l + 2\sum_{l=0}^{n-1} 1 \\ = 2n(n-1) + 2n = 2n^2.$$

Ex 5.6 (a) Sulphur will have the larger first ionization energy. Sulphur and magnesium belong to the same period. Sulphur has the higher Z, and more electrons in the valence shell. But electrons in the same shell are not very effective at screening the nuclear charge, so the valence electrons in sulphur will be more tightly bound.

(b) Barium will have the larger atomic radius. Magnesium and barium belong to the same group, with the same number of valence electrons, but the quantum number n is higher for barium, so its valence electrons are less tightly bound, leading to a larger radius.

Solutions to exercises

Ex 5.7 The 1s shell is closed, so we only need to consider the electron in the 2s shell. For this electron, $l_i = 0$ and $s_i = \frac{1}{2}$. So $L = 0$ and $S = \frac{1}{2}$.

Ex 5.8 $1s^2$ and $2s^2$ are closed shells, so we only need to consider the electrons in the 2p and 3p orbitals. We have $l_1 = l_2 = 1$, so the minimum and maximum values of L are

$$L_{\min} = |l_1 - l_2| = |1 - 1| = 0,$$
$$L_{\max} = l_1 + l_2 = 1 + 1 = 2.$$

The values of L increase in steps of 1, so the possible values are $L = 0$, $L = 1$ and $L = 2$. The corresponding values of M_L are

for $L = 0$: $M_L = 0$,
for $L = 1$: $M_L = -1, 0, 1$,
for $L = 2$: $M_L = -2, -1, 0, 1, 2$.

For spin, $s_1 = s_2 = \frac{1}{2}$, so the minimum and maximum values of S are

$$S_{\min} = |s_1 - s_2| = |\tfrac{1}{2} - \tfrac{1}{2}| = 0,$$
$$S_{\max} = s_1 + s_2 = \tfrac{1}{2} + \tfrac{1}{2} = 1.$$

The values of S increase in steps of 1, so the possible values are $S = 0$ and $S = 1$. When $S = 0$, M_S is 0. When $S = 1$, M_S is equal to $-1, 0$ or 1.

Ex 5.9 The 1s 2s configuration gives $L = 0$ and $S = 0$ or $S = 1$, so the possible atomic terms are ^{1}S and ^{3}S. For fixed L and S, the degeneracy of a term is $(2L + 1)(2S + 1)$, so the degeneracies of these terms are $1 \times 1 = 1$ and $1 \times 3 = 3$, respectively.

The 1s 2p configuration gives $L = 1$ and $S = 0$ or $S = 1$, so the possible atomic terms are ^{1}P and ^{3}P. The degeneracies of these terms are $3 \times 1 = 3$ and $3 \times 3 = 9$, respectively.

All of these terms are marked in Figure 5.2.

Ex 5.10 (a) For $L = S = 0$, we can only have $J = 0$ and hence $M_J = 0$.

(b) For $L = 2$, $S = 0$, the minimum value of J is $J_{\min} = |2 - 0| = 2$, and the maximum value of J is $J_{\max} = 2 + 0 = 2$, so $J = 2$ and $M_J = -2, -1, 0, 1, 2$.

(c) For $L = 1$, $S = 1$, we have $J_{\min} = |1 - 1| = 0$ and $J_{\max} = 1 + 1 = 2$, so the possible values of J are 0, 1 or 2. The corresponding values of M_J are: for $J = 0$, $M_J = 0$; for $J = 1$, $M_J = -1, 0, 1$; and for $J = 2$, $M_J = -2, -1, 0, 1, 2$.

Ex 5.11 (a) The term ^{3}P has $L = 1$, $S = 1$, so J takes the values $0, 1, 2$. Hence the levels are: ^{3}P$_0$, ^{3}P$_1$ and ^{3}P$_2$. The degeneracy of a level is $(2J + 1)$, so these three levels have degeneracies $(2 \times 0 + 1) = 1$, $(2 \times 1 + 1) = 3$ and $(2 \times 2 + 1) = 5$.

(b) The term ^{1}F has $L = 3$, $S = 0$, so we have only $J = 3$ and only one level: ^{1}F$_3$, which has degeneracy $(2 \times 3 + 1) = 7$.

(c) The term ^{2}D has $L = 2$, $S = \frac{1}{2}$, so J takes the values $\frac{3}{2}, \frac{5}{2}$. The levels are: ^{2}D$_{3/2}$ and ^{2}D$_{5/2}$, and these have degeneracies $(2 \times \frac{3}{2} + 1) = 4$ and $(2 \times \frac{5}{2} + 1) = 6$.

Ex 5.12 Hund's first rule shows that the quartets ^{4}P and ^{4}F are lower in energy than doublets ^{2}P, ^{2}D, ^{2}F, ^{2}G and ^{2}H. Hund's second rule tells us that, among the quartets, F ($L = 3$) is lower in energy than P ($L = 1$). Among the doublets, H ($L = 5$) is lower in energy than G ($L = 4$), which is lower in energy than D ($L = 2$), which is lower in energy than P ($L = 1$). So the required energy order is

$$^4\text{F} < {}^4\text{P} < {}^2\text{H} < {}^2\text{G} < {}^2\text{F} < {}^2\text{D} < {}^2\text{P}.$$

Ex 6.1 Yes because each electronic state is associated with a different electronic energy $E_{\text{el}}(\mathbf{R}_{\text{AB}})$, and therefore provides a different effective potential energy function for the nuclei.

Ex 6.2 (a) When R tends to zero, the two atomic orbitals become centred on the same point. We then have

$$S = \langle \phi_{1s}^{\text{A}} | \phi_{1s}^{\text{B}} \rangle \to \langle \phi_{1s}^{\text{A}} | \phi_{1s}^{\text{A}} \rangle = 1.$$

(b) When R tends to infinity, the overlap between the two atomic orbitals becomes vanishingly small, so $S \to 0$.

Ex 6.3 Substituting $\langle E \rangle = (H_{\text{AA}} + H_{\text{AB}})/(1 + S)$ into Equation 6.19 gives

$$c_1 \left(H_{\text{AA}} - \frac{H_{\text{AA}} + H_{\text{AB}}}{1 + S} \right) + c_2 \left(H_{\text{AB}} - S \frac{H_{\text{AA}} + H_{\text{AB}}}{1 + S} \right) = 0.$$

Multiplying through by $1 + S$ and rearranging, we obtain

$$c_1 (S H_{\text{AA}} - H_{\text{AB}}) + c_2 (H_{\text{AB}} - S H_{\text{AA}}) = 0,$$

which gives $c_1 = c_2$. We get the same result by starting with Equation 6.20, because of symmetry between c_1 and c_2.

Ex 6.4 Substituting $\langle E \rangle = (H_{AA} - H_{AB})/(1-S)$ into Equation 6.19 gives

$$c_1\left(H_{AA} - \frac{H_{AA} - H_{AB}}{1-S}\right) + c_2\left(H_{AB} - S\frac{H_{AA} - H_{AB}}{1-S}\right) = 0.$$

Multiplying through by $1-S$ and rearranging, we obtain

$$c_1(-SH_{AA} + H_{AB}) + c_2(H_{AB} - SH_{AA}) = 0,$$

which gives $c_1 = -c_2$, so the linear combination of atomic orbitals is of the form $c_1\left(\phi_{1s}^A(\mathbf{r}) - \phi_{1s}^B(\mathbf{r})\right)$.

Ex 6.5 With

$$\psi_{exc}(\mathbf{r}) = \frac{1}{\sqrt{2(1-S)}}\left(\phi_{1s}^A(\mathbf{r}) - \phi_{1s}^B(\mathbf{r})\right),$$

we have

$$\langle\psi_{exc}|\psi_{exc}\rangle = \frac{1}{2(1-S)}\big[\langle\phi_{1s}^A|\phi_{1s}^A\rangle + \langle\phi_{1s}^B|\phi_{1s}^B\rangle$$
$$- \langle\phi_{1s}^A|\phi_{1s}^B\rangle - \langle\phi_{1s}^B|\phi_{1s}^A\rangle\big]$$
$$= \frac{1}{2(1-S)}[1 + 1 - S - S] = 1,$$

so $\psi_{exc}(\mathbf{r})$ is normalized.

Since

$$\psi_{gs}(\mathbf{r}) = \frac{1}{\sqrt{2(1+S)}}\left(\phi_{1s}^A(\mathbf{r}) + \phi_{1s}^B(\mathbf{r})\right),$$

we have

$$\langle\psi_{exc}|\psi_{gs}\rangle = \frac{1}{2\sqrt{1-S^2}}\left(\langle\phi_{1s}^A| - \langle\phi_{1s}^B|\right)\left(|\phi_{1s}^A\rangle + |\phi_{1s}^B\rangle\right)$$
$$= \frac{1}{2\sqrt{1-S^2}}\big[\langle\phi_{1s}^A|\phi_{1s}^A\rangle - \langle\phi_{1s}^B|\phi_{1s}^B\rangle$$
$$+ \langle\phi_{1s}^A|\phi_{1s}^B\rangle - \langle\phi_{1s}^B|\phi_{1s}^A\rangle\big]$$
$$= \frac{1}{2\sqrt{1-S^2}}[1 - 1 + S - S] = 0,$$

so $\psi_{exc}(\mathbf{r})$ is orthogonal to $\psi_{gs}(\mathbf{r})$.

Ex 6.6 For $n = 3$, the allowed values of l are 0, 1 or 2, corresponding to 3s, 3p and 3d atomic orbitals.

For $l = 0$, m can only take the value 0. The corresponding 3s atomic orbitals produce bonding and antibonding σ molecular orbitals, each of which is non-degenerate.

For $l = 1$, m can take the values 0 and ± 1, corresponding to $3p_0$, $3p_{+1}$ and $3p_{-1}$ atomic orbitals. The $3p_0$ atomic orbitals produce bonding and antibonding σ molecular orbitals, each of which is non-degenerate. The $3p_{+1}$ and $3p_{-1}$ orbitals produce bonding and antibonding π molecular orbitals, and each is doubly-degenerate.

For $l = 2$, m can take the values 0, ± 1 and ± 2, corresponding to $3d_0$, $3d_{+1}$, $3d_{-1}$, $3d_{+2}$ and $3d_{-2}$ atomic orbitals. The $3d_0$ atomic orbitals produce bonding and antibonding σ molecular orbitals, each of which is non-degenerate. The $3d_{+1}$ and $3d_{-1}$ orbitals produce bonding and antibonding π molecular orbitals, and each is doubly-degenerate. The $3d_{+2}$ and $3d_{-2}$ orbitals produce bonding and antibonding δ molecular orbitals, and each is doubly-degenerate.

In total, there are $1 + (1+2) + (1+2+2) = 9$ bonding orbitals, and the same number of antibonding orbitals. Three of the bonding orbitals are non-degenerate, and the other six are in three doubly-degenerate pairs. The same is true for the antibonding orbitals.

Ex 6.7 There are 10 electrons in bonding orbitals ($1\sigma_g$, $2\sigma_g$, $3\sigma_g$ and $1\pi_u$), and 6 in antibonding orbitals, so the formal bond order of O_2 is $(10-6)/2 = 2$.

Ex 6.8 O_2^- has the electronic configuration

$$1\sigma_g^2\, 1\sigma_u^2\, 2\sigma_g^2\, 2\sigma_u^2\, 3\sigma_g^2\, 1\pi_u^4\, 1\pi_g^3.$$

It therefore has 10 electrons in bonding orbitals and 7 in antibonding orbitals giving a formal bond order of 1.5. This is less than the formal bond order for O_2, which is 2. Since O_2^- and O_2 belong to the same row of the Periodic Table, the dissociation energy of O_2^- is less than that for O_2.

Ex 6.9 (a) The ground-state electron configuration is

$$1\sigma_g^2\, 1\sigma_u^2\, 2\sigma_g^2\, 2\sigma_u^2\, 1\pi_u^4.$$

The C_2 molecule contains 12 electrons and in the ground state, these electrons occupy the 12 states with lowest energies. The σ orbitals corresponding to the lowest four energy levels in Figure 6.14b can each accommodate two electrons and the degenerate $1\pi_u$ orbitals can accommodate the other four electrons.

(b) The σ_g and π_u orbitals are bonding, and the σ_u orbitals are antibonding. So for C_2, there are 8 electrons in bonding orbitals and 4 electrons in antibonding orbitals, and the formal bond order is $(8-4)/2 = 2$.

Ex 7.1 Non-directional metallic bonding encourages atoms to pack closely together, leading to dense structures with twelve nearest neighbours. This does not happen in covalently-bonded solids because covalent bonds generally have preferred angles relative to one another. This leads to open structures in which each atom has only a few nearest neighbours.

Ex 7.2 In each period, the cohesive energy per atom is large when the d band (capacity 10 electrons per atom) is roughly half full. Let us assume that all the atoms in a given period have similar d bands in the solid. Then, elements with around 5 d-electrons per atom are strongly bound because all their electrons occupy bonding orbitals with energies below the d level in an isolated atom. Elements with a small number of d electrons are less strongly bound because some of their bonding orbitals are unoccupied, while elements with nearly 10 electrons have electrons in antibonding orbitals, cancelling to some extent the bonding effect of other electrons.

Comment: It is also noticeable that transition elements from Period 6 tend to be more strongly bound than transition elements from Period 5. This can be explained by the fact that the 5d atomic orbitals in Period 6 are more extended than the 4d atomic orbitals in Period 5. This leads to a greater overlap of wave functions, and hence a greater band width, in Period 6.

Ex 7.3 In a Bloch wave, $\psi(\mathbf{r}) = e^{i\mathbf{k}\cdot\mathbf{r}} u_{\mathbf{k}}(\mathbf{r})$, so the electron probability density is given by

$$|\psi(\mathbf{r})|^2 = \psi^*(\mathbf{r})\,\psi(\mathbf{r})$$
$$= \left[e^{-i\mathbf{k}\cdot\mathbf{r}} u_{\mathbf{k}}^*(\mathbf{r})\right] \times \left[e^{i\mathbf{k}\cdot\mathbf{r}} u_{\mathbf{k}}(\mathbf{r})\right]$$
$$= |u_{\mathbf{k}}(\mathbf{r})|^2.$$

The required result follows because Bloch's theorem guarantees that $u_{\mathbf{k}}(\mathbf{r})$ has the periodicity of the lattice.

Ex 7.4 For any function $f(\mathbf{r})$ we have

$$\widehat{T}(\mathbf{R})\big[\widehat{H}f(\mathbf{r})\big] = \widehat{H}f(\mathbf{r}+\mathbf{R}) = \widehat{H}\big[\widehat{T}(\mathbf{R})f(\mathbf{r})\big]$$

because $\widehat{H}$ is not affected by translations through lattice vectors. Since this equation is true for any $f(\mathbf{r})$, we can write $\widehat{T}(\mathbf{R})\widehat{H} = \widehat{H}\widehat{T}(\mathbf{R})$ and say that the lattice translation operator commutes with the Hamiltonian. We also have

$$\widehat{T}(\mathbf{R}_1)\widehat{T}(\mathbf{R}_2)f(\mathbf{r}) = f(\mathbf{r}+\mathbf{R}_1+\mathbf{R}_2)$$
$$= \widehat{T}(\mathbf{R}_2)\widehat{T}(\mathbf{R}_1)f(\mathbf{r}).$$

Since this is true for any $f(\mathbf{r})$, we can write $\widehat{T}(\mathbf{R}_1)\widehat{T}(\mathbf{R}_2) = \widehat{T}(\mathbf{R}_2)\widehat{T}(\mathbf{R}_1)$ and say that the lattice translation operators commute with one another.

Ex 7.5 Taking the square of the modulus of both sides of Equation 7.7 gives

$$|\psi(\mathbf{r}+\mathbf{R})|^2 = \left|e^{i\mathbf{k}\cdot\mathbf{R}}\psi(\mathbf{r})\right|^2 = |\psi(\mathbf{r})|^2,$$

as required.

Ex 7.6 For nearest neighbours i and j in the arrangement of atoms shown in Figure 7.8, the possible values of $\mathbf{R}_i - \mathbf{R}_j$ are $\pm a\mathbf{e}_x$, $\pm a\mathbf{e}_y$ and $\pm a\mathbf{e}_z$, and the corresponding values of $\mathbf{k}\cdot(\mathbf{R}_i - \mathbf{R}_j)$ are $\pm k_x a$, $\pm k_y a$ and $\pm k_z a$. So

$$f(\mathbf{k}) = \sum_{j=\text{ nn of } i} e^{-i\mathbf{k}\cdot(\mathbf{R}_i-\mathbf{R}_j)}$$
$$= e^{-ik_x a} + e^{ik_x a} + e^{-ik_y a} + e^{ik_y a}$$
$$+ e^{-ik_z a} + e^{ik_z a}$$
$$= 2\cos(k_x a) + 2\cos(k_y a) + 2\cos(k_z a),$$

where we have used the fact that $\cos\theta = (e^{i\theta} + e^{-i\theta})/2$. Hence

$$E(\mathbf{k}) = E_0 - \beta f(\mathbf{k})$$
$$= E_0 - 2\beta\big[\cos(k_x a) + \cos(k_y a) + \cos(k_z a)\big],$$

as required.

Ex 7.7 As mentioned in Subsection 7.2.1, different values of $\mathbf{k}$ can label the same state. According to Equation 7.13, the values of k_x, k_y and k_z should all be in the range between $-\pi/a$ and $+\pi/a$. Any change by $2\pi/a$ will take us outside this range, and will only lead to another way of labelling the same state.

Ex 7.8 As β decreases, the width of the energy band decreases, and the effective mass increases, so the effective mass is expected to be larger near the bottom of a narrow band than near the bottom of a wide band.

Ex 7.9 Because of thermal expansion, the average distance between neighbouring atoms increases with temperature. According to Figure 7.13, a larger interatomic separation implies a smaller band gap for silicon and germanium.

Ex 7.10 The number density of holes in the valence band is the same as the number density of electrons in the conduction band, 2.4×10^{19} m^{-3}. This is because each electron in the conduction band of a pure semiconductor arises from the thermal excitation of an electron in the valence band, and leaves a hole behind. For future reference, note that this applies only to a pure semiconductor (one that contains no impurities).

Ex 7.11 Ignoring small effects associated with reduced mass, the Bohr radius is

$$a_0 = \frac{4\pi\varepsilon_0\hbar^2}{m_e e^2} = 5.29 \times 10^{-11}\,\text{m}.$$

The corresponding scaled Bohr radius for an electron bound to a donor impurity in germanium is found by replacing m_e by $m_e^* = 0.12 m_e$, and replacing ε_0 by $\varepsilon\varepsilon_0$, where $\varepsilon = 15.8$. We therefore find a scaled Bohr radius of

$$(15.8/0.12) \times 5.29 \times 10^{-11}\,\text{m} = 6.97 \times 10^{-9}\,\text{m}.$$

Comment: The large value of the scaled Bohr radius justifies our implicit assumption that germanium can be treated as a continuous background medium, characterized by a relative permittivity ε.

Ex 7.12 The calculation is exactly the same as for the ionization energy of a donor atom in germanium, except that we must use the reduced mass μ of an exciton, which is a bound state of an electron and a hole. Since both particles in this system are assumed to have the same effective mass, m_e^*, we have

$$\mu = \frac{m_e^* m_e^*}{m_e^* + m_e^*} = 0.5 m_e^*.$$

The binding energy is therefore estimated to be $0.5 \times 6.5\,\text{meV} = 3.3\,\text{meV}$.

Ex 7.13 There is no band gap between the valence band and the conduction band of a semimetal. Unlike a semiconductor, the electrical conductivity of a semimetal would be expected to fall with temperature and be fairly insensitive to small concentrations of any impurity.

Ex 8.1 Each photon has wavelength λ, frequency $f = c/\lambda$ and energy

$$E_{\text{photon}} = \frac{hc}{\lambda}$$
$$= \frac{6.63 \times 10^{-34}\,\text{J s} \times 3 \times 10^8\,\text{m s}^{-1}}{5.0 \times 10^{-7}\,\text{m}}$$
$$= 4.0 \times 10^{-19}\,\text{J}.$$

The power of the transmitter is the number of photons emitted per second times the energy of each photon.

The number of photons emitted per second is therefore

$$\frac{\mathrm{d}N}{\mathrm{d}t} = \frac{100\,\text{W}}{4.0 \times 10^{-19}\,\text{J}} = 2.5 \times 10^{20}\,\text{s}^{-1}.$$

This vast rate of photon production would be even higher if the source were more powerful or if it operated at longer wavelengths.

Ex 8.2 No. In first-order time-dependent perturbation theory, Equation 8.20 shows that $a_j^{(1)}(t) = 0$ for all t if $V_{ji}(t) = 0$ and $j \neq i$.

Comment: This result is only true in first-order perturbation theory. Referring back to the exact result in Equation 8.13, we see that a_j can depend on time (and hence become non-zero) even if $V_{ji} = 0$. This is possible because, once some of the other coefficients a_k on the right-hand side of Equation 8.13 become non-zero, there are contributions to $a_j(t)$ from terms like $\mathrm{e}^{\mathrm{i}\omega_{jk} t} V_{jk} a_k$, with $k \neq i$. These contributions are usually very small because they are the product of two small terms; they are neglected in first-order perturbation theory, and in the remainder of this chapter.

Ex 8.3 From Equation 8.22, the required probability is

$$P_{i \to f} = \frac{U^2}{\hbar^2} \left| \int_0^t \mathrm{e}^{\mathrm{i}\omega_{fi} t'}\,\mathrm{d}t' \right|^2.$$

The integral inside the modulus signs is just

$$\int_0^t \mathrm{e}^{\mathrm{i}\omega_{fi} t'}\,\mathrm{d}t' = \left[\frac{\mathrm{e}^{\mathrm{i}\omega_{fi} t'}}{\mathrm{i}\omega_{fi}}\right]_0^t$$
$$= \frac{1}{\mathrm{i}\omega_{fi}}[\mathrm{e}^{\mathrm{i}\omega_{fi} t} - 1]$$
$$= \frac{2\mathrm{e}^{\mathrm{i}\omega_{fi} t/2}}{\omega_{fi}}\sin(\omega_{fi} t/2),$$

where we have used Equation 8.23 in the last step. Hence the probability of finding the system in state f is

$$P_{i \to f} = \frac{U^2}{\hbar^2} \left[\frac{\sin(\omega_{fi} t/2)}{(\omega_{fi}/2)}\right]^2.$$

Ex 8.4 Since $Y_{lm}(\theta,\phi)$ is proportional to $\mathrm{e}^{\mathrm{i}m\phi}$, the integral over ϕ in Equation 8.28 is

$$\int_0^{2\pi} [\mathrm{e}^{\mathrm{i}m_f \phi}]^* \mathrm{e}^{\mathrm{i}m_i \phi}\,\mathrm{d}\phi = \int_0^{2\pi} \mathrm{e}^{-\mathrm{i}m_f \phi}\mathrm{e}^{\mathrm{i}m_i \phi}\,\mathrm{d}\phi$$
$$= \int_0^{2\pi} \mathrm{e}^{\mathrm{i}(m_i - m_f)\phi}\,\mathrm{d}\phi$$
$$= 2\pi\delta_{m_f,m_i},$$

so the dipole matrix element in Equation 8.28 will vanish unless $m_f = m_i$.

Ex 8.5 The dipole matrix elements in Equation 8.31 include the following integral over ϕ:

$$\int_0^{2\pi} [e^{im_f \phi}]^* e^{\pm i\phi} e^{im_i \phi} \, d\phi$$

$$= \int_0^{2\pi} e^{-im_f \phi} e^{\pm i\phi} e^{im_i \phi} \, d\phi$$

$$= \int_0^{2\pi} e^{i(m_i - m_f \pm 1)\phi} \, d\phi,$$

which is equal to zero unless $m_i - m_f \pm 1 = 0$. Hence the transition is forbidden unless $m_f = m_i \pm 1$.

Ex 8.6 (a) According to our theory, no transitions should take place between singlet and triplet states because the dipole matrix element connecting singlet and triplet states is equal to zero. For example, suppose that the two electrons are initially in a singlet state described by the total wave function

$$\Psi_i = \psi_i(\mathbf{r}_1, \mathbf{r}_2) \, |S = 0, M_S = 0\rangle$$

and are finally in a triplet state described by

$$\Psi_f = \psi_f(\mathbf{r}_1, \mathbf{r}_2) \, |S = 1, M_S = 0\rangle.$$

We shall see that the nature of $\psi_i(\mathbf{r}_1, \mathbf{r}_2)$ or $\psi_f(\mathbf{r}_1, \mathbf{r}_2)$ doesn't enter the discussion. The relevant matrix element for this transition is $\langle \Psi_f | \widehat{V} | \Psi_i \rangle$, where $\widehat{V} = e\,\boldsymbol{\mathcal{E}}(t) \cdot (\mathbf{r}_1 + \mathbf{r}_2)$ is the electric dipole interaction. The operator $\widehat{V}$ depends only on the positions of the particles and is independent of their spin, so we have

$$\langle \Psi_f | \widehat{V} | \Psi_i \rangle$$
$$= \langle S = 1, M_S = 0 | S = 0, M_S = 0 \rangle \langle \psi_f | \widehat{V} | \psi_i \rangle,$$

and this is equal to zero because the singlet and triplet spin vectors are orthogonal.

(b) Excited triplet states can be produced in a number of ways. The electric dipole interaction is not the only interaction between light and matter; the magnetic field also contributes, and this can convert singlet states into triplet states and vice versa. Moreover, transitions need not be radiative (i.e. associated with the emission of photons). Atoms in a gas can bump into one another, exchanging energy as they do so. Non-radiative transitions are not limited by selection rules.

Ex 8.7 (a) The range of angular frequencies is

$$\delta\omega = \frac{4\pi}{t} = \frac{4\pi}{3 \times 10^{-14}\,\mathrm{s}} = 4.2 \times 10^{14}\,\mathrm{s}^{-1}.$$

(b) For the longer interaction time, the range of angular frequencies is smaller:

$$\delta\omega = \frac{4\pi}{t} = \frac{4\pi}{3 \times 10^{-3}\,\mathrm{s}} = 4.2 \times 10^{3}\,\mathrm{s}^{-1}.$$

Ex 8.8 The minimum time duration that maximizes the transition probability is

$$\frac{\pi}{|\omega_{21} - \omega|} = \frac{\pi}{1 \times 10^9\,\mathrm{s}^{-1}} = 3\,\mathrm{ns}.$$

Ex 8.9 'Everything else the same' implies that B_{12} does not change, but the ω_{21}^3-dependence in Equation 8.47 shows that A_{21} increases by a factor of a thousand. That in turn means that, since the average lifetime is inversely proportional to the probability of emitting a photon, it will be reduced by a factor of a thousand.

Ex 8.10 Yes, because the spontaneous emission rate is directly proportional to the stimulated emission rate.

Index

Items that appear in the Glossary have page numbers in **bold type**. Ordinary index items have page numbers in Roman type.

absorption of light   195, **196**, 211–5, 216
absorption spectrum   64, 145–6, 192, 198
acceptor atoms   **191**
acceptor energy level   191
A-coefficient   216–8
alkali atoms   **129**, 130
amorphous solid   **169**
Anderson, Carl   109
anharmonic oscillator   **78**
antibonding orbitals   **157**–60, 162–4, 171, 174
    electron probability density in   156
antiparticles   9, 89, **109**–10
antiproton   **99**
antiprotonic atom   99
antisymmetric
    molecular orbitals   158
    spatial function   118–20, 123, 165
    spin ket   119, 120
    total wave function   118, 133
arsenic donors   190
associated Legendre function   19
atomic core   171, 172, 188
atomic levels   **137**–8
    degeneracy   137
    Hund's rules for   138
    spectroscopic notation for   137
atomic mass   101
atomic number   91, 95, 101, 115
atomic orbitals   114, **117**
    degeneracy   125
    energy ordering   125, 127
    in central field approximation   124
    in diatomic molecules   148–9
    in independent particle model   117
    in tight-binding method   180–3
    radial functions   124
    spectroscopic notation for   124
    spherical harmonics for   124
atomic radius   130–1
atomic terms   **134**–8
    degeneracy   134, 136
    Hund's rules for   138
    multiplicity   136
    spectroscopic notation for   135–6

azimuthal angle   **12**, 147–8
azimuthal quantum number   10, 148

Balmer series   36, 38, 89, 91, 215
    for deuterium   51
Balmer, Johann   36
Balmer's formula   36, 38
band gap   **174**, 183, **189**
    in diamond   187
    in germanium   185, 187, 189
    in silicon   187, 189
band structure   **184**–5
B-coefficient   216–8
binomial expansion   104, 182
Bloch wave   **176**–80
    labelling of 179–80
    probability density of   176–7
Bloch, Felix   175–6
Bloch's theorem   169, **175**–80, 188
    and electrical conduction   176, 188
    and tight-binding method   180–1
    proof   177–79
Bohr model   35, **37**–8, 89, 93
    and Heisenberg's uncertainty principle   38
    and helium   38
Bohr orbit   **37**, 61
Bohr radius   **37**, 42, 90, 91–2, 117
Bohr, Neils   37, 56, 89, 93, 195, 197
Boltzmann distribution law   **217**
Boltzmann, Ludwig   217
Boltzmann's constant   185, 187, 217
bond order   163–5
bonding orbitals   **157**–60, 162–4, 171, 174
bonding
    covalent   171–2, 173
    ionic   171–2
    metallic   171–2
Born, Max   143
Born–Oppenheimer approximation   143–**144**, 145, 147, 153, 172
Born's rule
    in Cartesian coordinates   13
    in spherical coordinates   **14**
boron as acceptor   191

Index

Brackett series   38
braking radiation   99
bremsstrahlung   **99**
bronze age   33

calculus of variations   75
carbon atom energy levels   132, 138
cathode rays   141
central field   124
central force   12
central-field approximation   114, **124**–6, 130, 132, 133, 181
   self-consistent solutions   126
centre of mass   39, 143
centre of symmetry   **158**
centre-of-mass frame   39, 143
centrifugal barrier   **18**, 41, 125
ceramic   170
charmonium   **112**
chemical bond   141, 146
chemical bonding   152, 155
Clebsch, Alfred   29
closed shell   **127**, 134, 171
cohesive energy in a solid   174–5
commuting operators   28, 133, 134, 147, 178
commutation relations
   for orbital angular momentum operators   10
   for total angular momentum operators   27
compatible observables   11
conduction band   **189**–92
conductivity   see *electrical conductivity*
conductors   **186**–8
   in Periodic Table   187
configuration   see *electronic configuration*
conservation of momentum   61
constructive interference   152
copper
   band structure   184–5
   crystal structure   170
   electrical conductivity   188
   light absorption   184
Coulomb integral   **122**–3
Coulomb interaction in quantum field theory   111
Coulomb model
   of a hydrogen atom   **35**, 36–63, 95
   failure for muonic atoms   96
Coulomb potential energy   77
   due to electron–electron repulsion   115–7, 133, 143
   due to electron–nuclear attraction   115, 143
   due to proton–proton repulsion   154
   in a hydrogen atom   41
Coulomb's electrostatic force law   11, 36
covalent bonding   **171**
   and LCAO method   173
crystal defects   188
crystalline grains   170, 188
crystalline solid   **169**
crystal structure
   copper   170
   sodium chloride   171
   tungsten   170
current density   **189**

Darwin term   **106**, 109
de Broglie, Louis   36
degeneracy
   in hydrogen atom   46–7, 51–2
   of atomic levels   137
   of atomic orbitals   125
   of atomic terms   136
   of molecular orbitals   159–60, 163
   remaining with spin–orbit interaction   30
destructive interference   156
deuterium   50–1, 90, 91
diamond   169
   as crystalline solid   169
   as insulator   187
   as semiconductor   191
   electrical conductivity   186
diatomic molecule   141–66
   as anharmonic oscillator   78
Dirac equation   **108**–10
   and Darwin term   109
   and electron magnetic dipole moment   109
   and hydrogen atom energy levels   109
   and muonic atoms   97
   and relativistic kinetic energy correction   109
   and spin   109
   and spin–orbit interaction   109
Dirac, Paul   38, 108, 111, 195, 202
dissociation energy   155, 161–2, 165
donor atoms   **191**
donor energy level   191
doping   **190**, 192
double bond   **163**
doublet term   136
Drude, Paul   188

effective Hamiltonian   124
effective mass   **184**
    of electron   184, 188, 191
    of hole   192
effective potential energy function   41, 47
Einstein, Albert   103, 196, 216, 218
Einstein coefficients   **216**–8
electric current   186
electric dipole approximation   195, **201**, 207
electric dipole moment   **201**
electric field   201
electrical conductivity   186, 188
    of semiconductors   188–9, 192
    temperature dependence   188, 192
electrical resistance   176, 188
electromagnetic spectrum   145–6, 199–200
electromagnetic waves   199–200
electron–electron interaction   115, 117, 124, 132, 133, 143, 181
electronic configuration
    of atoms   **126**–9
    of molecules   **161**–5
electronic energy eigenfunction   **144**
electronic energy in a molecule   **144**, 146
electronic time-independent Schrödinger equation   **144**, 146–7
electron–positron annihilation   110
electron–positron pair   110–1
emission spectrum   35, 198
energy band width   174, 183–4
    in LCAO method   174
    in tight-binding method   183–4
energy bands   **174**
    for quantum dots   192
    in LCAO method   174
    in tight-binding method   183–4
    labelling of   174
energy curve   153, **154**, 155–9, 162
energy eigenfunctions
    as complete orthonormal set   68, 76, 203
    in crystals   173, 175–81
    in helium atom   120–1
    in hydrogen atom   51–8, 90
    in hydrogen molecule ion   152, 156, 166
    in many-electron atoms   116
energy eigenvalues
    in carbon atom   132
    in crystals   174, 181–4
    in helium atom   121–3
    in hydrogen atom   37, 44, 46, 52, 90, 106, 109
    in many-electron atoms   116
    in muonic atoms   94
energy zero   37, 118, 154, 159
equilibrium nuclear separation   **154**, 161, 162, 164, 165
equivalent electrons   **136**
even parity   **22**–3, 158
Ewen, Harold   108
exchange   114, 122, **123**
exchange integral   **122**–3
exciton   **190**
expectation value   58–9, 68, 83

Faraday, Michael   188
Fermi energy   **185**, 186
fermion   114, 118
Feynman, Richard   195
fine structure   **90**
    in hydrogen atom   64, 89, **102**–7
    relativistic kinetic energy   103–5
    spin–orbit perturbation   105–6
fine structure constant   **104**
first ionization energy   **130**–1
first-order approximation   **80**, 83, 205
first-order approximation for energy eigenvalue   **83**
first-order correction   80–2
first-order perturbation theory   see *perturbation theory*
first-order transition probability   206
fluorescent lighting tubes   197
Foley, Henry   111
forbidden transition   **209**
formal bond order   **163**–5
Fourier transform in three dimensions   63
Fowler, Alfred   93
full energy band, inability to conduct electricity   186

gamma rays   199–200
Gauss's law   95, 124
gerade orbital   **158**
germanium
    band gap   185, 189
    band structure   184–5
    effective electron mass   191
    electron number density   189, 190
    relative permittivity   191
glass, as amorphous solid   169
good quantum numbers   **28**, 132–**133**, 134, 137, 147
Gordan, Paul   29

Index

grains   170, 188
graphite, as semimetal   192
group in Periodic Table   **128**

half-life   93, 96
halogen atoms   **129**
Hamiltonian   **67**
Hamiltonian function   12, 39
Hamiltonian operator
    as Hermitian operator   82
    for atom in electromagnetic field   200–2
    for diatomic molecules   142–3
    for hydrogen atom   39
    for lithium atom   115–6
    in Cartesian coordinates   12
    in spherical coordinates   13–4
    including spin–orbit interaction   25
harmonic oscillator   45, 72–3, 78
Heisenberg uncertainty principle   38
Heisenberg, Werner   9, 175, 176, 188, 195
helium atom   114, 118–23
    and Bohr model   38
    and perturbation theory   120–3
    energy levels   123
    excited states   121–3
    ground state   120–1
    selection rules   211
    singlet states   123, 211
    spectrum   131, 211
    terms   123
    triplet states   123, 211
helium ion   92–3
helium molecule   162
hole   **189**–90
hole energy   190
homonuclear diatomic molecule   **158**, 161–5
Hund's rules   **138**
hydrogen atom
    and Darwin term   **106**
    and Dirac equation   109
    and finite size of proton   86
    and relativistic kinetic energy   103–5
    and spin–orbit interaction   77, 105–6
    Bohr model   **35**, 37–9
    centrifugal barrier   41
    Coulomb model   **35**, 36–63, 95
    degeneracy   46–7, 51–2
    effective potential energy function   41, 47
    energy eigenfunctions   51–8, 90,
    energy levels   37, 44, 46, 52, 90, 106, 109
    energy quantization   46
    expectation values   58–9, 90
    fine structure   64, 89, **102**–7
    Hamiltonian operator   39
    hyperfine structure   **107**–8, 111
    ionization energy   37
    momentum probability density   61–3
    probability density   53–5, 57
    quantum numbers   51–2
    radial equation   40
    radial functions   40, 47–50
    radial probability density   54–6
    reduced radial equation   40
    reduced radial function   40
    Rydberg states   **60**–1
    spectrum   9, 35, 36, 38, 64, 131, 198
    stationary states   39
    time-independent Schrödinger equation   39
    uncertainties   60
hydrogen chloride molecule
    absorption spectrum   145–6, 198
    electronic energy   146
    rotational energy   146, 198
    vibrational energy   146
hydrogen-like atom   **91**, 190
    scaled Bohr radius   92
    scaled Rydberg energy   92
hydrogen molecule   162
hydrogen molecule ion   141, **146**–57, 165–6
    dissociation energy   155, 165
    electronic time-independent Schrödinger equation   146–7
    energy curves   153–5, 157, 159
    equilibrium proton–proton separation   155, 165
    excited states   155–7
    good quantum numbers   147–8
    ground state   148–55, 153, 155
    stability   153
    total static energy   154
    trial function   148–9
    variational method   148–57
hyperfine structure in hydrogen atom   **107**–8, 111

incompatible observables   11
independent-particle model   114, **116**–7, 118, 120, 121
infrared light   145–6, 198
insulator   **186**
integrals in three dimensions, limits of   14

interatomic overlap integral 149
interference 152, 156
inversion **22**, 158
ionic bonding **171**
ionization energy 37, 191

jj-coupling scheme **138**

Kusch, Polycarp 111

Lamb shift **111**, 216
Lamb, Willis 111
Laplacian operator **13**
    in Cartesian coordinates 13
    in spherical coordinates 13–4
lasers 195, 197, 215, 218
lattice **170**
lattice point **170**
lattice translation operator **177**–8
lattice vector **170**, 175, 177–80
LCAO approximation **148**
    and tight binding 180
    for molecules 148–65
    for solids 172–4, 175, 177
    improvements on in molecules 165–6
    problems in crystals 174, 177
lead 169
levels see *atomic levels* or *molecular levels*
linear combination of atomic orbitals see *LCAO*
lithium atom 115–6, 128
lithium molecule 163–4
LS coupling scheme **137**
Lyman series 38

magnesium as conductor 187
magnetic dipole moment
    of electron 105, 107, 109, 111, 216
    of nucleus 200–1
    of proton 107
magnetic properties of oxygen molecule 141, 165
magnetic quantum number **10**, 18, 20
    in hydrogen atom 51
    in hydrogen molecule ion 148
    in molecular orbitals 157
mass number **95**
matrix elements
    in LCAO method 150
    in tight-binding method 181–2
    in time-dependent perturbation theory **204**, 208
    in time-independent perturbation theory **87**

mean radius of nucleus 97
Mendeleev, Dmitri 128
metallic bonding **171**–2, 173
microwaves 145–6, 199–200
modulated plane wave 176
molecular levels 166
molecular orbitals **148**, **157**–61
    degeneracy 159–60, 163
    energy ordering 163
    spectroscopic notation 157–8
molecular terms 166
molybdenum 100
momentum amplitude 63
momentum probability density in hydrogen 61–3
monochromatic light 199
Moseley, Henry 101
multiple integrals, limits of 14
multiplicity of terms **136**
muon **93**–6
muonic atom 93, **94**, 95–8
mutually-commuting operators 17, 25, 27–8, 51, 147, 178

neutron mass 92
neutron distribution in nuclei 99
Newton, Isaac 76
nitrogen molecule 141, 164
noble gases **129**
nodal surfaces of hydrogen atom energy eigenfunctions **57**–8
nodeless radial functions for hydrogen atom 43–4
non-equivalent electrons 136
non-radiative transitions 197
non-separable equation 13, 206
normalization
    in perturbation theory 82
    in spherical coordinates 14
    of hydrogen atom radial functions 48
    of molecular orbital 152
    of spherical harmonics 22
nth-order approximation **80**
n-type semiconductor **191**
nuclear eigenfunctions 145
nuclear time-independent Schrödinger equation **145**
nucleus
    charge density 97, 99
    charge-to-mass ratio 200
    distribution of neutrons 99
    magnetic dipole moment 200–1

Index

    mean radius 97
    size 89, 95–9
    skin thickness 97
    spin–orbit interaction in 32–3
number density
    of electrons in 189
    of holes 190

odd parity **22**, 158
open shell **127**, 134–5, 171
Oppenheimer, Robert 143
orbital angular momentum
    classical 10
    commutation relations for 10
    components in spherical coordinates 10, 15
    eigenvalues 10, 17, 40
    eigenvectors see *spherical harmonics*
    in Bohr model 37
    in hydrogen molecule 39–41
    in hydrogen molecule ion 147–48
    operators 10, 15, 16
    square of magnitude of 10, 16
orbital angular momentum quantum number **10**, 17, 20, 40, 46, 124, 198
orbital see *atomic orbital* or *molecular orbital*
order of approximation **79**
orthonormality
    of hydrogen atom radial functions 48
    of spherical harmonics 22
overlap of atomic orbitals 174, 183
overlap rule 206
oxygen molecule 141, 164–5

pair production **110**
parametric dependence 144
parity
    even 22, 158
    odd 22, 158
    of molecular orbitals 158
    of spherical harmonics 22–3
partly full band, ability to conduct electricity 186–7
Paschen series 38
Pauli exclusion principle 95, 100, 109, 114, 127, 136, 185, 186
Pauli, Wolfgang 30, 188
period in Periodic Table **128**
periodic boundary conditions **179**
Periodic Table 30–1, 101, 114, **128**–31
periodicity of the lattice **175**, 177, 179

permittivity of free space 11, 36
perturbation **77**–8
    for helium atom 120
    for hydrogen atom 77, 86, 103–6
perturbation theory, time-dependent 195, **202**–7
    first-order transition probability 206, 207
    first-order wave function 205
    higher-order effects 209
    initial conditions 204–5
    matrix elements of perturbation 204
perturbation theory, time-independent **75**–87, 89, 96, 105, 106, **202**
    and anharmonic oscillator 78
    and finite size of proton 86
    and helium atom 120–3
    and hydrogen atom 77, 86, 103–6
    and one-dimensional infinite square well 79, 84–5
    and spin–orbit interaction 77
    compared with variational method 83
    first-order correction 82
    first-order energy eigenvalue **83**
    notation 76–7
    second-order approximation **87**
    second-order energy eigenvalue 87
    zeroth-order approximation 82
perturbed Hamiltonian **76**
PET 110
Pfund series 38
Pickering, Edward 93
pion **99**
pionic-atoms 99
Planck distribution law **217**
Planck, Max 217
Planck's constant 36, 217
plane wave, modulated 176
plane-polarized light 199
plasma 61
Poisson's law 106
polar angle **12**
polar diagram 23–4
population inversion 215
positron **109**–10, 190
positron emission tomography 110
positronium **110**, 190
principal quantum number **46**, 90, 95
probability density
    in π molecular orbital 160
    in σ molecular orbital 160

in crystals   172, 176–7
in hydrogen atom   53–5, 57
in hydrogen molecule ion   152–3, 156, 166
proton mass   92
proton magnetic dipole moment   107
p-type semiconductor   **191**
Purcell, Edward   108

QED   111, 195, 216
quantum chemistry   141
quantum dot   **192**
quantum electrodynamics   9, **111**, 195, 216
vacuum fluctuations   111, 197, 216
quantum field theory   89, **111**
quantum-mechanical interference   122, 152, 156
quarkonium   **112**
quarks   9, **112**
quartet term   136

radial coordinate   **12**
radial equation   **40**
solutions   42–51
radial functions   **40**, 47–9, 124
nodeless   43–4
normalization of   48
table of   49
radial probability density   54, **55**, 56
radiative transition   **197**
driven by electric field in light   201
driven by magnetic field in light   209
selection rules for   208–11
radio waves   199–200
Rayleigh, Lord   67
recurrence relation   45
reduced mass   27, 39, 50–1, 90
for deuterium atom   50–1, 91
for diatomic molecule   78
for exciton   235
for helium ion   92–3
for hydrogen atom   37, 90
for muonic atom   94
reduced radial equation   **40**
reduced radial function   **40**
relative atomic mass   **101**
relative permittivity   191
relativistic quantum mechanics   89, 108–10
relaxation time   **188**
residual electron–electron interaction   133, 136
resonance   212

rest energy   104
rest frame   105
rest mass   110
Retherford, Robert   111
rotational energies of hydrogen chloride molecule   146
rotational states, selection rules for   199
Rydberg energy   **37**, 42, 90–1 117
Rydberg states   **60**–1, 96

scaled Bohr radius   **92**, 192
scaled Rydberg energy   **92**
Schrödinger, Erwin   36, 89
Schrödinger's equation
acceptance of   35
for hydrogen atom   39
non-separable   202, 206
Schwinger, Julian   195
screening   124–**125**, 130, 171, 181
second-order approximation for energy eigenvalue   **87**
second-order correction   87
second-order perturbation theory
for degenerate unperturbed states   87
secular determinant   **151**, 173
secular equation   **151**
selection rules   196, **208**–11
for helium atom   211
for one-electron atom   209–11
self-consistent solution   **126**
semiconductor   169, **187**, 188–92
doped   190–2
electrical conductivity   188, 189, 192
pure   189–90
purification   188
semimetal   192
separable equation   17, 116, 124
separation constant   17, 19
separation of variables   17, 19, 40, 117, 120, 124
shell   **126**
energy-ordering   127
silicon   169
band gap   189
electron number density   189
silver, electrical conductivity   186
simultaneous eigenfunctions   11, 17, 21, 25, 51, 147, 178
single bond   **163**
singlet state   119, 122–3, 211
singlet term   136
single-valuedness condition   19

Index

skin thickness in nuclei   97
sodium atom, and spin–orbit interaction   31–2, 103
sodium as conductor   187
sodium chloride   130
    bonding   171–2
    crystal structure   171
solid state physics   169
Sommerfeld, Arnold   89
special relativity   9, 97, 103–4, 108
spectral energy density function   **214**, 216–7
spectroscopic dissociation energy   154–**155**, 161
spectroscopic notation
    for atomic levels   **137**
    for atomic orbitals   **117**, 124
    for atomic terms   **135**–6
    for hydrogen atom   **52**
    for molecular orbitals   **157**–8
spectrum
    deuterium atom   50–1
    electromagnetic   145–6, 199–200
    helium atom   9, 38, 131, 211
    hydrogen atom   9, 35, 36–8, 64, 131, 198
    hydrogen chloride   145–6, 198
speed of light   199
spherical coordinates   **12**–5
    and Born's rule   14
    and Hamiltonian operator   13–4
    and orbital angular momentum   15–6
    time-independent Schrödinger equation   12–4, 17–8
    volume element   15
spherical harmonics   **18**–24
    in atomic orbitals   117, 124, 210
    in hydrogen atom   40, 51
    normalization   22, 48, 55
    orthonormality   22
    parity   22–3
    table of   21
    visualization   23–4
spherical symmetry   11, **12**, 25
spin angular momentum operator   77, 107
spin ket   119
spin-$\frac{1}{2}$ particles and Dirac equation   109
spin–orbit interaction   **25**, 77, 105–6, 132–3, 136–8
    energy-level splitting due to   30
    Hamiltonian operator including   25
    in nuclei   32–3
    in sodium atom   31–2
    remaining degeneracy   30, 137
    strong   137
    weak   136–7
spinors   26, 109
spontaneous emission of light   195, 196–**197**, 216, 218
stationary states   39, 197, 215–6
stimulated emission of light   195, 196–**197**, 212, 216, 218
symmetric
    molecular orbitals   158
    spatial function   118–9, 123
    spin ket   119

Taylor expansion   79–80, 104
terms   see *atomic terms* or *molecular terms*
thermal energy   185
Thompson, J. J.   141
three-dimensional Fourier transform   **63**
tight-binding method   **180**–4
    and Bloch's theorem   180–1
    effective mass   184
    energy bands   183
    matrix elements   181–2
time-dependent Hamiltonian operator   201–2
time-dependent perturbation   202
time-dependent perturbation theory   see *perturbation theory, time-dependent*
time-independent Schrödinger equation
    for a diatomic molecule   142–3
    for a hydrogen atom   39
    for a many-electron atom   116
    for electrons in molecule   144
    for nuclei in a molecule   145
    in Cartesian coordinates   13
    in independent-particle model   116
    in spherical coordinates   14, 17
    in spherically-symmetric well   12–4, 16–7
    non-separable   13
    separable   17, 116, 124
time-independent perturbation theory   see *perturbation theory, time-independent*
tin as semimetal   192
total angular momentum   **26**–30, 136–8
    atomic   see *total atomic angular momentum*
    commutation relations   27
    eigenvalues   28
    eigenvectors   28–30
    operators   26
    square of magnitude of   27

Index

total angular momentum quantum number **28**, 137
total atomic angular momentum **136**–8
total kinetic energy operator 115
total magnetic quantum number **28**
total orbital angular momentum operator 134
total orbital angular momentum quantum number **134**
total spin operator 107, 134
total spin quantum number **134**
total static energy **145**, 154–5
total wave function 118–20, 122
transistor 7, 189
transition element, cohesive energy 174–5
transition probability 206–**207**, 211–5
translational symmetry **170**, 174
transverse wave 199
trial function **68**, 69–71, 74, 148–9, 172
triple bond **164**
triplet states 119, 122–3, 165, 211
triplet term 136
tritium 91
tungsten, crystal structure 170
twenty-one centimetre radiation **108**

ultraviolet light 145–6, 199–200
uncertainties in hydrogen atom 60
ungerade orbital **158**
unperturbed Hamiltonian **76**, 77, 78
    for helium atom 120
    in time-dependent perturbation theory 202
Urey, Harold 51

vacuum in quantum field theory 111
valence band **189**
valence electron 31, **127**, 128–9, 171, 172, 207
van de Hulst, Hendrik 108
variational method **67**–75
    accuracy 71
    and adjustable parameters 69, 72–4
    and excited states 74–5
    and ground state 68–73, 74
    and harmonic oscillator 72–3
    and hydrogen molecule ion 148–53, 155–6, 165
    and one-dimensional infinite square well 70–1
    and solids 173
    choice of trial function 69, 70, 74, 148, 172
    compared with perturbation theory 83
vibrating atomic cores, and electrical conductivity 188
vibrational energy 146
vibrational states, selection rules 199
virtual particles 197
visible light 199–200
volume element in spherical coordinates 14–5, 54

wave vector
    in a Bloch wave **176**, 180
    quantization in solids 179
    restrictions in crystal 179–80

X-ray photon 94, 99–101
X-ray spectrum 89, 99–101

zeroth-order approximation **80**, 82

Group	1	2	3	4	5	6	7	8	9	10	11	12	13	14	15	16	17	18
Period																		
1	1 H																	2 He
2	3 Li	4 Be											5 B	6 C	7 N	8 O	9 F	10 Ne
3	11 Na	12 Mg											13 Al	14 Si	15 P	16 S	17 Cl	18 Ar
4	19 K	20 Ca	21 Sc	22 Ti	23 V	24 Cr	25 Mn	26 Fe	27 Co	28 Ni	29 Cu	30 Zn	31 Ga	32 Ge	33 As	34 Se	35 Br	36 Kr
5	37 Rb	38 Sr	39 Y	40 Zr	41 Nb	42 Mo	43 Tc	44 Ru	45 Rh	46 Pd	47 Ag	48 Cd	49 In	50 Sn	51 Sb	52 Te	53 I	54 Xe
6	55 Cs	56 Ba	57 La	72 Hf	73 Ta	74 W	75 Re	76 Os	77 Ir	78 Pt	79 Au	80 Hg	81 Tl	82 Pb	83 Bi	84 Po	85 At	86 Rn
7	87 Fr	88 Ra	89 Ac	104 Rf	105 Db	106 Sg	107 Bh	108 Hs	109 Mt	110 Ds	111 Rg	112 Uub						

s shells filling; p shells filling; d shells filling; 4f shell filling; 5f shell filling

Period														
6	58 Ce	59 Pr	60 Nd	61 Pm	62 Sm	63 Eu	64 Gd	65 Tb	66 Dy	67 Ho	68 Er	69 Tm	70 Yb	71 Lu
7	90 Th	91 Pa	92 U	93 Np	94 Pu	95 Am	96 Cm	97 Bk	98 Cf	99 Es	100 Fm	101 Md	102 No	103 Lr

1 hydrogen; 2 helium; 3 lithium; 4 beryllium; 5 boron; 6 carbon; 7 nitrogen; 8 oxygen; 9 fluorine; 10 neon; 11 sodium; 12 magnesium; 13 aluminium; 14 silicon; 15 phosphorus; 16 sulphur; 17 chlorine; 18 argon; 19 potassium; 20 calcium; 21 scandium; 22 titanium; 23 vanadium; 24 chromium; 25 manganese; 26 iron; 27 cobalt; 28 nickel; 29 copper; 30 zinc; 31 gallium; 32 germanium; 33 arsenic; 34 selenium; 35 bromine; 36 krypton; 37 rubidium; 38 strontium; 39 yttrium; 40 zirconium; 41 niobium; 42 molybdenum; 43 technetium; 44 ruthenium; 45 rhodium; 46 palladium; 47 silver; 48 cadmium; 49 indium; 50 tin; 51 antimony; 52 tellurium; 53 iodine; 54 xenon; 55 caesium; 56 barium; 57 lanthanum; 58 cerium; 59 praseodymium; 60 neodymium; 61 promethium; 62 samarium; 63 europium; 64 gadolinium; 65 terbium; 66 dysprosium; 67 holmium; 68 erbium; 69 thulium; 70 ytterbium; 71 lutetium; 72 hafnium; 73 tantalum; 74 tungsten; 75 rhenium; 76 osmium; 77 iridium; 78 platinum; 79 gold; 80 mercury; 81 thallium; 82 lead; 83 bismuth; 84 polonium; 85 astatine; 86 radon; 87 francium; 88 radium; 89 actinium; 90 thorium; 91 protoactinium; 92 uranium; 93 neptunium; 94 plutonium; 95 americium; 96 curium; 97 berkelium; 98 californium; 99 einsteinium; 100 fermium; 101 mendelevium; 102 nobelium; 103 lawrencium; 104 rutherfordium; 105 dubnium; 106 seaborgium; 107 bohrium; 108 hassium; 109 meitnerium; 110 darmstadtium; 111 roentgenium; 112 ununbium

Physical constants

Planck's constant	h	6.63×10^{-34} J s		Planck's constant/2π	$\hbar$	1.06×10^{-34} J s
vacuum speed of light	c	3.00×10^{8} m s^{-1}		Coulomb law constant	$\frac{1}{4\pi\varepsilon_0}$	8.99×10^{9} m F^{-1}
permittivity of free space	ε_0	8.85×10^{-12} F m^{-1}		permeability of free space	μ_0	$4\pi \times 10^{-7}$ H m^{-1}
Boltzmann's constant	k	1.38×10^{-23} J K^{-1}		Avogadro's constant	N_m	6.02×10^{23} mol^{-1}
electron charge	$-e$	-1.60×10^{-19} C		proton charge	e	1.60×10^{-19} C
electron mass	m_e	9.11×10^{-31} kg		proton mass	m_p	1.67×10^{-27} kg
Bohr radius	a_0	5.29×10^{-11} m		atomic mass unit	u	1.66×10^{-27} kg